MOLECULAR SYSTEMATICS AND PHYLOGEOGRAPHY OF MOLLUSKS

SMITHSONIAN SERIES IN COMPARATIVE EVOLUTIONARY BIOLOGY

Douglas H. Erwin, Smithsonian Institution
Ted R. Schultz, Smithsonian Institution

The intent of this series is to publish innovative studies in the field of comparative evolutionary biology, especially by authors willing to introduce new ideas or to challenge or expand views now accepted. Within this context, and with some preference toward the organismic level, a diversity of viewpoints is sought.

Also in the series

Parascript

Parasites and the Language of Evolution
Daniel R. Brooks and Deborah A. McLennan

The Development and Evolution of Butterfly Wing Patterns

H. Frederick Nijhout

Hawaiian Biogeography

Evolution on a Hot Spot Archipelago
Edited by Warren L. Wagner and V. A. Funk

The Origin and Early Diversification of Land Plants

A Cladistic Study
Paul Kenrick and Peter R. Crane

Phylogenetic Analysis of Morphological Data

Edited by John J. Wiens

Molecular Systematics and Phylogeography of
Mollusks

Edited by

Charles Lydeard and
David R. Lindberg

Foreword by Geerat J. Vermeij

SMITHSONIAN BOOKS
Washington and London

Copy Editor: Fran Aitkens

Production Editor: Ruth G. Thomson

Library of Congress Cataloging-in-Publication Data

  Molecular systematics and phylogeography of mollusks / edited by Charles Lydeard and
  David Lindberg.
      p.   cm. — (Smithsonian series in comparative evolutionary biology)
  Includes bibliographical references and index.
  ISBN 1-58834-148-8 (alk. paper)
      1. Mollusks—Classification—Molecular aspects. I. Lydeard, Charles. II. Lindberg,
  David R., 1948– III. Series.
  QL406.M64   2003
  594′.012—dc21   2003042459

British Library Cataloging-in-Publication Data available

Manufactured in the United States of America

10 09 08 07 06 05 04 03   5 4 3 2 1

∞ The paper used in this publication meets the minimum requirements of the American National
Standard for Information Sciences—Permanence of Paper for Printed Library Materials ANSI
Z39.48-1984.

CONTENTS

FOREWORD
Molecules, Markers, and Mollusks

Classifying objects and facts—imposing order on a chaotic jumble of data—is a central, if often maligned, activity of scholars. Why do we engage in it? What do we really want to know that requires classification?

For evolutionary biologists, answers to these questions might seem obvious. We classify to infer the branching order in the evolutionary tree, on which we place species and lineages and clades according to the characteristics they share by virtue of inheritance. The characters we use have expanded in range over the centuries, beginning with the features of adult form, supplemented in time with characters from embryology and, most recently, with molecular sequences within genes and with the order in which genes are situated on chromosomes and organelles. With each new source of evidence, practitioners hope that previously unresolved patterns of evolutionary branching will finally be clarified. Every new type of character is viewed as a marker, an historical indicator of descent, a conserved trait. Everyone worries about whether the marker is truly inherited from the ancestor or whether it evolved multiple times, but always it remains an indicator, an atomized isolate that happens to be part of an organism's phenotype or genome.

However, there is more to classification than just identifying clades and the order in which they appear in the tree of life. As some contributors to this book realize, there are fascinating questions to be asked. When, how, where, and how often do functionally useful traits or adaptive syndromes arise? How are new characters introduced into the established pattern of development and gene regulation? How do clades and their members come to have the geographic and ecological distributions they now have or which they had in the past? How have the limits of range changed through time? Why are certain plausible functional types unattainable given established developmental pathways, genetic architectures, and ecological realities?

In trying to answer such questions, we find ourselves facing a central evolutionary conundrum: the markers on which we rely to show us pathways of descent with modification exist within a context of genes, chemical environments, physical forces, physiological processes, and interactions with other organisms. Collectively, the parts and properties of organisms form an adapted whole that works. At every scale of inclusion, a unit of life affects—and is affected by—its internal and external context; it is an entity incomprehensible when taken out of context. By abstracting characters of sequences or properties or bits of code, we not only eliminate essential evidence of feedback and other interactions, but we accept the implicit and untested assumption that the attributes we isolate are indeed functionally neutral, that they are merely helpful signs pointing to the correct, historically unique, evolutionary pathway we are trying to reconstruct.

To be sure, many observed changes at the molecular level—and some at the phenotypic level of emergent morphology as well—are effectively neutral, untouched by selection, impervious to the molecular environment and the forces that mold phenotypes from genetically coded sequences. To a large and mostly unknown extent, however, our reliance on such markers rests more on hope and experience than on established fact. We already know from years of observation and experiment that many aspects of form are adaptive, and that the expression of organic form is affected not just by genetic instructions, but by the interaction-rich environment within and outside the organism's body. Surely this dependence on context—this adaptive and constructional aspect—applies with equal force to base-pair sequences within genes and to the order of genes in the genome. There exists a natural history of organisms, of morphology, of developing parts of an embryo, and of the coding and constructional molecules that provide the machinery of life. Understanding this natural history is, I believe, key to transforming classification and phylogenetic inference from essentially descriptive and phenomenological pursuits to more explicitly theory-driven disciplines.

What we need is a predictive framework that enables us to distinguish truly reliable historical markers from characters and codes that are under strong selection. Of course, characters are integral parts of adapted, selection-driven, physically molded complexes that can be conserved just as neutral characters are, because contexts are subject to evolution as much as their components are. My point is, however, that decisions about which characters or genes "work" and which ones do not provide useful historical information about the order of evolutionary branching, and these decisions must be informed by an intimate knowledge of natural history, of the context in which organisms and their parts function and evolve. In short, we must try to understand why or whether some parts of the genome are informative while others are not.

Scientists outside our discipline often accuse us of stamp-collecting, of zealous description rather than theory-driven analysis. Cladists and molecular evolutionists would hotly contest such a caricature, but the critics have a point. Yes, we test hypotheses of relationships with new characters, new statistical tools, new computer-assisted analytical protocols, and estimates of how confident we should be in our results, but in the end it all remains descriptive and inductive. A theory of context is one of the missing pieces that could transform phenomenology into a more genuinely predictive enterprise.

Mollusks can play a crucial role in showing us how such a predictive framework can be built. Because of their extremely rich fossil record and a treasure trove of past and present characters and character combinations, shell-bearing mollusks allow us to reconstruct ancient and modern contexts ranging from the ecology within the developing and mature body to the interactions of the whole animal with its surroundings. They already give us insights into how the environment molds form and how form molds environment, and there is no reason why this perspective on macroscopic form cannot be integrated with a similar framework for microscopic molecules and developmental pathways.

No one source of data and no single methodology is a panacea. Authors in this book tell us time and time again that much remains unknown and unresolved despite the latest advances. Some of this surely reflects reality: events of branching are so closely spaced in the distant past that inferring the order of branching—including the sprouting of many branches simultaneously—is difficult to infer; there is rampant, iterative evolution of the same traits; and signals are lost in the mists of time. But there are also methodological shortcomings. We need to think deeply about the assumptions underlying the analytical procedures, the alignments of molecular sequences, the difficulties inherent in inferring pattern from only living representatives of clades. We need to be much more open to other lines of evidence, especially to data on the order of appearance in the fossil record, a source of data too little appreciated or trusted by cladists and other biologists. Above all, we need to think about whether the intermediates we invent as ancestors could really work, and whether the gains and losses of traits we infer are really plausible. Again, we need a theory, an understanding of assumptions and of context.

Historical sciences such as evolutionary biology pose challenges that experimental scientists do not have to confront. We cannot modify the past at will or invent worlds in which history is played out differently; but we can and must test hypotheses, and above all test assumptions, with as many independent sources of evidence as we can uncover, informed by an overarching theory of historical change. I believe with the contributors of this book that mollusks will provide us with the tools and insights to establish an evolutionary tree from which we can harvest a cornucopia of answers to the fundamental questions of history.

Geerat J. Vermeij
Department of Geology
University of California at Davis

PREFACE

Fifty years have passed since James Watson and Francis Crick won the race to describe the three-dimensional molecular structure of deoxyribonucleic acid (DNA). Since that time, extraordinary advances have been made in molecular biology, including the sequencing of the human genome and the cloning of complete organisms, such as the lamb named Dolly. But in a less public arena, the same DNA technology is also allowing investigators to address evolutionary questions that have perplexed natural historians since Charles Darwin provided his seminal treatise *On the Origin of Species by Means of Natural Selection*. Darwin envisioned a world in which the life of our planet would be understood and organized in terms of its genealogical relationships, but in practice this has proven harder than Darwin could have imagined.

Mollusks are one of those groups that have proven quite difficult. To begin with, they are an extremely diverse group of animals with more living species than birds, mammals, reptiles, amphibians, and fishes combined. But despite their diversity, abundance, and a fossil record that dates back more than 500 million years, as well as their ecological and economical importance, our knowledge of their evolutionary relationships remains surprisingly incomplete.

Over the past decade, the revolution in molecular biological techniques has blossomed into a discipline referred to as molecular phylogenetics, which holds great promise for resolving relationships among mollusks where morphology has been unsuccessful. Here, for the first time, we have assembled a cast of molluscan experts from around the world to contribute to this book. We asked each contributor to synthesize what is known about the molecular systematics or phylogeography of their favorite group of mollusks and/or to provide novel data or a fresh perspective or approach for analyzing data in a phylogenetic framework. We did not attempt to cover every major group of mollusk; some groups have been relatively little researched and some have been recently treated elsewhere (e.g., Cephalopoda). We hope this volume serves as a useful primer and valuable resource for those interested in the rapidly developing field of molluscan systematics or phylogeography and that it will inspire evolutionary biologists to consider the use of mollusks in their research programs. Like Darwin, we envision a time when the molluscan "Tree of Life" will be reconstructed, in part with the help of their molecules.

We begin the volume (Chapter 1) with an introduction to molluscan molecular systematics and phylogeography. Medina and Collins (Chapter 2) present an excellent overview of molluscan phylogenetic relationships and suggest future directions of systematic research, including evolutionary developmental biology (evo-devo). Giribit and Distel (Chapter 3) provide an impressive dataset to examine relationships among all of the major lineages of the Bivalvia. Roe and Hoeh (Chapter 4) provide an overview of one particular group of bivalves, the freshwater mussels (Unionoida). This group is particularly important because of its extremely imperiled status globally (Bogan 1993; Neves 1999). Steiner and Reynolds (Chapter 5) present the latest molecular-based phylogeny of the Scaphopoda based on their own considerable research effort on the group. McArthur and Harasewych (Chapter 6) examine the molecular systematics of the major gastropod lineages using a variety of brilliant approaches that include meta- and Bayesian analyses of published and novel data. Dayrat and Tillier (Chapter 7) present an informative summary of phylogenetic studies of the Euthyneura (a group referred to informally as "heterostropha," pulmonates + opisthobranchs) using a taxonomic congruence approach. Wägele and coauthors (Chapter 8) generate novel data and exhaustively exploit available data from the Internet (i.e., GenBank) to focus on phylogenetic relationships within the Nudibranchia (which includes the sea hares, popular model organisms for examining the molecular and cellular basis of basic neural processes). The last two chapters focus on the phylogeography of mollusks. Intraspecific DNA-based phylogeographic studies have only recently been conducted on mollusks, so we decided to divide the chapters into two major categories: aquatic and terrestrial molluscan phylogeography. Wares and Turner (Chapter 9) provide an admirable introduction to phylogeography in general and illustrate the utility of using freshwater and marine mollusks to explore the interactions among intrinsic life history traits and extrinsic forces acting on genealogical history. The phylogeography of terrestrial mollusks, the youngest of the DNA-based molluscan areas of research, is covered by Hugall and coauthors (Chapter 10), who present an extraordinary approach linking phylogeography of a rainforest-inhabiting group of gastropods from Queensland, Australia, with explicit paleo- and current bioclimatic modeling.

ACKNOWLEDGMENTS

We would like to extend our thanks to the authors for agreeing to participate in the production of this volume and to the many reviewers who took the time to improve the quality of each chapter: David Campbell, Donald Colgan, Kevin Cummings, Benoît Dayrat, M. G. Harasewych, Andrew McArthur, Mónica Medina, Paula Mikkelsen, Kathryn Perez, Winston Ponder, Kevin Roe, Jeanne Serb, Heike Wägele, and Patricia West. Special thanks to Steve Savarese Jr. for help standardizing and formatting all of the tables, Patricia West for editing the penultimate draft of each chapter, and Vincent Burke for guidance and seeing the value of such a scholarly volume. Our work on molecular phylogenetics of mollusks has been generously funded by grants from the National Science Foundation, the National Geographic Society, United States Fish and Wildlife Service, Bureau of Land Management, the University of Alabama, and the University of California at Berkeley. A special thanks to our spouses, Dixie Lindberg and Patricia West, and children, Jason Lindberg and Andrew, Emily, and Catherine Lydeard, for their continued support and patience as we pursue a marvelous and enriching career in invertebrate evolutionary biology.

CHARLES LYDEARD AND DAVID R. LINDBERG

1

CHALLENGES AND RESEARCH OPPORTUNITIES IN MOLLUSCAN MOLECULAR PHYLOGENETICS

One of the primary goals of the scientific discipline referred to as systematics is to describe, classify, and name the world's taxa on the basis of evolutionary relationships (Systematics Agenda 2000 1994; Eshbaugh 1995). Reconstructing the "Tree of Life" is a primary goal of systematists and has been identified as a high priority for the National Science Foundation of the United States. Systematists rely on shared morphological attributes of organisms to reconstruct an evolutionary tree or phylogeny of a given taxon. However, since the late 1960s, a revolution in molecular biology has resulted in the availability of methods for directly examining molecular variation of proteins and nucleic acids among taxa. Because the "stream of heredity makes phylogeny" (Simpson 1945), only genetically transmitted traits are informative to phylogenetic estimation (Avise 1994). That is not to say that molecular data are "better" than morphological data; indeed, the evolution of an organism's morphological *bauplan* is often what we are trying to understand and is typically what draws us to work on a particular group of creatures in the first place. Nevertheless, the birth of molecular systematics has provided an additional dataset for reconstructing the Tree of Life and has enabled workers to investigate phylogenetic relationships among widely disparate taxa where morphological homologies are wanting or remain dubious or debatable.

Molecular systematics encompasses a variety of molecular techniques that include isozyme electrophoresis, immunological techniques, molecular cytogenetics, DNA–DNA hybridization, restriction site analysis, and nucleic acid sequencing. Two excellent books on the topic entitled *Molecular Systematics*

1

provide detailed accounts on each molecular method, as well as the technical and theoretical information for properly designing and implementing a study (Hillis and Moritz 1990; Hillis et al. 1996).

Phylogeography, a term and discipline founded largely by Avise and his colleagues (Avise et al. 1987; Avise 1994, 2000), is a subset of molecular systematics and is primarily concerned with the spatial analysis of gene lineages within and among closely related species. It attempts to build empirical and conceptual bridges between macro- and microevolutionary patterns and processes (Avise 2000). Early, seminal phylogeographic studies in the late 1970s and early 1980s revealed mitochondrial restriction site variation among populations of animals, which often reflected historical barriers to dispersal within a species (Avise 1994, 2000). An excellent account of the historical development of phylogeography and the broader history of molecular phylogenetics can be found elsewhere (Avise et al. 1987; Avise 1994, 2000), and an entire issue of the journal *Molecular Ecology* has been dedicated to phylogeography, which also figures prominently in conservation genetics (Bermingham and Moritz 1998).

The phylum Mollusca is a diverse group of animals that includes such familiar organisms as scallops, clams, snails, slugs, squids, octopuses, and less familiar ones such as monoplacophorans, chitons, and tusk shells. Molluscan diversity extends from before the Cambrian explosion and also includes numerous, diverse extinct groups such as the bellerophonts and ammonites. Mollusks are second only to the phylum Arthropoda in living diversity, with more than 90,000 known species and an estimated 100,000 to 200,000 living species (the molluscan class Gastropoda is the second-most diverse animal class after the arthropod class Insecta). Mollusks live in almost every habitat, including deep-sea vents, coral reefs, estuaries, rocky and sandy shorelines, freshwater lakes and rivers, and almost everywhere on land from deserts to rainforests. Ecologically, mollusks play a fundamental role in a variety of ecosystems and often are the most abundant benthic invertebrate members of a particular ecosystem. Regrettably, many molluscan species are now considered extinct, endangered, or threatened because of the direct or indirect effects of human activities—habitat destruction or modification, introduction of exotic species, pollution, and overexploitation of commercially important or rare species.

Economically, mollusks have figured prominently in fisheries and mariculture, being used as food (e.g., clams, scallops, abalone, calamari, and conch) and commercial pearl production (Landman et al. 2001). In contrast, considerable economic loss is caused by the role of some mollusks as obligatory intermediate hosts of digenic trematodes and helminth parasites that infect livestock (e.g., liver flukes, *Fasciola* species) and humans (e.g., schistosomiasis). In addition, many terrestrial snails and slugs cause considerable damage to ornamental and agricultural plants, and introduced taxa often harm the native molluscan fauna (Cowie 1998, 2001; Carlton 1999; Pointier 1999; Strayer 1999). Mollusks are fascinating creatures and make excellent study organisms. They have provided a wealth of information following the methodological trajectory of new laboratory techniques of molecular systematics and phylogeography. The first molecular approach widely used in the field was protein electrophoresis to study allozyme and isozyme variation within and among species and/or populations. Several early seminal electrophoretic studies examined the population genetic structure and breeding systems in terrestrial snails, including the microgeographic structure of nonnative *Helix aspersa* colonies within a city block (Selander 1975) and the nominally diverse *Cerion* (Gould et al. 1975). Allozyme studies have provided useful data for studying gene flow (Dillon 1988), mating behavior (Jarne et al. 1993; Wethington and Dillon 1996, 1997), hybridization (Bianchi et al. 1994), and systematics (Chambers 1978, 1980; Murphy 1978; Davis et al. 1981; Buth and Suloway 1983; Hillis et al. 1991; Colgan and Ponder 1994; Ponder et al. 1994, 1995; Falniowski et al. 1996). Allozyme studies have also provided useful data on population genetic structure, particularly in commercially important bivalve species (e.g., Buroker 1982, 1983; Jordaens et al. 2000) and in gastropod species from South America, Asia, and Africa that serve as intermediate hosts to a variety of parasites that affect humans (e.g., Mimpfoundi and Greer 1989; Bandoni et al. 1990, 1995; Davis 1990, 1992; Mascara and Morgante 1991; Brown and Rollinson 1996; Brown et al. 1996; Mukaratirwa et al. 1996). Reviews of allozyme-based population genetic studies of mollusks can be found elsewhere (Davis 1978; Selander and Ochman 1983; Jarne and Delay 1991; Backeljau et al. 2001).

Other techniques that have been used in molecular systematics of mollusks include immunological comparisons of proteins (Davis and Fuller 1981; Davis 1984) and molecular cytogenetic methods (Burch 1960, 1963; Dillon 1989, 1991; Choudhury and Pandit 1997). The discovery of restriction enzymes followed by the development and study of restriction fragment-length polymorphisms (RFLPs), particularly of mitochondrial DNA (mtDNA), began in the 1970s and reached fruition in the 1980s (Avise 1994). However, relatively few studies used these methods on mollusks (e.g., Liu and Mitton 1993; Liu et al. 1996), apparently because of the problems associated with the presence of mucopolysaccharides that made isolation, purification, and digestion of mtDNA difficult.

Today the primary method of choice relies on the polymerase chain reaction (PCR) for in vitro amplification of specific DNA fragments (Palumbi 1996), followed by the actual determination of the nucleotide sequence. Several journals developed in response to the explosion of empirical molecular phylogenetic and phylogeographic studies (e.g., *Molecular Ecology, Molecular Phylogenetics and Evolution,* and *Conservation Genetics*). Recent molecular technological breakthroughs have resulted in a steady increase during the past decade of published articles on mollusks emphasizing systematics and phylogeography.

Technological advances of DNA sequencing have led to the ability to generate large nucleotide sequence datasets that include entire genomes. Although the genome of a mollusk is not yet completed, one for a squid is underway and nearly a dozen complete or near complete molluscan mitochondrial genomes have been generated; others are in the works. Unlike vertebrate mtDNA genomes, those of mollusks exhibit extraordinary variation in gene order, which provides the opportunity to consider these presumably conservative or rare genomic changes (RGCs; Rokas and Holland 2000) as characters for phylogenetic reconstruction (e.g., Boore and Brown 1994; Hatzoglou et al. 1995; Sasuga et al. 1999; Wilding et al. 1999; Kurabayashi and Ueshima 2000; Rawlings et al. 2001; Chapter 2, this volume). Comparative mitochondrial genomics is also providing opportunities to test hypotheses regarding the molecular mechanisms that shape the evolution of gene rearrangement (Boore 1999, 2000). Other RGCs that are being explored for phylogenetic reconstruction include variation in rRNA secondary structure (Thollesson 1999; Lydeard et al. 2000, 2002), which, when used in conjunction with morphological characters, provides evidence for recognizing groups such as the gastropod subclass Heterobranchia.

Biotechnological breakthroughs are providing opportunities to explore the phylogeography of mollusks using a battery of molecular tools and techniques. Although mitochondrial haplotype data provide a wealth of useful genetic data, other genetic markers such as random amplified polymorphic DNA (RAPD; Adamkewicz and Harasewych 1994) and particularly microsatellites (Jarne et al. 1994; Jarne and Lagoda 1996; Winnepenninckx and Backeljau 1998; Davison 1999; Samadi et al. 1999; Charbonnel et al. 2000; Mavárez et al. 2002) are sure to become more commonly used in genetic surveys in the future (e.g., Dowling et al. 1996). The importance of using several independent molecular markers has been exemplified in an extraordinary case of the American oyster (*Crassostrea virginica*). Allozyme data revealed little genetic differentiation among populations of the species, including populations from the Gulf of Mexico and the Atlantic coast of southeastern United States. This pattern was initially attributed to high interpopulational gene flow due to the long planktonic larval stage (Buroker 1983). However, subsequent mtDNA haplotype (Reeb and Avise 1990) and nuclear RFLP (Karl and Avise 1992) data revealed an unambiguous genetic split between Atlantic and Gulf oyster populations. Some authors hypothesize that the discrepancy in results is due to several of the allozyme loci being under uniform balancing selection and thus not recording the population subdivision evidenced by the neutral genetic markers (Avise 1994). We suspect that future studies will routinely use multiple, neutral genetic markers.

Evolutionary developmental biology, or "evo-devo," has yet to make significant contributions to molluscan molecular systematics, but the early signs are there (Degnan and Morse 1993, 1995; Callaerts et al. 2002). Developmental genes and their expression patterns are often extraordinarily stable. For example, *Hox* gene sequences appear to have remained relatively unchanged for more than half a billion years. Moreover, these genes are ultimately responsible for the continuum of morphology from embryo through larva to adult. Although the conserved sequences of these genes are not likely to resolve molluscan relationships, the documentation of gene cascades and expression patterns in molluscan taxa can provide fundamental insights into molluscan relationships and the evolution of molluscan morphology. Such an approach is also likely to have important implications for extinct molluscan morphologies in the fossil record. The potential of these approaches has been previously demonstrated in studying the formation of appendages (Shubin et al. 1997), but remains untapped for most other structures, especially outside of model organisms (Shubin and Marshall 2000).

In addition to evo-devo studies, the excellent molluscan fossil record is important in molecular systematic and phylogeographic studies as an aid in interpreting molecular phylogenetic patterns, as well as in estimating divergence times and morphological evolution (Collins et al. 1996; Hellberg et al. 2001).

Hand-in-hand with the development of molecular datasets has been the parallel development of methodologies and algorithms to analyze the new data. Analyses of early allozyme data used existing statistical techniques such as regression analysis (e.g., Mantel 1967) and a variety of techniques based on principal component analyses (PCA), Euclidean distances, and geometric distances (e.g., Cavalli-Sforza and Edwards 1967). As the population genetic models underlying our understanding of diversification became more robust, new distance measures to incorporate parameters such as drift and mutation appeared, including Nei's (1972) distance and coancestry distance (Reynolds et al. 1983). Today, more statistically rigorous methods are being used to test hypotheses about gene flow and population history using nested clade analysis (Templeton et al. 1995; Templeton 1998, 2001). Analysis of organelle gene order is particularly problematic, and currently there is no preferred method for coding and analyzing these data. Phenetic breakpoint analysis has been used (Blanchette et al. 1999) and new parsimony techniques are being developed (Cosner et al. 2000).

The cladistic methods used today for reconstructing molecular phylogenies also date from the early 1970s (e.g., Farris 1970) and rely on maximum parsimony to resolve relationships among taxa. Rather than using any of the previously developed distance measures and overall similarity to group taxa, cladistic methods use only Manhattan distances and shared, special similarities or genomic signatures known as synapomorphies (= shared derived characters) to construct trees or cladograms. Unfortunately, the reconstruction of the most parsimonious tree from the universe of all trees is a member of the statistical problem set known as NP-complete, meaning that the run time is polynomial to the size of the input. Thus, for large datasets, heuristic methods have been developed to subsample from the universe of trees and to converge relatively quickly on (one of) the most parsimonious solutions.

New methods in phylogenetic analysis have appeared that incorporate additional assumptions and models of nucleotide substitution. Foremost among these methods is maximum likelihood, in which the best model or tree consistent with the data is preferred. In phylogenetic analysis, the best tree under the maximum likelihood criterion is the tree that is the most likely to have occurred given the dataset and an assumed model of evolution (Swofford et al. 1996). Recently, Bayesian analysis has joined maximum likelihood and maximum parsimony as a tree-building method. Bayesian approaches are based on posterior probabilities, which include prior probabilities (informative, diffused, or uniform) and a likelihood model that describes the probability of the data given a specific topology and model of evolution. Bayesian methods commonly use the Metropolis–Hastings algorithm to generate tree topologies, and the rate at which specific areas of tree space are visited corresponds to the probability of those trees—some of which are higher than others and are therefore to be preferred under Bayesian assumptions.

Currently, the literature is contentious over the use of newer techniques such as maximum likelihood and Bayesian probabilities in phylogenetic reconstruction. Many of the arguments are philosophical, others are methodological. Most likely they reflect a healthy discourse in an active and growing field in which dogma and tradition have yet to become entrenched, and predict a rich future where intellectual inquiry has yet to become clichéd. Moreover, by-products of the emerging field of bioinformatics resulting from the human and other genomic sequencing activities can only benefit phylogenetic reconstruction and new analytical approaches.

We encourage readers to explore web-based resources (see below) or to join the Mollusca discussion listserver (e-mail subscription request to mollusclist@uclink.berkeley.edu). Software and literature bibliographies are available online through several "clearing-houses," such as:

University of California Museum of Paleontology:
    http://www.ucmp.berkeley.edu/subway/phylo/phylosoft.html

Felsenstein Laboratory, University of Washington, Phylogeny Programs:
http://evolution.genetics.washington.edu/phylip/software.html
Hennig Society–Parsimony Analysis Software:
http://www.cladistics.org/education.html
Dr. Keith Crandall Laboratory, Computer Programs, Brigham Young
University: http://zoology.byu.edu/crandall_lab/programs.htm
Cummings, K. S., A. E. Bogan, G. T. Watters, and C. A. Mayer. 2001, Freshwater mollusk bibliography database:
http://ellipse.inhs.uiuc.edu/mollusk/
Dr. Elizabeth Boulding, Mollusc Molecular Newsletter:
http://www.webapps.ccs.uoguelph.ca/mmnw/index.htm

REFERENCES

Adamkewicz, S. L., and M. G. Harasewych. 1994. Use of random amplified polymorphic DNA (RAPD) markers to assess relationships among beach clams of the genus *Donax*. The Nautilus, Supplement 2:51–60.

Avise, J. C. 1994. Molecular Markers, Natural History and Evolution. Chapman and Hall, New York, NY.

Avise, J. C. 2000. Phylogeography. Harvard University Press, Boston, MA.

Avise, J. C., J. Arnold, R. M. Ball Jr., E. Bermingham, T. Lamb, J. E. Neigel, C. A. Reeb, and N. C. Saunders. 1987. Intraspecific phylogeography: The mitochondrial DNA bridge between population genetics and systematics. Annual Review of Ecology and Systematics 18:489–522.

Backeljau, T. A. Baur, and B. Baur. 2001. Population and conservation genetics. Pp. 383–412 in The Biology of Terrestrial Molluscs (G. M. Barker, ed.). CAB International, Oxon, UK.

Bandoni, S. M., M. Mulvey, D. K. Koech, and E. S. Loker. 1990. Genetic structure of Kenyan populations of *Biomphalaria pfeifferi* (Gastropoda: Planorbidae). Journal of Molluscan Studies 56:383–392.

Bandoni, S. M., M. Mulvey, and E. S. Loker. 1995. Phylogenetic analysis of eleven species of *Biomphalaria* Preston, 1910 (Gastropoda: Planorbidae) based on comparisons of allozymes. Biological Journal of the Linnean Society 54:1–27.

Bermingham, E., and C. Moritz. 1998. Comparative phylogeography: Concepts and applications. Molecular Ecology 7:367–369.

Bianchi, T. S., G. M. Davis, and D. Strayer. 1994. An apparent hybrid zone between freshwater gastropod species *Elimia livescens* and *E. virginica*. American Malacological Bulletin 11:73–78.

Blanchette, M., T. Kunisawa, and D. Sankoff. 1999. Gene order breakpoint evidence in animal mitochondrial phylogeny. Journal of Molecular Evolution 49:193–203.

Boore, J. L. 1999. Animal mitochondrial genomes. Nucleic Acids Research 27:1767–1780.

Boore, J. L. 2000. The duplication/random loss model for gene rearrangement exemplified by mitochondrial genomes of deuterostome animals. Pp. 133–147 in Computational Biology Series. I: Comparative Genomics (D. Senkoff and J. Nadeau, eds.). Kluwer Academic Publ., Dordaecht, The Netherlands.

Boore, J. L., and W. M. Brown. 1994. Complete DNA sequence of the mitochondrial genome of the black chiton, *Katharina tunicata*. Genetics 138:423–443.

Brown, D. S., B. A. Curtis, and D. Rollinson. 1996. The freshwater snail *Bulinus tropicus* (Planorbidae) in Namibia, characterised according to chromosome number, enzymes and morphology. Hydrobiologia 317:127–139.

Brown, D. S., and D. Rollinson. 1996. Aquatic snails of the *Bulinus africanus* group in Zambia identified according to morphometry and enzymes. Hydrobiologia 324:163–177.

Burch, J. B. 1960. Chromosomes of *Gyraulus circumstriatus*, a freshwater snail. Nature 186:497–498.

Burch, J. B. 1963. Cytotaxonomic studies of freshwater limpets (Gastropoda: Basommatophora). I: The European lake limpet, *Acroloxus lacustris*. Malacologia 1:55–72.

Buroker, N. E. 1982. Allozyme variation in three nonsibling *Ostrea* species. Journal of Shellfish Research 2:157–163.

Buroker, N. E. 1983. Genetic differentiation and population structure of the American oyster *Crassostrea virginica* (Gmelin) in Chesapeake Bay. Journal of Shellfish Research 3:153–167.

Buth, D. G., and J. J. Suloway. 1983. Biochemical genetics of the snail genus *Physa:* A comparison of populations of two species. Malacologia 23:351–359.

Callaerts, P., P. N. Lee, B. Hartmann, C. Farfan, D. W. Y. Choy, K. Zkeo, K.-F. Fischbach, W. J. Gehring, and H. G. de Couet. 2002. *Hox* genes in the sepiolid squid *Euprymna scolopes:* Implications for the evolution of complex body plans. Proceedings of the National Academy of Sciences USA. 99:2088–2093.

Carlton, J. T. 1999. Molluscan invasions in marine and estuarine communities. Malacologia 41:439–454.

Cavalli-Sforza, L.L., and A. W. F. Edwards. 1967. Phylogenetic analysis: Models and estimation procedures. Evolution 21:550–570.

Chambers, S. M. 1978. An electrophoretically detected sibling species of *Goniobasis floridensis* (Mesogastropoda: Pleuroceridae). Malacologia 17:157–162.

Chambers, S. M. 1980. Genetic divergence between populations of *Goniobasis* (Pleuroceridae) occupying different drainage systems. Malacologia 20:63–82.

Charbonnel, N., B. Angers, R. Razatavonjizay, P. Bremond, and P. Jarne. 2000. Microsatellite variation in the freshwater snail *Biomphalaria pfeifferi*. Molecular Ecology 9:1006–1007.

Choudhury, R. C., and R. K. Pandit. 1997. Chromosomes of three prosobranch gastropods from Viviparidae, Pilidae and Cyclophoridae (Order: Mesogastropoda). Caryologia 50:341–353.

Colgan, D. J., and W. F. Ponder. 1994. The evolutionary consequences of restriction in gene flow: Examples from hydrobiid snails. The Nautilus, supp. 4:25–43.

Collins, T. M., K. Frazer, A. R. Palmer, G. J. Vermeij, and W. M. Brown. 1996. Evolutionary history of northern hemisphere *Nucella* (Gastropoda, Muricidae): Molecular, morphological, ecological, and paleontological evidence. Evolution 50:2287–2304.

Cosner, M. E., R. K. Jansen, B. M. E. Moret, L. A. Raubeson, L. S. Wang, T. Warnow, and S. Wyman. 2000. A new fast heuristic for computing the breakpoint phylogeny and a phylogenetic analysis of a group of highly rearranged chloroplast genomes. Pp. 104–115 in Proceedings of the 8th International Conference on Intelligent Systems for Molecular Biology (ISMB-2000), San Diego, CA.

Cowie, R. H. 1998. Patterns of introduction of non-indigenous non-marine snails and slugs in the Hawaiian Islands. Biodiversity and Conservation 7:349–368.

Cowie, R. H. 2001. Can snails ever be effective and safe biocontrol agents? International Journal of Pest Management 47:23–40.

Davis, G. M. 1978. Experimental methods in molluscan systematics. Pp. 99–169 in Pulmonates, vol. IIA (V. Fretter and J. Peake, eds.). Academic Press, London.

Davis, G. M. 1984. Genetic relationships among some North American Unionidae (Bivalvia): Sibling species, convergence, and cladistic relationships. Malacologia 25:629–648.

Davis, G. M. 1990. The genus *Wuconchona* of China (Gastropoda: Pomatiopsidae Triculinae): Anatomy, systematics, cladistics and transmission of *Schistosoma*. Proceedings of the Academy of Natural Sciences of Philadelphia 142:119–142.

Davis, G. M. 1992. Evolution of prosobranch snails transmitting Asian Schistosoma; coevolution with *Schistosoma:* A review. Progress in Clinical Parasitology 3:145–204.

Davis, G. M., and S. L. H. Fuller. 1981. Genetic relationships among recent Unionacea (Bivalvia) of North America. Malacologia 20:217–253.

Davis, G. M., W. H. Heard, S. L. H. Fuller, and C. Hesterman. 1981. Molecular genetics and speciation in *Elliptio* and its relationships to other taxa of North American Unionidae (Bivalvia). Biological Journal of the Linnean Society 15:131–150.

Davison, A. 1999. Isolation and characterization of long compound microsatellite repeat loci in the land snail, *Cepaea nemoralis* L. (Mollusca, Gastropoda, Pulmonata). Molecular Ecology 8:1760–1761.

Degnan, B. M., and D. E. Morse. 1993. Identification of eight homeobox-containing transcripts expressed during larval development and at metamorphosis in the gastropod mollusc *Haliotis rufescens*. Molecular Marine Biology and Biotechnology 2:1–9.

Degnan, B. M, and D. E. Morse. 1995. Developmental and morphogenetic gene regulation in *Haliotis rufescens* larvae at metamorphosis. American Zoologist 35:391–398.

Dillon, R. T., Jr. 1988. Evolution from transplants between genetically distinct populations of freshwater snails. Genetica 76:111–120.

Dillon, R. T., Jr. 1989. Karyotypic evolution in pleurocerid snails. I. Genomic DNA estimated by flow cytometry. Malacologia 31:197–203.

Dillon, R. T., Jr. 1991. Karyotypic evolution in pleurocerid snails. II. *Pleurocera, Goniobasis,* and *Juga*. Malacologia 33:339–344.

Dowling, T. E., C. Moritz, J. D. Palmer, and L. H. Rieseberg. 1996. Nucleic acids. III. Analysis of fragments and restriction sites. Pp. 249–320 in Molecular Systematics (D. M. Hillis, C. Moritz, and B. K. Mable, eds.). Sinauer Associates, Sunderland, MA.

Eshbaugh, W. H. 1995. Systematics Agenda 2000: An historical perspective. Biodiversity and Conservation 4:455–462.

Falniowski, A., A. Kozik, M. Szarowska, W. Fialkowski, and K. Mazan. 1996. Allozyme and morphology evolution in European Viviparidae (Mollusca: Gastropoda: Architeanioglossa). Journal of Zoological Systematics and Evolutionary Research 34:49–62.

Farris, J. S. 1970. Methods for computing Wagner trees. Systematic Zoology 19:83–92.

Gould, S. J., D. S. Woodruff, and J. P. Martin. 1975. Genetics and morphometrics of *Cerion* at Pongo Carpet: A new systematic approach to this enigmatic land snail. Systematic Zoology 23:518–535.

Hatzoglou, E., G. C. Rodakis, and R. Lecanidou. 1995. Complete sequence and gene organization of the mitochondrial genome of the land snail *Albinaria coerulea*. Genetics 140:1353–1366.

Hellberg, M. E., D. P. Balch, and K. Roy. 2001. Climate-driven range expansion and morphological evolution in a marine gastropod. Science 292:1707–1710.

Hillis, D. M., M. T. Dixon, and A. L. Jones. 1991. Minimal genetic variation in a morphologically diverse species (Florida tree snail, *Liguus fasciatus*). Journal of Heredity 82:282–286.

Hillis, D. M., and C. Moritz. 1990. Molecular Systematics. Sinauer Associates, Sunderland, MA.

Hillis, D. M., C. Moritz, and B. K. Mable (eds.). 1996. Molecular Systematics. Sinauer Associates, Sunderland, MA.

Jarne, P., and B. Delay. 1991. Population genetics of freshwater snails. Trends in Ecology and Evolution 6:383–386.

Jarne, P., and P. J. L. Lagoda. 1996. Microsatellites, from molecules to populations and back. Trends in Ecology and Evolution 11:424–429.

Jarne, P., M. Vianey-Liaud, and B. Delay. 1993. Selfing and outcrossing in hermaphroditic freshwater gastropods (Basommatophora): Where, when, and why. Biological Journal of the Linnean Society 49:99–125.

Jarne, P., F. Viard, B. Delay, and G. Cuny. 1994. Variable microsatellites in the highly selfing snail *Bulinus truncatus* (Basommatophora: Planorbidae). Molecular Ecology 3:527–528.

Jordaens, K., H. De Wolf, T. Willems, S. Van Dongen, C. Brito, A. M. F. Martins, and T. Backeljau. 2000. Loss of genetic variation in a strongly isolated Axorean population of the edible clam, *Tapes decussatus*. Journal of Shellfish Research 19:29–34.

Karl, S. A., and J. C. Avise. 1992. Balancing selection at allozyme loci in oysters: Implications from nuclear RFLPs. Science 256:100–102.

Kurabayashi, A., and R. Ueshima. 2000. Complete sequence of the mitochondrial DNA of the primitive opisthobranch gastropod *Pupa strigosa:* Systematic implication of the genome organization. Molecular Biology and Evolution 17:266–277.

Landman, N. H., P. M. Mikkelsen, R. Bieler, and B. Bronson. 2001. Pearls: A Natural History. H. A. Abrams, New York, NY.

Liu, H. P., and J. B. Mitton. 1993. A technique to reveal restriction fragment length polymorphisms (RFLPs) in the genus *Physa*. Malacological Review 26:89–90.

Liu, H. P., J. B. Mitton, and S. K. Wu. 1996. Paternal mitochondrial DNA differentiation far exceeds maternal mitochondrial DNA and allozyme differentiation in the
freshwater mussel, *Anodonta grandis grandis*. Evolution 50:952–957.

Lydeard, C., W. E. Holznagel, M. N. Schnare, and R. R. Gutell. 2000. Phylogenetic
analysis of molluscan mitochondrial LSU rDNA sequences and secondary structures.
Molecular Phylogenetics and Evolution 15:83–102.

Lydeard C., W. E. Holznagel, R. Ueshima, and A. Kurabayashi. 2002. Systematic implications of extreme loss or reduction of mitochondrial LSU rRNA helical-loop structures in gastropods. Malacologia 44:333–336.

Mantel, N. A. 1967. The detection of disease clustering and a generalized regression approach. Cancer Research 27:209–220.

Mascara, D., and J. S. Morgante. 1991. Enzyme polymorphism and genetic structure of
Biomphalaria tenagophila (Gastropoda: Planorbidae) populations founder effect.
Brazilian Journal of Genetics 14:631–644.

Mavárez, J., M. Amarista, J. Pointier, and P. Jarne. 2002. Fine-scale population structure
and dispersal in *Biomphalaria glabrata,* the intermediate snail host of *Schistoma mansoni,* in Venezuela. Molecular Ecology 11:879–889.

Mimpfoundi, R., and G. J. Greer. 1989. Allozyme comparisons among species of the *Bulinus forskalii* group (Gastropoda: Planorbidae) in Cameroon. Journal of Molluscan
Studies 55:405–410.

Mukaratirwa, S., H. R. Siegismund, T. K. Kristensen, and S. K. Chandiwana. 1996. Genetic structure and parasite compatibility of *Bulinus globosus* (Gastropoda: Planorbidae) from two areas of different endemicity of *Schistosoma haematobium* in Zim-
babwe. International Journal for Parasitology 26:269–280.

Murphy, P. G. 1978. *Collisella austrodigitalis* sp. nov.: A sibling species of limpet (Ac-
maeidae) discovered by electrophoresis. Biological Bulletin 155:193–206.

Nei, M. 1972. Genetic distance between populations. American Naturalist 106:283–292.

Neves, R. J. 1999. Conservation and commerce: Management of freshwater mussel (Bivalvia: Unionoidea) resources in the United States. Malacologia 41:461–474.

Palumbi, S. 1996. Nucleic acids. II. The polymerase chain reaction. Pp. 205–247 in
Molecular Systematics (D. M. Hillis, C. Moritz, and B. K. Mable, eds.). Sinauer Associates, Sunderland, MA.

Pointier, J. P. 1999. Invading freshwater gastropods: Some conflicting aspects for public health. Malacologia 41:403–411.

Ponder, W. F., D. J. Colgan, G. A. Clark, A. C. Miller, and T. Terzis. 1994. Microgeographic genetic and morphological differentiation of freshwater snails: A study on the
Hydrobiidae of Wilsons Promontory, Victoria, south eastern Australia. Australian Journal of Zoology 42:557–678.

Ponder, W. F., P. Eggler, and D. J. Colgan. 1995. Genetic differentiation of aquatic snails
(Gastropoda: Hydrobiidae) from artesian springs in arid Australia. Biological Journal
of the Linnean Society 56:553–596.

Rawlings, T. A., T. M. Collins, and R. Bieler. 2001. A major mitochondrial gene rearrangement among closely related species. Molecular Biology and Evolution 18:1604–1609.

Reynolds, J., B. S. Weir, and C. C. Cockerham. 1983. Estimation of the coancestry coefficient: Basis for a short-term genetic distance. Genetics 105:767–779.

Reeb, C. A., and J. C. Avise. 1990. A genetic discontinuity in a continuously distributed species: Mitochondrial DNA in the American oyster, *Crassostrea virginica*. Genetics 124:397–406.

Rokas, A., and P. W. H. Holland. 2000. Rare genomic changes as a tool for phylogenetics. Trends in Ecology and Evolution 15:454–459.

Saitou, N., and M. Nei. 1987. The neighbor-joining method: A new method for reconstructing phylogenetic trees. Molecular Biology and Evolution 4:406–425.

Samadi, S., J. Mavaarez, J. P. Pointier, B. Delay, and P. Jarne. 1999. Microsatellite and morphological analysis of population structure in the parthenogenetic freshwater snail *Melanoides tuberculata:* Insights into the creation of clonal variability. Molecular Ecology 8:1141–1153.

Sasuga, J., S. Yokobori, M. Kaifu, T. Ueda, K. Nishikawa, and K. Watanabe. 1999. Gene contents and organization of a mitochondrial DNA segment of the squid *Loligo bleekeri*. Journal of Molecular Evolution 48:692–702.

Selander, R. K. 1975. Genetic structure of populations of the brown snail (*Helix aspersa*). I. Microgeographic variation. Evolution 29:385–401.

Selander, R. K., and H. Ochman. 1983. The genetic structure of populations as illustrated by molluscs. Pp. 93–123 in Isozymes: Current Topics in Biological and Medical Research 10: Genetics and Evolution. Alan R. Liss, New York.

Shubin, N. and C. R. Marshall. 2000. Fossils, genes, and the origin of novelty. Paleobiology, supp. 26:324–340.

Shubin, N., C. Tabin, and S. Carroll. 1997. Fossils, genes and the evolution of animal limbs. Nature 388:639–648.

Simpson, G. G. 1945. The principles of classification and a classification of mammals. Bulletin of the American Museum of Natural History 85:1–350.

Strayer, D. L. 1999. Effects of alien species on freshwater mollusks in North America. Journal of the North American Benthological Society 18:74–98.

Swofford, D. L., G. J. Olsen, P. J. Waddell, and D. M. Hillis. 1996. Phylogenetic inference. Pp. 407–514 in Molecular Systematics (D. M. Hillis, C. Moritz, and B. K. Mable, eds.). Sinauer Associates, Sunderland, MA.

Systematics Agenda 2000. 1994. Systematics Agenda 2000: Charting the biosphere. Technical Report, Society of Systematic Biologists, American Society of Plant Taxonomists, Willi Hennig Society, Association of Systematics Collections, New York.

Templeton, A. R. 1998. Nested clade analyses of phylogeographic data: Testing hypotheses about gene flow and population history. Molecular Ecology 7:381–397.

Templeton, A. R. 2001. Using phylogeographic analyses of gene trees to test species status and processes. Molecular Ecology 10:779–791.

Templeton, A. R., E. Routman, and C. A. Phillips. 1995. Separating population structure from population history: A cladistic analysis of the geographical distribution of mitochondrial DNA haplotypes in the tiger salamander, *Ambystoma tigrinum*. Genetics 140:767–782.

Thollesson, M. 1999. Phylogenetic analysis of Euthyneura (Gastropoda) by means of the 16S rRNA gene: Use of a "fast" gene for "higher-level" phylogenies. Proceedings of the Royal Society of London B 266:75–83.

Wethington, A. R., and R. T. Dillon, Jr. 1996. Gender choice and gender conflict in a nonreciprocally mating simultaneous hermaphrodite, the freshwater snail, *Physa*. Animal Behavior 51:1107–1118.

Wethington, A. R., and R. T. Dillon, Jr. 1997. Selfing, outcrossing, and mixed mating in the freshwater snail *Physa heterostropha:* Lifetime fitness and inbreeding depression. Invertebrate Biology 116:192–199.

Wilding, C. S., P. J. Mill, and J. Grahame. 1999. Partial sequence of the mitochondrial genome of *Littorina saxatilis:* Relevance to gastropod phylogenetics. Journal of Molecular Evolution 48:348–359.

Winnepenninckx, B., and T. Backeljau. 1998. Isolation and characterization of microsatellite markers in the periwinkle *Littorina striata* King and Broderip, 1832 (Mollusca, Gastropoda, Prosobranchia). Molecular Ecology 7:1253–1254.

MÓNICA MEDINA AND ALLEN G. COLLINS

2

THE ROLE OF MOLECULES IN UNDERSTANDING MOLLUSCAN EVOLUTION

Few, if any, invertebrate groups of animals have received as much study as Mollusca, the second largest animal phylum after Arthropoda. Cuvier (1795) used several molluscan taxa to illustrate the use of internal characters (soft body parts) in comparative anatomy. Subsequently, he defined the phylum Mollusca in the form it is now accepted (Cuvier 1797). Mollusks are coelomate protostomes and their anatomy is characterized in most cases by a well-developed head, a ventral foot, and a mantle layer covering a visceral mass. The mantle glands secrete calcareous skeletal elements, which can be spicules or shells that enclose much of the body. Ctenidia (the respiratory organs) are plesiomorphically present in pairs. The feeding organ of mollusks is the radula, a chitinous structure with numerous curved teeth often used in rasping. This complex structure is considered one of the strongest morphological synapomorphies of the Mollusca, although it has been secondarily lost in bivalves and derived members of other groups. Molluscan ontogeny involves spiral cleavage and, in many cases, a triphasic life cycle composed of trochophore, veliger, and adult.

Four main sources of data can be used to reconstruct the evolutionary history of Mollusca: anatomy, fossils, molecular sequences, and genomic characters. By genomic characters, we refer to rare genomic changes such as mitochondrial gene-order data and large insertions or deletions of DNA, as well as secondary and tertiary structures of biomolecules and gene-expression patterns of developmental genes. The continued use and exploration of this latter type of data will likely enhance understanding of both the phylogenetic relationships among mollusks and the associations between genotype and phenotype in the different molluscan body plans.

An ideal appreciation for molluscan evolution will come from the integration of all sources of evidence, including fossils and anatomy. In this chapter, however, we concentrate on reviewing recent evidence stemming from advances in molecular biotechnology that enlightens the phylogenetic history of Mollusca. Specifically, we review what molecular sequence data and genomic characters suggest about the phylogenetic position of Mollusca within Metazoa, the identity of the closest living relatives of mollusks, and the relationships among the major groups of mollusks that have traditionally been called classes. We also discuss new directions that molecular techniques are taking research in molluscan evolution and phylogenetics.

THE PHYLOGENETIC POSITION OF MOLLUSCA WITHIN METAZOA

Morphological and developmental traits such as segmentation and spiral cleavage have often led to what could be termed a "traditional" morphology-based view that Mollusca shares a relatively close phylogenetic association with annelids (including echiurans and pogonophorans), arthropods, onychophorans, sipunculans, and tardigrades (Schram 1991; Nielsen et al. 1996; Nielsen 2001). However, since Taylor's (1996) major review on different aspects of molluscan evolution, this general placement of mollusks in the metazoan tree has been challenged, to a certain extent, on the basis of molecular data. Specifically, a large number of analyses of nuclear small subunit ribosomal DNA (rDNA) (18S) data confirm that mollusks share a relatively recent history with other animals that possess a trochophore larval stage, such as annelids (including echiurans and pogonophorans) and sipunculids. However, 18S studies also suggest that mollusks are more closely related to ectoprocts, nemertines, and *Xenoturbella,* as well as to groups whose members have a lophophore (brachiopods, phoronids, and ectoprocts), than they are to arthropods, onychophorans, and tardigrades (Halanych et al. 1995; Winnepenninckx et al. 1995, 1996; Kim et al. 1996; Aguinaldo et al. 1997; Littlewood et al. 1998; Zrzavy et al. 1998; Ruiz-Trillo et al. 1999; Adoutte et al. 2000).

An early such 18S study defined the clade stemming from the last common ancestor of mollusks, annelids, brachiopods, bryozoans, and phoronids as Lophotrochozoa (Halanych et al. 1995). A related phylogenetic hypothesis based on 18S data is that arthropods are members of a clade, Ecdysozoa, that includes all animals that molt their cuticles (Aguinaldo et al. 1997). Some confusion about the definition of Lophotrochozoa has arisen in the literature of metazoan phylogenetics, where the name is often applied to all or most nonecdysozoan protostomes (Aguinaldo et al. 1997; Ruiz-Trillo et al. 1999; Adoutte et al. 2000; Peterson and Eernisse 2001). However, a review of the many studies of metazoan phylogeny based on 18S reveals little evidence for the assertion that Platyhelminthes (with or without Acoela), Gnathostomulida, Gastrotricha, or Rotifera, and others are descended from the common ancestor on which the definition of Lophotrochozoa is based. Relationships among Mollusca and the other lophotrochozoan taxa, however, are not clearly delineated by 18S data.

Although 18S studies have provided important insights into animal phylogeny, they are based on a single set of nucleotide characters and must be assessed in light of other data sources. A number of genes have been suggested as potentially useful for revealing relationships among major metazoan groups, for instance, elongation factor–1α, RNA polymerase II, and large subunit (28S) rDNA. However, at present, only 28S has been applied to a broad diversity of metazoan groups (Medina et al. 2001b; Mallatt and Winchell 2002; Winchell et al. 2002). Like 18S data, 28S sequences appear to support the Ecdysozoa and Lophotrochozoa hypotheses by suggesting that mollusks are more closely related to annelids, brachiopods, echiurans, nemertines, phoronids, and sipunculans than they are to Arthropoda and its molting allies (Mallatt and Winchell 2002).

Most studies using molecular data to investigate the phylogenetics of Metazoa have relied on comparisons of aligned sets of nucleotides or amino acids. Over time, better theoretical models of nucleotide evolution have been applied to the analysis of nucleotide and amino acid data. In addition, as computer speeds and the amounts of molecular data increase, the methods used to analyze them have become more sophisticated. Nevertheless, just as for morphological characters, well-documented limitations exist for nucleotide data. They are subject to homoplasy, as well as single-site insertions and deletions (indels), nucleotide biases, different substitution rates, site-dependent saturation, and other problems. As a consequence, researchers have been increasingly interested in finding new nontraditional phylogenetic markers (Rokas and Holland 2000). Advances in genome research have made it easier to identify large genomic changes (i.e., intron indels, ribosomal secondary structure, organelle gene-order data), which are starting to be examined for their potential use for phylogeny (reviewed in Rokas and Holland 2000). We will refer to these new types of markers as *genomic signatures*. If they do occur at a rate that is appropriate for the phylogenetic divergences being addressed, then genomic signatures should be powerful phylogenetic characters that can be readily analyzed by the cladistic method of phylogenetic reconstruction (Manuel et al. 2000; Rokas and Holland 2000). For instance, *Hox* genes and neural expression of horseradish peroxidase immunoreactivity are consistent with the Ecdysozoa and Lophotrochozoa hypotheses (de Rosa et al. 1999; Haase et al. 2001).

Mitochondrial DNA (mtDNA) gene-order data also play a potentially important role in understanding the position of Mollusca within Metazoa. Animal mitochondria generally have a circular genome, which usually is less than 20 kb in size and is characterized by a highly conserved gene content (13 protein-coding genes, 2 ribosomal genes, and 22 transfer RNA genes; Westenholme 1992; Boore 1999). Gene rearrangements in animal mtDNA genomes occur more slowly than single nucleotide substitutions, and models of evolution of its gene order are appearing in the literature (Boore 2000). Additionally, despite the few animal lineages sampled, gene-order data have already been useful in reconstructing broad animal phylogenetic relationships (Boore and Brown 1998; Boore 1999). As this chapter goes to print, 260 complete mitochondrial genomes are available either in the literature or in GenBank. Most of these genomes are chordates (178), but a rapidly growing number represent invertebrate phyla, including 11 from mollusks, 2 from annelids, and 3 from brachiopods. Several partial mitochondrial genomes are also known that include gene-order data. A comprehensive review of all published gene-order datasets in mollusks is available online at http://www.jgi.doe.gov/programs/comparative/ MGA_Source_Guide.html.

Although the relatively small number of genomes limits how much one can infer from gene-order data at this point, these data have contributed something to our understanding of metazoan phylogeny. For instance, gene-order data supports a lophotrochozoan clade that includes mollusks (Boore and Staton 2002; Jeffrey Boore, Joint Genome Institute, personal communication). Unfortunately, although gene order comparisons within phyla tend to be conservative (e.g., within vertebrates), a major exception is Mollusca (Boore and Brown 1994a, 1994b; Boore 1999; Jeffrey Boore, Joint Genome Institute, unpublished data). Major rearrangements are observed among all of the molluscan mtDNA genomes available (see "Mitochondrial Gene Order Comparisons among Molluscan Classes"). The rapid evolution of gene order in mollusks is particularly well illustrated by the recent finding of major rearrangements among mitochondrial coding genes in several species of the gastropod genus *Dendropoma* (Rawlings et al. 2001).

THE SISTER GROUP TO MOLLUSCA

Asking the question "What is the sister group to Mollusca?" assumes that Mollusca is monophyletic. In fact, at present, molecular data have provided little evidence to support such an assertion. Ghiselin (1988) and Winnepenninckx and coauthors (1994, 1995) did pioneering work in molluscan systematics using 18S. These studies included few molluscan taxa (<10) as part of a larger metazoan dataset, which was enough to weakly recover monophyly of mollusks but not to address relationships between the different major lineages. Analyses of large 18S invertebrate datasets do not always find optimal trees that contain a monophyletic Mollusca (Winnepenninckx et al. 1996; Giribet and Wheeler 1999; Giribet et al. 2000; Peterson and Eernisse 2001). In fact, this lack of resolution is broadly true for all of the lophotrochozoan groups. Initially investigators suggested that the lack of lophotrochozoan resolution in 18S phylogenies was due to poor taxon sampling. The addition of more taxa has yet to resolve the relationships among the lophotrochozoan lineages (Adoutte et al. 2000). Adoutte et al. proposed that this probably reflects a burst of rapid speciation in the Cambrian within three major bilaterian lineages. On the one hand, if this is the case, new data from other molecular markers will also probably fail to resolve lophotrochozoan relationships or accurately reveal the phylogenetic status of the phyla that comprise Lophotrochozoa. On the other hand, new molecular data may help elucidate some of those questions for which 18S have failed to provide insight. Despite the poor resolution of the available data, no molecular evidence strongly contradicts the monophyly of mollusks. Additionally, historical work on morphology certainly suggests that Mollusca is monophyletic.

A number of potential sister groups to Mollusca have been suggested in print. One potential sister group is Sipuncula, a phylum of vermiform organisms whose larval morphology at the 64-cell stage is similar to that seen in mollusks (i.e., the molluscan cross; Scheltema 1993). However, it should be noted that some annelids also exhibit the same developmental pattern (Ronald Jenner, Cambridge University, personal communication). Combined analysis of 18S data and morphology produced a similar conclusion (Zrzavy 1998). Partial mitochondrial genome data for a sipunculan, however, supports a phylogenetic affinity to annelids rather than to mollusks (Boore and Staton 2002). In many instances, mollusks are thought to have a close association with annelids (Ghiselin 1988), or form an assemblage that includes both annelids and arthropods (Schram 1991; Nielsen et al. 1996; Nielsen 2001). Turbellarian flatworms have also been considered as the sister group to mollusks (Salvini-Plawen 1991). Potential synapomorphies, such as similarities in larval morphology and circulatory systems, have also been proposed for a grouping of Entoprocta plus Mollusca (Haszprunar 1996). Although 18S data contradicts the hypothesis that Arthropoda or any flatworm clade is part of a sister group to Mollusca, they do not contradict a sister-group relationship with any combination of the lophotrochozoan taxa, including Sipuncula, Annelida, or Entoprocta.

Using complete 28S sequences, Mallatt and Winchell (2002) obtained some support for the idea that brachiopods and/or phoronids may be the sister group to mollusks. Although this hypothesis has been little discussed in the literature in recent years, early analysis of the secondary structure of 5.8S also suggested this relationship (Hendriks et al. 1986). There is, therefore, reason to reevaluate the anatomical traits used previously to unite these groups. For instance, pores formed from mantle extensions into the shell are present in both phyla. These structures are known as caeca in brachiopods and the shell-bearing mollusks, and as aesthetes in polyplacophorans (Reindl et al. 1995). According to ultrastructural observations and immunocytochemistry, these organs closely resemble each other (Reindl et al. 1995, 1997). However, Reindl and colleagues attributed this similarity to convergence caused by similar function rather than to a shared phylogenetic history. If additional molecular evidence shows these two groups to be sister taxa, this character may be homologous.

RELATIONSHIPS WITHIN MOLLUSCA

Investigators generally agree on seven primary extant molluscan groups, correspondingly designated as classes of the phylum Mollusca. These classes are Aplacophora, Bivalvia, Cephalopoda, Gastropoda, Monoplacophora, Polyplacophora, and Scaphopoda. As shown in Figure 2.1, a general consensus view holds that the univalved and bivalved groups, collectively termed Conchifera, constitute a clade with Monoplacophora as the earliest diverging lineage (Brusca and Brusca 1990; Scheltema 1993; Haszprunar 1996; Salvini-Plawen and Steiner 1996; Haszprunar 2000; Nielsen 2001). Within Conchifera, Gastropoda and Cephalopoda are thought to be sister groups. Competing hypotheses suggest that Scaphopoda is either the sister group to Bivalvia (Scheltema 1996) or Gastropoda plus Cephalopoda (Lindberg and Ponder 1996; Haszprunar 2000). Additional controversy exists about whether the aplacophorans are monophyletic (Ivanov 1996; Scheltema 1996) or a paraphyletic grade representing the two earliest diverging lineages within Mollusca (Figure 2.1A). Polyplacophora is also thought to be a relatively early divergence within Mollusca, although it has alternatively been hypothesized as the sister group to Aplacophora or Conchifera (Figure 2.1B).

Not surprisingly, most of the molecular data brought to bear on these questions comes from 18S sequences. Nevertheless, taxon sampling for this gene is still limited for some groups, so strong conclusions are premature. Most studies to date have focused on the relationships within each class, or on a particular class and its suspected sister groups. Gastropoda and Bivalvia are well-

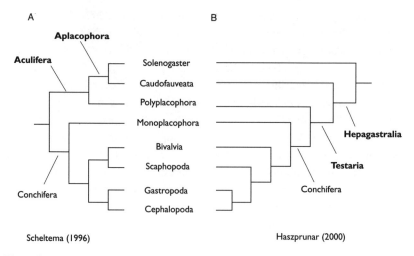

Figure 2.1. Two hypotheses of phylogenetic relationships among major groups within Mollusca. (A) Placement of the two aplacophoran subclasses in one clade with Polyplacophora as the sister taxon. This grouping is traditionally named Aculifera. (B) The gradist hypothesis renders the Aculifera paraphyletic, placing the two aplacophoran lineages at the base of the molluscan tree and Polyplacophora as the sister taxon to the crown lineages. The latter grouping is known as the Testaria.

represented in GenBank for 18S rDNA gene sequences, whereas other classes have fewer than a handful or no sequences. Investigators have evaluated the potential for phylogenetic reconstruction of several other genes traditionally used in molecular systematics. These studies have involved partial sequences of the 28S rRNA gene (Rosenberg et al. 1994, 1997) and several mitochondrial genes such as 16S and cytochrome oxidase I (Bonnaud et al. 1994, 1997; Harasewych et al. 1997; Carlini and Graves 1999; Thollesson 2000; Wollscheid-Lengeling et al. 2001; Giribet and Wheeler, 2002). These studies also suffer from poor representation of all major molluscan lineages. Complete 28S sequences have been shown to elucidate the relationships within at least one metazoan phylum (Medina et al. 2001a), and similar success can be hoped for in the case of Mollusca.

MITOCHONDRIAL GENE ORDER COMPARISONS AMONG MOLLUSCAN CLASSES

The chiton *Katharina tunicata* was one of the first mtDNA genomes sequenced (Boore and Brown 1994a, 1994b). This genome has more features in common with other metazoan genomes (Figure 2.2A, only lophotrochozoan genomes depicted). These observations suggest that *Katharina* has retained several features that were present in the ancestral bilaterian mitochondrial genome, a characteristic that has been used to polarize the evolution of molluscan rearrangements. A partial sequence (11 kb) from the aplacophoran *Epimenia verrucosa* supports the primitive status of this class based on the number of gene junctions shared with *Katharina* (Jeffrey Boore, Joint Genome Institute, personal communication). The only cephalopod in the database (*Loligo bleekeri*) shares some gene junctions with *Katharina,* but several unique rearrangements are present (Figure 2.2B). A second cephalopod (*Nautilus* sp.) has been sequenced and, although it is conservative in its gene order, it has some rearrangements relative to *Loligo* (Boore, personal communication). The scaphopod *Dentalium eboreum* is highly rearranged relative to the other molluscan genomes (Boore, personal communication). Gastropods have been the most widely sampled group to date, but the sampling has been biased toward the crown lineages (heterobranchs), and only a substantial partial sequence is available for a caenogastropod (*Littorina saxatilis;* Figure 2.2B). *Littorina* (Wilding et al. 1999) exhibits a conservative gene order, whereas the heterobranch gastropods share a widely rearranged gene order (Hatzoglou et al. 1995; Terret et al. 1996; Yalazaki et al. 1997; Kurabayashi and Ueshima 2000a, 2000b).

In the case of the bivalve genomes, not only are they highly rearranged, but they have also lost or kept additional gene copies, which is rare in metazoan mtDNA evolution. The ATPase 8 (atp8) gene is lost in *Mytilus* (Hoffman et al. 1992), which has previously been observed only in nematodes (Okimoto et al. 1991), whereas *Crassostrea* has a second copy of the small subunit ribosomal gene (Kim et al. 1999; Figure 2.2B). Bivalve mitochondrial genomes are also particularly interesting because some species exhibit a special mode of mtDNA inheritance in which both male and female mtDNAs are passed on to the offspring. Females usually have a unique mitochondrial genome, whereas males exhibit heteroplasmy. The male somatic tissue contains mainly female mtDNA, but the germline is characterized by a dominant male-type mtDNA (Sibinski et al. 1994; Zouros et al. 1994). The male and female mitochondrial genomes show up to 30% sequence divergence depending on the species, and they are distinct to the point that there are major rearrangements between the two sexes (Okazaki and Ueshima 2001). This sex-associated heteroplasmy has been termed *doubly uniparental inheritance* (DUI) (Zouros et al. 1994). Initially identified in marine mussels of the family Mytilidae (Sibinski et al. 1994; Zouros et al. 1994), it was subsequently detected in freshwater mussels of the family Unionidae (Hoeh et al. 2002; Okazaki and Ueshima 2001). Recently, DUI has been characterized from a venerid clam of genus *Tapes,* a phylogenetically divergent group of bivalves (Passamonti and Scali 2001). Although

A — Terebratulina retusa (Brachiopoda, Articulata) AJ245743

Lumbricus terrestris (Annelida, Oligochaeta) U24570

Katharina tunicata (Polyplacophora, Ischnochitonida) U09810

B — Katharina tunicata (Polyplacophora, Ischnochitonida) U09810

Mytilus edulis (Bivalvia, Lamellibranchia) M83756-62

Crassostrea gigas (Bivalvia, Pteriomorphia) AF177226

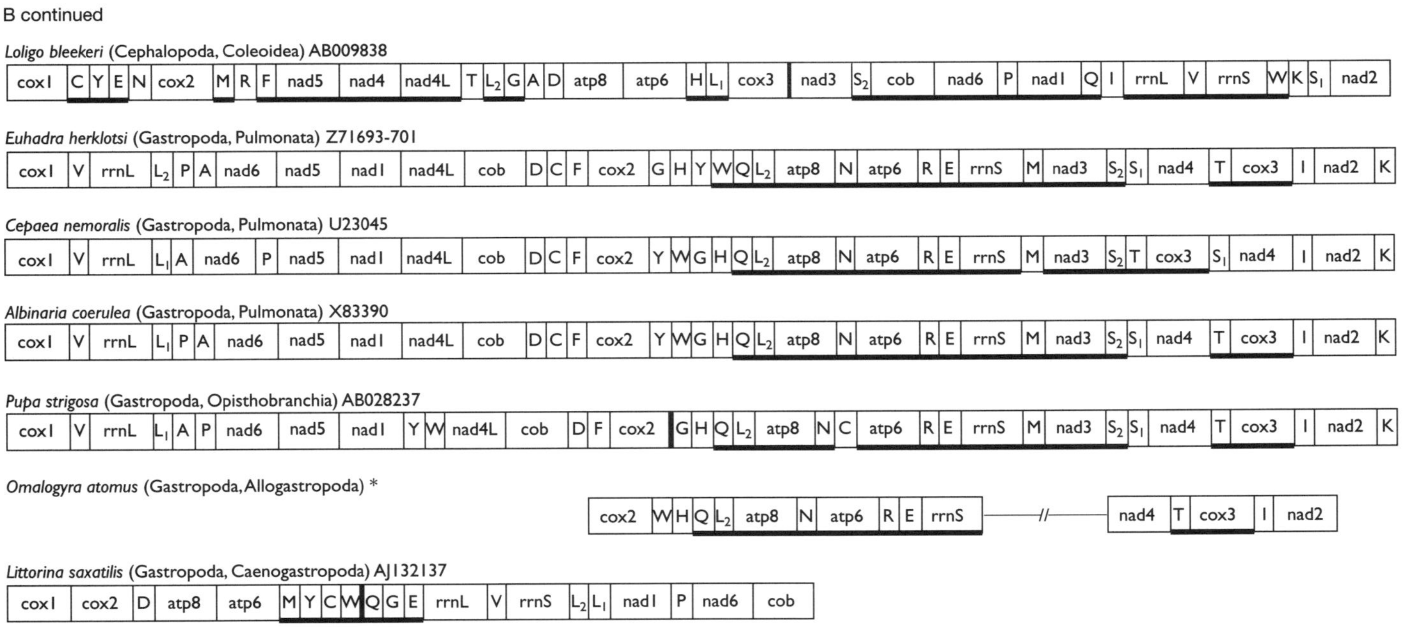

B continued

Figure 2.2. Mitochondrial gene arrangements of lophotrochozoan and molluscan taxa. Genomes are graphically linearized at cox1. All genes are transcribed left to right except those with a heavy underline, indicating the opposite orientation. NCBI accession numbers are next to each genome; * is from Kurabayashi and Ueshima (2000b). Gene designations: Cytochrome oxidase subunit I, II, III (cox1, cox2, cox3), cytochrome b (cob), NADH dehydrogenase subunits 1–6, 4L (nad1–6, 4L), ATP synthase subunits 6,8 (atp6, atp8), large ribosomal subunit (rrnL), small ribosomal subunit (rrnS), 18 transfer RNAs specifying a single amino acid (trnX), two transfer RNAs specifying leucine L₁ = L(CUN) and L₂ = L(UUR), two transfer RNAs specifying serine S₁ = S(AGN) and S₂ = S(UCN). (Modified from Boore 1999). (A) Comparison of three conserved lophotrochozoan mitochondrial genomes. (B) Complete and partial molluscan mitochondrial gene arrangements.

other venerid clams have been examined and no heteroplasmy is present (Marco Passamonti, University of Bologna, personal communication), this finding suggests that this type of inheritance may have been present in the ancestral bivalve that gave rise to the clade containing pteriomorph and heterodont bivalves (Passamonti and Scali 2001). Bivalve lineages that are considered to have diverged earlier than the pteriomorph/heterodont split, as well as other molluscan groups, should now be sampled for DUI. It remains to be evaluated whether the occasional presence of DUI in bivalve lineages may account for the high number of rearrangements observed in them. With the help of molecular approaches, we are beginning to understand this unique mode of inheritance, which previously could not be detected by other methods of observation. Molecules can therefore be considered ideal markers to understand not only phylogenetic relationships, but also biological phenomena at the organismal level.

With improvements in sequencing capabilities, more molluscan mtDNA genomes are being sequenced (Collins et al. 2001; Grande et al. 2001; Medina et al. 2001a). The availability of multiple mitochondrial genomes will benefit studies in molluscan molecular systematics in two ways. First, the gene-order data, being considered the product of rare genomic rearrangements, will help establish major groups with confidence by revealing conserved gene junctions that can be used as synapomorphies. Second, a more thorough traditional molecular phylogenetic analysis, between and within classes, will be possible. Analysis of concatenated protein and transfer RNA (tRNA) datasets has already been a fruitful undertaking in vertebrate groups (Macey et al. 1997; Miya et al. 2001), and this method will be applicable to investigating molluscan relationships.

rRNA SECONDARY STRUCTURE

Although most molecular systematists are aware of the additional information available in the secondary structure of ribosomal genes, most studies limit the use of this type of data to improving the alignment (Kjer 1995; Hickson et al. 1996). In many cases, no secondary structure model exists for a group of interest or its related taxa, and this lack affects the assessment of nucleotide homology. Additionally, nuclear ribosomal genes (28S and 18S) have been more widely studied and larger datasets are available for the secondary structure (Van de Peer et al. 2000; Wuyts et al. 2001) than for their mitochondrial counterparts (16S and 12S). There is a growing interest, however, in developing new methods to ease the comparison and assembly of secondary structure models of new taxa for phylogenetic reconstruction (Billoud et al. 2000; Hickson et al. 2000;

Page 2000; Parsch et al. 2000). Only the studies of Lydeard et al. (2000, 2002) have incorporated secondary structure into phylogenetic reconstruction of molluscan relationships. These studies include representatives from four classes (Polyplacophora, Bivalvia, Gastropoda, and Cephalopoda) and a consensus molluscan secondary structure was folded. The models developed for each taxon used are available online at http://www.rna.icmb.utexas.edu/. Lydeard et al. (2002) identified three structural characters as informative for phylogenetic reconstruction and coded them as binary characters for a cladistic analysis, which appears to unite the gastropod subclass Heterobranchia. As well as the large number of 28S and 18S sequences appearing in the literature, new 16S and 12S data will also become available as a result of the increasing number of complete mitochondrial genomes being sequenced. With the new advances in secondary structure reconstruction, the use of this type of data should become standard practice in phylogenetic reconstruction when analyzing ribosomal data. This is therefore another example of the potential of genomic signatures for molluscan phylogeny.

GENE FAMILY EVOLUTION IN MOLLUSKS

Although most molecular systematists have concentrated their efforts on both ribosomal genes and mitochondrial genes for their phylogenetic studies, other nuclear gene data are available that could be used to address phylogenetic questions. Particular cases are discussed below.

Hemocyanin

This gene encodes for a blue copper protein that acts as an oxygen carrier freely dissolved in the hemolymph. Hemocyanin has been studied in several molluscan taxa (reviewed in van Holde and Miller 1995). Although hemocyanins are also present in arthropods, there seems to be no evidence of common ancestry (Burmester 2001). Partial sequences are available for several species, but only recently has the gene been completely sequenced in a cephalopod (*Octopus dofleini;* Miller et al. 1998). One isoform has been sequenced for the gastropod *Haliotis tuberculata,* and a second isoform has been partially sequenced in this species (Lieb et al. 2000). In *Octopus,* hemocyanin consists of 10 subunits, which in turn contain seven oxygen binding functional units (FUs) (Miller et al. 1998). In *Haliotis,* this protein is a didecamer with eight FUs instead of seven (Lieb et al. 2000). It reveals several interesting features that are shared only among the two mollusks when compared with related proteins from out-group taxa: the placement of some introns—a linker intron between each pair of FUs, and introns separating the two exons of the signal peptide, the signal peptide from the first FU, and the two halves of the untranslated region of the 3' end (3' UTR) (Lieb et al. 2001: Fig. 1). Other introns show no correspondence between the two taxa. A phylogenetic analysis including most FUs from *Octopus* and *Haliotis,* in addition to the few FUs available from other mollusks, shows that they are homologous units and that they are probably the result of ancient multiple duplication events before the cephalopod–gastropod divergence (Lieb et al. 2000, 2001). The comparison of the gene structure from just these two species, combined with X-ray crystallography (Cuff et al. 1998) and three-dimensional electron microscopy (Meissner et al. 2000), has allowed for a fairly sophisticated analysis of the evolution of this protein in mollusks. This initial effort has suggested that hemocyanin is a useful phylogenetic marker for understanding phylogenetic relationships at the class level, and other taxa such as chitons and bivalves are now being sequenced (Lieb and Markl 2001; Markl et al. 2001). Studies of hemocyanin will give insight not only into the relationships among different classes, but also into gene family evolution in mollusks.

Arginine Kinase (AK)

This usually monomeric molecule is a member of the highly conserved protein family phosphagen (guanidino) kinases. It catalyzes the reversible transfer of phosphate from a phosphagen (arginine in this case) to ADP:

$$\text{phosphoarginine} + Mg^{+2}\, \text{ADP} \overset{AK}{\Leftrightarrow} \text{arginine} + MG^{+2}\, \text{ATP}$$

Invertebrates have at least six phosphagen kinases as opposed to one in vertebrates (creatine kinase); arginine kinase is the most widely distributed (Suzuki et al. 1997a, 1997b). A slightly different comparative approach has been undertaken to study the evolution of this protein in mollusks than for hemocyanins. In this case, a combination of direct protein sequencing and amplification from arginine kinase cDNAs was used to obtain the amino acid sequence for several mollusks: a chiton (*Liolophura japonica*) and a gastropod (*Battilus cornutus;* Suzuki et al. 1997a); a clam (*Pseudocardium sachalinensis;* Suzuki et al. 1998); three cephalopods (*Nautilus pompilius, Octopus vulgaris,* and *Sepioteuthis lessoniana;* Suzuki et al. 2000a); and two more gastropods (*Cellana grata* and *Aplysia kurodai;* Suzuki et al. 2000b). These studies identified a gene duplication and fusion so far found only in heterodont bivalves (*Pseudo-*

cardium, Solen, and *Corbicula*—the two latter are unpublished sequences discussed only in a manuscript; Suzuki et al. 2000a). Although this can be considered a synapomorphy (a genomic signature) for this group of bivalves, when either of the two domains is used for phylogenetic reconstruction, the topologies conflict with the traditional view by having bivalves branching earlier than chitons. Suzuki et al. (2000a) conclude that bivalves are the most basal molluscan lineage. Their topology, however, may be artifactual because of limited taxon sampling and faster rates in the bivalve lineage, which in turn can result in saturation and consequently incorrect reconstructions.

One important aspect of protein functional studies is identifying the substrate recognition region. In the case of molluscan arginine kinases, a four–amino-acid region was recognized by site-directed mutagenesis. Introduced mutations that altered the amino acid composition resulted in reduced substrate affinity (Suzuki et al. 2000a). This amino acid region is not conserved in the two-domain arginine kinases from clams, which suggests that bivalve arginine kinases may have a different substrate-binding system (Suzuki et al. 2000a).

A phylogenetic analysis of several phosphagen kinases suggests an early divergence of arginine kinases in metazoans. An early branching of cnidarian arginine kinases, followed by a protostome clade containing a well-supported arthropod lineage and a lophotrochozoan clade (mollusks and annelids), also are strongly supported (Suzuki et al. 1997a: Fig. 3). We can conclude that arginine kinase sequence data from three different molluscan classes and a few outgroups can have potential as a phylogenetic marker for molluscan systematics.

Shell Proteins

The molluscan shell is a mineralized structure composed of layers of calcium carbonate crystals and organic polymers. The crystals are of two types: a prismatic layer formed of calcite and a nacreous layer formed of aragonite. The organic matrices secreted by the mantle may play a crucial role in shell formation. Consequently, several shell matrix proteins have recently been sequenced from different mollusks, and protein secretion has been characterized to be exclusive to the mantle. In gastropods, two distinct shell proteins have been identified from the red abalone, *Haliotis rufescens* (Bowen and Tang 1996; Shen et al. 1997), and several from the oyster pearl, *Pinctada fucata* (Miyamoto et al. 1996; Miyashita et al. 2000; Sudo et al. 1997). Some of these different extracellular matrix proteins are likely present in many mollusks and could be used as traditional phylogenetic markers or examined for genomic signatures if the gene organization is determined.

MOLLUSCAN EVO-DEVO

Comprehensive reviews of molluscan comparative embryology from reproduction to organogenesis can be found in Verdonk et al. (1983). For an updated
summary, refer to van den Biggelaar et al. (1994), which describes the pros and
cons of mollusks as developmental models. In this section, we provide some background information, followed by a short overview of recent molecular findings.

The study of metazoan evolution traditionally has been based on comparative embryology, anatomy, and paleontology. The development of both phylogenetic theory and molecular systematics were the first fundamental steps to
revolutionize the field. However, the most recent and rapid advances have come
from molecular developmental biology. New molecular tools finally allow the
analysis of patterns of gene expression, but most importantly, these tools are
crucial in producing a better understanding of how different genes affect biological processes in the overall functioning of the cell and the organism. Differential gene expression, both during development and in the adult stage, is
what makes an organism unique in phenotype and physiology, and use of these
approaches in a comparative framework has produced a new synthesis in the
study of animal body plan evolution. This is now known as the rapidly growing field of evolutionary developmental biology or "evo-devo."

With the exception of cephalopods, molluscan development starts with spiral cleavage, followed by blastula and gastrula stages. Although some groups
exhibit direct development, most groups possess a lecithotrophic trochophore
larval stage (Nielsen 2001). Gastropods and bivalves usually have a feeding
veliger larval stage as well. Whereas larvae are mobile, adults are usually slow
moving. Although both stages use the shell and the operculum as a protective
shield, their feeding habits are modified according to their different living environments. Thus, the diverse life history strategies adopted by many species,
combined with extreme body plan modifications, make mollusks an ideal group
to investigate the plasticity of the underlying mechanisms of development. In
the particular case of mollusks, evo-devo studies are in their infancy, yet a growing interest is already reflected in the recent literature. The groundwork for modern developmental studies was laid in the late nineteenth and early twentieth
centuries when embryologists carefully described the early development of several spiralian taxa (Kofoid 1894; Wierzejski 1905; Smith 1935). Later work
involved embryological manipulations (i.e., deletion experiments) that helped
unravel some of the mechanisms underlying spiralian development (Guerrier
1970a, 1970b, 1970c). For instance, it was shown that the D quadrant (in particular, the macromere 3D and the mesentoblast or 4d), acts as an organizer of
dorsoventral patterning by carrying cytoplasmic determinants that signal the de-

velopmental fates of other blastomeres (Clement 1976; van den Biggelaar 1977; Damen 1994). Additionally, cell lineage studies demonstrated that most molluscan embryos develop in one of two ways: (1) unequal cleavage (this includes polar lobe–forming taxa) yields different-size quadrants from the first cell division, or (2) equal cleavage produces similar-size blastomeres in the first two cell divisions (Boyer and Henry 1998; Freeman and Lundelius 1992). In molluscan equal cleavers, cell fate is induced late by the D quadrant, whereas in unequal cleavers, blastomere specification occurs early (Freeman and Lundelius 1992). Because a reliable phylogeny for lophotrochozoans is not available, no strict character state reconstruction has actually been performed on developmental modes in spiralians; however, a loose attempt suggests that equal cleavage with late specification is the ancestral state (Freeman and Lundelius 1992). As cell division proceeds to later stages, cell lineages tend to differ between taxa (Freeman and Lundelius 1992). Character mapping of cell lineages on an accepted gastropod phylogeny indicates an evolutionary trend toward an early formation of the mesentoblast in more derived lineages (van den Biggelaar 1996; van den Biggelaar and Haszprunar 1996). Guralnick and Lindberg (2001) presented the first attempt at a rigorous phylogenetic reconstruction of molluscan relationships based on cell lineage data. To be able to use cell lineages in a modern phylogenetic analysis (maximum parsimony and distance-based), these authors developed a relative measure of cell formation that allowed them to code homologous characters. Because taxon sampling was limited and the dataset contained some homoplasy, the resulting topologies recovered only a few nodes. These data, however, suggest that gathering more cell lineage data for molluscan phylogenetics may be worthwhile.

Most of the work on molluscan development has been carried out with gastropod embryos, although development of taxa from other classes has occasionally been described. Despite their key taxonomic placement, aplacophorans have been neglected. The development of two aplacophoran taxa, however, has recently been described by using light and scanning electron microscopy (SEM). The development of the neomenid aplacophoran *Epimenia babai* occurs by unequal cleavage, and no indication of metamerism is observed (Okusu 2002). In the caudofaveate *Chaetoderma nitidulum*, early cleavage and early larval stages have not been studied, but it is now known that this species exhibits epibolic gastrulation. Eight transverse dorsal ridges of late larvae may indicate structural homology with metameric polyplacophorans (Claus Nielsen, University of Copenhagen, personal communication). Additional detailed descriptions of early development will likely be described for new taxa in the near future. Both advanced microscopy and fluorescent lineage tracing techniques will greatly improve these observations.

MOLECULAR DEVELOPMENTAL DATA

Homeobox Genes (*Hox*)

This gene family has been shown to play a key role in anterior–posterior patterning in bilaterian animals. These genes are usually organized in clusters and are expressed in colinear patterns. Although just a few taxa representative of key protostome lineages have been sampled, the observations are consistent with findings from phylogenetic studies using 18S (de Rosa et al. 1999). Many phyla, including mollusks, are still poorly sampled, and additional information is needed. However, putative homeobox domains have been identified in the heterodont bivalve *Donax texianus* (Adamkewicz et al. 2001), the abalone *Haliotis rufescens* (Degnan and Morse 1993; Degnan et al. 1995), and the limpet *Patella vulgata* (de Rosa et al. 1999). The expression of five *Hox* genes has been better characterized in *Haliotis asinina* to the trochophore through post-larval stages (Giusti et al. 2000; Degnan 2001). Expression of most of these genes is usually restricted to the neuroectoderm as in annelids, interpreted as playing a role in the development of the central nervous system in lophotrochozoans. Two *Hox* genes, however, are expressed in the mantle margin of the larvae, suggesting involvement in shell formation. This expression pattern has been hypothesized as co-option of these *Hox* genes for a phyletic specialization in mollusks (Degnan 2001). Expression studies in representatives from other molluscan classes will help test these potential new functions.

Otx and *Otp*

Orthodenticle/otx and *orthopedia/otp* are homeobox gene families present in all bilaterians. Orthologs to these two genes (*Pv-otx* and *Pv-otp*) were cloned from *Patella vulgata* (Nederbragt et al. 2002b). The authors used the expression of *Pv-otx* around the stomodeum to argue in favor of homology of the larval mouth in bilaterians. In addition, *Pv-otx* is involved in the formation of ciliary bands, which are also found in other bilaterian taxa. Thus, this gene seems to maintain an ancestral function in multiple lineages. Finally, *Pv-otp* expression was found to be associated with the development of the larval nervous system, and in particular the larval apical sensory organ.

Engrailed

Engrailed is a homeobox gene shown to be involved in segmentation and neurogenesis in arthropods and annelids (Patel 1994). Because some mollusks exhibit metamerism (i.e., polyplacophorans and monoplacophorans), this gene has been an early point of interest for evo-devo studies. Degenerate PCR in five different molluscan classes (Wray et al. 1995) initially identified homologues of the gene. Expression studies by immunocytochemistry and in-situ hybridization have been performed in representatives from four different classes: (1) the chiton *Lepidochitona caverna* (Jacobs et al. 2000); (2) the bivalve *Transennella tantilla* (Jacobs et al. 2000); (3) the scaphopod *Antalis entalis* (Wanninger and Haszprunar 2001); and (4) the gastropods *Ilyanassa obsoleta* (Moshel et al. 1998) and *Patella vulgata* (Nederbragt et al. 2002d). In all cases, *engrailed* expression was associated with the embryonic shell gland in later embryonic stages or in larvae. Expression in early embryonic stages was examined in *Ilyanassa,* whereas only later stages were examined in the other taxa. The results are therefore not totally comparable and new comparative studies using similar experimental designs will likely be useful in understanding the role of *engrailed* in molluscan shell formation and synthesis.

Dpp-BMP2/4

In vertebrates and insects, these genes play a role in the specification of the dorsoventral axis. In *Patella vulgata,* however, the ortholog of this gene appears to have a different function, being expressed in ectodermal cells surrounding the cells expressing *engrailed* (Nederbragt et al. 2002d). By comparing these genes with other *dpp/engrailed* interactions on other experimental organisms (i.e., *Drosophila*), the authors hypothesize an ancient involvement of *engrailed* and *dpp* in setting up compartment boundaries between embryonic domains.

Snail and *Twist*

These genes are involved in mesoderm formation in chordates and ecdysozoans. To assess the function of *snail* in a member of the lophotrochozoan clade, Lespinet et al. (2002) studied the spatiotemporal expression of two *snail* homologues (*Pv-Sna1* and *Pv-Sna2*) in the gastropod *Patella vulgata*. Although some small mesodermal expression is present in *Patella,* the two homologues are most often expressed in ectodermal derivatives. This expression pattern led the authors to propose a new ancestral function for *snail* in the bilaterian ancestor. In the case of *Twist,* this gene (*Pv-twist*) was shown to express in ectomesoderm of trochophore larvae in *Patella vulgata* (Nederbragt et al. 2002a), but not in endomesoderm. The authors suggest that other genes are likely involved in mesoderm formation, and therefore the homology of this germ layer in Bilateria still needs to be evaluated.

Esther32

Esther32 has been identified as a putative RNA-binding protein in *Patella vulgata* (Klerkx et al. 2001). Spatiotemporal expression patterns show that this gene is not expressed in cleavage-arrested or differentiated cells, which led the authors to suggest that *Esther32* may be involved in maintenance of undifferentiated cells where expressed.

Mitogen-Activated Protein Kinase

As mentioned previously, the D quadrant acts as an embryonic organizer for axial patterning in mollusks. The first attempt to uncover the molecular mechanisms behind D quadrant specification was performed in embryos from an unequal cleaver, the gastropod snail *Ilyanassa obsoleta* (Lambert and Nagy 2001). Activation of the mitogen-activated protein kinase *(MAPK)* signal transduction cascade was found to be a key player in cell fate specification of early development in *Ilyanassa*. By inhibiting activation of the *MAPK* cascade in a time series, Lambert and Nagy (2001) demonstrated that a progressive specification of *MAPK* correlates with a progressive activation of cell fates. In the equal cleaver *Patella vulgata,* using a similar experimental procedure, Lartillot et al. (2002) were able to show that the inhibition of *MAPK* signaling also leads to an equalized cell-division pattern in this limpet snail. Although alterations are more marked in *Patella,* the role of the *MAPK* cascade in the regulation of cell cleavage appears to be conserved in gastropods.

Brachyury

A homologue of this gene *(PvuBra)* has been identified in *Patella vulgata* (Lartillot et al. 2002). *PvuBra* is expressed in the macromere (3D) as soon as its fate is determined, consequently spreading to neighboring cells that give rise to the posterior edge of the blastopore during gastrulation. *PvuBra* expression is maintained at the posterior pole and along the developing anterior–posterior axis until the completion of gastrulation. This broad pattern of anterior–posterior axis development induced by the activity of a posterior growth zone is observed across Bilateria, which led the authors to suggest that *Brachyury* is a key component of a conserved developmental process already present in the bilaterian ancestor. Lartillot et al. successfully showed that *PvuBra* is a target of the *MAPK* signaling cascade in the macromere, and propose that this new gene–gene interaction may be conserved within gastropod mollusks, but more likely among more divergent lineages.

Hedgehog

Protostomes show a ventrally located nervous system, whereas deuterostomes have a dorsally located nervous system. There has been controversy about whether these two systems are developmental homologues. *Hedgehog* plays a key role in midline patterning in deuterostomes and has recently been shown to express in the ventral midline ectodermal cells of the one-day-old larva of *Patella vulgata* (Nederbragt et al. 2002c). These cells usually give rise to ciliated structures in mollusks, which in turn are thought to give rise to the nervous system. This observation led the authors to conclude that *hedgehog* had an ancient role in the early patterning of the nervous system in the ancestral bilaterian.

Morphogenesis

One of the fascinating questions about molluscan body plans is the evolution of gastropod torsion, and many hypotheses have been proposed to explain this extreme anatomical modification. Larval musculature plays a key role in early torsion, and it is important to determine if larval muscles give rise to adult muscle structures before models for the evolution of torsion can be improved. New higher-resolution techniques (i.e., SEM, transmission electron microscopy [TEM], spatiotemporal expression of tropomyosin, and phalloidin staining) have demonstrated that larval retractor muscles and the adult shell muscles are not, as previously thought, homologous structures in basal gastropods (*Patella* and *Haliotis;* Degnan et al. 1997a, 1997b; Page 1997; Wanninger et al. 1999a, 1999b).

Potential for Genomic Approaches

The availability of powerful molecular and genomic techniques has triggered an immense interest in comparative developmental biology. Most of these studies have included only model organisms developed initially as genetic systems (two ecdysozoans: the fruitfly and the nematode; and several vertebrates: the zebrafish, frog, mouse, and chick). Before the new animal phylogeny, *C. elegans* (a pseudocoelomate) was thought to represent a lineage basal to coelomate bilaterians. Strong support for the ecdysozoan clade, which includes both the nematode and the fruitfly, leaves us with no lophotrochozoan organism represented in the modern developmental revolution. This clade, however, includes key phyla that exhibit extreme modification of their body plan, and Mollusca is one of the clearest examples. With several genomes from model organisms either complete or well underway, there is now room for using these modern approaches to understand the body plan architecture of nonmodel organisms.

Mollusks, because of their diverse body plan architecture, are obvious candidates as targets of this new revolution.

Easy access to sequence data was clearly just a turning point in the whole process of understanding how organisms function and evolve. As new, non-model organisms become targets of the new technological advances, we predict that comparative thinking will have more effect on the understanding of metazoan evolution.

ACKNOWLEDGMENTS

We are grateful to Jeffrey Boore, Kevin Helfenbein, Marty Shankland, Marco Passamonti, Benoît Dayrat, and Bernhard Lieb for helpful comments on different aspects of molecular evolution and molluscan phylogeny. We also thank Olivier Lespinet, Alexander Nederbragt, Ronald Jenner, Jon Mallatt, Gonzalo Giribet, Claus Nielsen, Jeffrey Boore, Kevin Helfenbein, and Heike Wägele for sharing manuscripts in press and unpublished data. A. G. Collins is supported by the University of California Marine Council. Part of this work was performed under the auspices of the U.S. Department of Energy, Office of Biological and Environmental Research, by the University of California, Lawrence Berkeley National Laboratory under contract DE-AC03-76SF00098.

REFERENCES

Adamkewicz, S. L., M. K. Arterburn, and D. F. Bailey, D. F. 2001. *Hox* domains in a heterodont bivalves. Abstracts, World Congress of Malacology, Vienna, Austria, Unitas Malacologica.

Adoutte, A., Balavoine, G., Lartillot, N., Lespinet, O., Prud'homme, B. and de Rosa, B. 2000. The new animal phylogeny. Proceedings of the National Academy of Sciences U.S.A. 97:4453–4456.

Aguinaldo, A. M., J. M. Turbeville, L. S. Linford, M. Rivera, J. Garey, R. Raff, and J. A. Lake. 1997. Evidence for a clade of nematodes, arthropods and other molting animals. Nature 387:489–493.

Billoud, B., M. A. Guerrucci, M. Masselot, and J. S. Deutsch. 2000. Cirripede phylogeny using a novel approach: Molecular morphometrics. Molecular Biology and Evolution 17:1435–1445.

Bonnaud, L., R. Boucher-Rodoni, and M. Monnerot. 1994. Phylogeny of decapod cephalopods based on partial 16S rDNA nucleotide sequences. C.R. Academy of Sciences III 317:581–588.

Bonnaud, L., R. Boucher-Rodoni, and M. Monnerot. 1997. Phylogeny of cephalopods inferred from mitochondrial DNA sequences. Molecular Phylogenetics and Evolution 7:44–54.

Boore, J. L. 1999. Survey and summary. Animal mitochondrial genomes. Nucleic Acids Research 27:1767–1780.

Boore, J. L. 2000. The duplication/random loss model for gene rearrangement exemplified by mitochondrial genomes of deuterostome animals. Pp. 133–147 in Comparative Genomics (D. Sankoff and J. H. Nadeau, eds.). Kluwer Academic Publishers, Dordrecht, The Netherlands.

Boore, J. L., and W. M. Brown. 1994a. Complete DNA sequence of the mitochondrial genome of the black chiton, *Katharina tunicata*. Genetics 138:423–443.

Boore, J. L., and W. M. Brown. 1994b. Mitochondrial genomes and the phylogeny of mollusks. The Nautilus, supp. 2:61–78.

Boore, J. L., and W. M. Brown. 1998. Big trees from little genomes: Mitochondrial gene order as a phylogenetic tool. Current Opinion in Genetics and Development 8:668–674.

Boore, J. L., and J. L. Staton. 2002. The mitochondrial genome of the sipunculid *Phascolopsis gouldii* supports its association with Annelida rather than Mollusca. Molecular Biology and Evolution 19:127–137.

Bowen, C. E., and H. Tang. 1996. Conchiolin-protein in aragonite shells of mollusks. Comparative Biochemistry and Physiology Part A 115:269–275.

Boyer, B. C., and J. Q. Henry. 1998. Evolutionary modifications of the spiralian development program. American Zoologist 38:621–633.

Brusca, R. C., and G. J. Brusca. 1990. Invertebrates. Sinauer Associates, Sunderland, MA.

Burmester, T. 2001. Molecular evolution of the arthropod hemocyanin superfamily. Molecular Biology and Evolution 18:184–195.

Carlini, D. B., and J. E. Graves. 1999. Phylogenetic analysis of cytochrome oxidase I sequences to determine higher-level relationships within the coleoid cephalopods. Bulletin of Marine Science 64:57–76.

Clement, A. C. 1976. Cell determination and organogenesis in molluscan development: A reappraisal based on deletion experiments in *Ilyanassa*. American Zoologist 16:447–453.

Collins, T. C., T. A. Rawlings, and R. Bieler. 2001. Learning about mechanisms of mtDNA gene-order change from the caenogastropods: Major gene order change and gene order homoplasy. Abstracts, World Congress of Malacology, Vienna, Austria, Unitas Malacologica.

Cuff, M. E., K. I. Miller, K. E. van Holde, and W. A. Hendrickson. 1998. Crystal structure of a functional unit from *Octopus* hemocyanin. Journal of Molecular Biology 278:855–870.

Cuvier, G. 1795. Helminthologie. Magazin Encyclopédique, Journal des Sciences, des Lettres, et des Arts 2:433–448.

Cuvier, G. 1797. Tableau Élémentaire de L'histoire Naturelle des Animaux. Baudouin, Paris.

Damen, P. 1994. Cell lineage of the prototroch of *Patella vulgata* (Gastropoda, Mollusca). Developmental Biology 162:364–383.

De Rijk, P., E. Robbrecht, S. de Hoog, A. Caers, Y. Van de Peer, and R. De Wachter. 1999. Database on the structure of large subunit ribosomal RNA. Nucleic Acids Research 27:174–178.

de Rosa, B., J. K. Grenier, T. Andreeva, C. E. Cook, A. Adoutte, M. Akam, S. Carroll, and G. Balavoine. 1999. *Hox* genes in brachiopods and priapulids and protostome evolution. Nature 399:772–776.

Degnan, B. M. 2001. Conserved developmental gene regulatory systems expressed in *Haliotis asinina* larvae: Insights into spiralian evolution and development. Abstracts, World Congress of Malacology, Vienna, Austria, Unitas Malacologica.

Degnan, B. M., S. M. Degnan, G. Fentenany, and D. E. Morse. 1997a. A *Mox* homeobox gene in the gastropod mollusc *Haliotis rufescens* is differentially expressed during larval morphogenesis and metamorphosis. FEBS Letters 411:119–122.

Degnan, B. M., S. M. Degnan, A. Giusti, and D. E. Morse. 1995. A *hox/hom* homeobox gene in sponges. Gene 155:175–177.

Degnan, B. M., S. M. Degnan, and D. E. Morse. 1997b. Muscle-specific regulation of tropomyosin gene expression and myofibrillogenesis differs among muscle systems examined at metamorphosis of the gastropod *Haliotis rufescens*. Development, Genes and Evolution 206:464–471.

Degnan, B. M., and D. E. Morse. 1993. Identification of eight homeobox-containing transcripts expressed during larval development and at metamorphosis in the gastropod mollusc *Haliotis rufescens*. Molecular Marine Biology and Biotechnology 2:1–9.

Freeman, G., and J. W. Lundelius. 1992. Evolutionary implications of the mode of D quadrant specification in coelomates with spiral cleavage. Journal of Evolutionary Biology 5:205–247.

Ghiselin, M. 1988. The origin of molluscs in the light of molecular evidence. Pp. 66–95 in Oxford Surveys in Evolutionary Biology (H. P. Partridge and L. Partridge, eds.). Oxford University Press, Oxford.

Giribet, G., D. L. Distel, M. Polz, W. Sterrer, and W. Wheeler. 2000. Triploblastic relationships with emphasis on the acoelomates and the position of Gnathostomulida, Cycliophora, Plathelminthes, and Chaetognatha: A combined approach of 18S sequences and morphology. Systematic Biology 49:539–562.

Giribet, G., and W. Wheeler. 1999. The position of arthropods in the animal kingdom: Ecdysozoa, islands, trees, and the "parsimony rachet". Molecular Phylogenetics and Evolution 13:619–623.

Giribet, G., and W. Wheeler. 2002. On bivalve phylogeny: A high-level phylogeny of the mollusc class Bivalvia on a combined analysis of morphology and DNA sequence data. Invertebrate Biology 121:271–324.

Giusti, A., V. F. Hinman, S. M. Degnan, B. M. Degnan, and D. E. Morse. 2000. Expression of a *Scr/Hox5* gene in the central nervous system of the gastropod *Haliotis*, a non-segmented spiralian lophotrochozoan. Evolution and Development 2:294–302.

Grande, C., J. Templado, J. L. Cervera, and R. Zardoya. 2001. Complete sequence of the mitochondrial genome of the nudibranch *Roboastra europaea* (Mollusca, Opisthobranchia). Abstracts, World Congress of Malacology, Vienna, Austria, Unitas Malacologica.

Guerrier, P. 1970a. Les caractères de la segmentation et de la détermination de la polarité dorsoventrale dans le développement de quelques Spiralia. I. Les formes à clivage égal. Journal of Embryology and Experimental Morphology 23:611–637.

Guerrier, P. 1970b. Characteristics of segmentation and determination of dorsoventral polarity in the development of Spiralia. II. *Sabellaria alveolata* (Annelid polychaete). Journal of Embryology and Experimental Morphology 23:639–665.

Guerrier, P. 1970c. Les caractères de la segmentation et de la détermination de la polarité dorsoventrale dans le développement de quelques Spiralia. III. *Pholas dactylus* et *Spisula subtruncata*. Journal of Embryology and Experimental Morphology 23:667–692.

Guralnick, R. P., and D. R. Lindberg. 2001. Reconnecting cell and animal lineages: What do cell lineages tell us about the evolution and development of Spiralia? Evolution 55:1501–1519.

Haase, A., M. Stern, K. Wächtler, and G. Bicker. 2001. A tissue-specific marker of Ecdysozoa. Development, Genes and Evolution 211:428–433.

Halanych, K. M., J. D. Bacheller, A. M. A. Aguinaldo, S. M. Liva, D. M. Hillis, and J. A. Lake. 1995. Evidence from 18S ribosomal DNA that the Lophophorates are protostome animals. Science 267:1641–1643.

Harasewych, M. G., S. L. Adamkewicz, J. A. Blake, D. Saudek, T. Spriggs, and C. J. Bult. 1997. Phylogeny and relationships of pleurotomarid gastropods (Mollusca: Gastropoda): An assessment based on partial 18S rDNA and cytochrome x oxidase I sequences. Molecular Marine Biology and Biotechnology 6:1–20.

Haszprunar, G. 1996. The Mollusca: Coelomate turbellarians or mesenchymate annelids? Pp. 1–28 in Origin and Evolutionary Radiation of the Mollusca (J. Taylor, ed.). Oxford University Press, Oxford.

Haszprunar, G. 2000. Is the Aplacophora monophyletic? A cladistic point of view. American Malacological Bulletin 15:115–130.

Hatzoglou, E., G. C. Rodakis, and L. Lecanidou. 1995. Complete sequence and gene organization of the mitochondrial genome of the land snail *Albinaria coerulea*. Genetics 140:1353–1366.

Hendriks, L., E. Huysmans, A. Vandenberghe, and R. De Wachter. 1986. Primary structures of the 5S ribosomal RNAs of 11 arthropods and applicability of 5S to RNA to the study of metazoan evolution. Journal of Molecular Evolution 24:103–109.

Hickson, R. E., C. Simon, A. Cooper, G. S. Spicer, J. Sullivan, and D. Penny. 1996. Conserved sequence motifs, alignment, and secondary structure for the third domain of animal 12S rRNA. Molecular Biology and Evolution 13:150–169.

Hickson, R. E., C. Simon, and S. W. Perrey. 2000. The performance of several multiple-sequence programs in relation to secondary structure for an rRNA sequence. Molecular Biology and Evolution 17:530–539.

Hoeh, W. R., D. T. Stewart, and S. I. Guttman. 2002. High fidelity of mitochondrial genome transmission under the doubly uniparental mode of inheritance in freshwater mussels (Bivalvia: Unionoidea). Evolution: International Journal of Organic Evolution 56:2252–2261.

Hoeh, W. R., D. T. Stewart, C. Saavedra, B. W. Sutherland, and E. Zouros. 1997. Phylogenetic evidence for role-reversals of gender-associated mitochondrial DNA in *Mytilus* (Bivalvia: Mytilidae). Molecular Biology and Evolution 14:959–967.

Hoffman, R. J., J. L. Boore, and W. M. Brown. 1992. A novel mitochondrial genome organization for the blue mussel, *Mytilus edulis*. Genetics 131:397–412.

Ivanov, D. L. 1996. Origin of Aculifera and problems of monophyly of higher taxa in molluscs. Pp. 59–65 in Origin and Evolutionary Radiation of the Mollusca (J. D. Taylor, ed.). Oxford University Press, Oxford.

Jacobs, D. K., C. G. Wray, C. J. Wedeen, R. Kostriken, R. DeSalle, J. L. Staton, R. D. Gates, and Lindberg, D. R. 2000. Molluscan engrailed expression, serial organization, and shell evolution. Evolution and Development 2:340–347.

Kim, S. H., E. Y. Je, and D. W. Park. 1999. *Crassostrea gigas* mitochondrial DNA. GenBank AF177226.

Kim, C. B., S. Y. Moon, S. R. Gelder, and W. Kim. 1996. Phylogenetic relationships of annelids, molluscs, and arthropods evidenced from molecules and morphology. Journal of Molecular Evolution 43:207–215.

Kjer, K. M. 1995. Use of rRNA secondary structure in phylogenetic studies to identify homologous positions: An example of alignment and data presentation from the frogs. Molecular Phylogenetics and Evolution 4:314–330.

Klerkx, A. H. E. M., E. de Boer, and A. E. van Loon. 2001. Spatio-temporal expression of a gene encoding a putative RNA-binding protein during the early larval development of the mollusc *Patella vulgata*. Development, Genes and Evolution 211:423–427.

Kofoid, C. A. 1894. One some laws of cleavage in *Limax*. Proceedings of the American Academy of Arts and Sciences 29:180–203.

Kurabayashi, A., and R. Ueshima. 2000a. Complete sequence of the mitochondrial DNA of the primitive opisthobranch gastropod *Pupa strigosa:* Systematic implication of the genome organization. Molecular Biology and Evolution 17:266–277.

Kurabayashi, A., and R. Ueshima. 2000b. Partial mitochondrial genome organization of the heterostrophan gastropod *Omalogyra atomus* and its systematic significance. Venus 59:7–18.

Lambert, J. D., and L. M. Nagy. 2001. MAPK signaling by the D quadrant embryonic organizer of the mollusc *Ilyanassa obsoleta*. Development 128:45–56.

Lartillot, N., O. Lespinet, M. Vervoot, and A. Adoutte. 2002. Expression pattern of *Brachyury* in the mollusc *Patella vulgata* suggests a conserved role in the establishment of the AP axis in Bilateria. Development 129:1411–1421.

Lespinet, O., A. J. Nederbragt, M. Cassan, W. J. A. Dictus, A. E. van Loon, and A. Adoutte. 2002. Characterization of two snail genes in the gastropod mollusc *Patella vulgata*. Implications for understanding the ancestral function of the snail-related genes in Bilateria. Development, Genes and Evolution 212:186–195.

Lieb, B., B. Altenhein, and J. Markl. 2000. The sequence of a gastropod hemocyanin (HtH1) from *Haliotis tuberculata*. Journal of Biological Chemistry 275:5675–5681.

Lieb, B., B. Altenhein, J. Markl, A. Vincent, E. van Olden, K. E. van Holde, and K. I.

Miller. 2001. Structures of two molluscan hemocyanin genes: Significance for gene evolution. Proceedings of the National Academy of Sciences U.S.A. 98:4546–4551.

Lieb, B., and J. Markl. 2001. Hemocyanin: Clues for molluscan evolution. Abstracts, World Congress of Malacology, Vienna, Austria, Unitas Malacologica.

Lindberg, D. R., and W. F. Ponder. 1996. An evolutionary tree for the Mollusca: Branches or roots? Pp. 135–154 in Origin and Evolutionary Radiation of the Mollusca (J. Taylor, ed.). Oxford University Press, Oxford.

Littlewood, D. T. J., M. J. Telford, K. A. Clough, and K. Rohde. 1998. Gnathostomulida: An enigmatic metazoan phylum from both morphological and molecular perspectives. Molecular Phylogenetics and Evolution 9:72–79.

Lydeard, C., W. E. Hoznagel, M. N. Schnare, and R. R. Gutell. 2000. Phylogenetic analysis of molluscan mitochondrial LSU rDNA sequences and secondary structures. Molecular Phylogenetics and Evolution 15:83–102.

Lydeard, C., W. E. Holznagel, R. Ueshima, and A. Kurabayashi. 2002. Systematic implications of extreme loss or reduction of mitochondrial LSU rRNA helical-loop structures in gastropods. Malacologia 44:333–336.

Macey, R. J., A. Larson, N. B. Ananjeva, Z. Fang, and T. J. Papenfuss. 1997. Two novel gene orders and the role of light-strand replication in rearrangement of the vertebrate mitochondrial genome. Molecular Biology and Evolution 14:91–104.

Mallatt, J., and C. J. Winchell. 2002. Testing the new animal phylogeny: First use of combined large-subunit and small-subunit rRNA gene sequences to classify the protostomes. Molecular Biology and Evolution 19:289–301.

Manuel, M., M. Kruse, W. E. G. Müller, and Y. Le Parco. 2000. The comparison of b-thymosin homologues among Metazoa supports an arthropod-nematode clade. Journal of Molecular Evolution 51:378–381.

Markl, J., B. Lieb, U. Meissner, W. Gebauer, and J. R. Harris. 2001. Blue blood: Structure and evolution of gastropod hemocyanin. Abstracts, World Congress of Malacology, Vienna, Austria. Unitas Malacologica.

Medina, M., A. G. Collins, J. D. Silberman, and M. L. Sogin. 2001a. Evaluating hypotheses of basal animal phylogeny using complete sequences of large and small subunit rRNA. Proceedings of the National Academy of Sciences U.S.A. 98:9707–9712.

Medina, M., Y. Valles, T. Gosliner, and J. L. Boore. 2001b. Mitochondrial evolution of crown gastropods: Insight from large subunit sequences and gene order data. Abstracts, World Congress of Malacology, Vienna, Austria, Unitas Malacologica.

Meissner, U., P. Dube, J. R. Harris, H. Stark, and J. Markl. 2000. Structure of a molluscan hemocyanin didecamer (HtH1 from *Haliotis tuberculata*) at 12 A resolution by cryoelectron microscopy. Journal of Molecular Biology 298:21–34.

Miller, K. I., G. Dimopoulus, M. E. Cuff, W. F. Lang, K. Varga-Weiz, G. Field, and K. E. van Holde. 1998. Sequence of the *Octopus doflenini* hemocyanin subunit: Structural and evolutionary implications. Journal of Molecular Biology 278:827–842.

Miya, M., A. Kawaguchi, and M. Nishida. 2001. Mitogenomic exploration of higher teleostean phylogenies: A case study for moderate-scale evolutionary genomics with

38 newly determined complete mitochondrial DNA sequences. Molecular Biology and Evolution 18:1993–1009.

Miyamoto, H., T. Miyashita, T. Okushima, S. Nakano, T. Morita, and A. Matsushiro. 1996. A carbonic anhydrase from the nacreous layer in oyster pearls. Proceedings of the National Academy of Sciences U.S.A. 93:9657–9660.

Miyashita, T., R. Tagaki, S. Nakano, H. Miyamoto, E. Nishikawa, and A. Matsushiro. 2000. Complementary DNA cloning and characterization of Perlin, a new class of matrix protein in the nacreous layer of oyster pearls. Marine Biotechnology 2:409–418.

Moshel, S. M., M. Levine, and J. R. Collier. 1998. Shell differentiation and engrailed expression in the *Ilyanassa* embryo. Development, Genes and Evolution 208:135–141.

Nederbragt, A. J., O. Lespinet, S. van Wageningen, A. E. van Loon, A. Adoutte, and W. J. A. G. Dictus. 2002a. A lophotrochozoan *twist* gene is expressed in the ectomesoderm of the gastropod mollusc *Patella vulgata*. Evolution and Development 4:334–343.

Nederbragt, A. J., P. te Welscher, S. van den Driesche, A. E. van Loon, and W. J. A. G. Dictus. 2002b. Novel and conserved roles for *orthodenticle/otx* and *orthopedia/otp* orthologs in the gastropod mollusc *Patella vulgata*. Development, Genes and Evolution 212:330–337.

Nederbragt, A. J., A. E. van Loon, and W. J. A. G. Dictus. 2002c. Evolutionary biology: *Hedgehog* crosses the snail's midline. Nature 417(6891):811–812.

Nederbragt, A. J., A. E. van Loon, and W. J. A. G. Dictus. 2002d. Expression of *Patella vulgata* orthologs of engrailed and *dpp-BMP2/4* in adjacent domains during molluscan shell development suggests a conserved compartment boundary mechanism. Developmental Biology 246:341–355.

Nielsen, C. 2001. Animal Evolution. Oxford University Press, New York.

Nielsen, C., N. Scharff, and D. Eibye-Jacobsen. 1996. Cladistic analysis of the animal kingdom. Biological Journal of the Linnean Society 57:385–410.

Okazaki, M., and R. Ueshima. 2001. Evolutionary diversity between the gender-associate mitochondrial DNA genomes of freshwater mussels. GenBank ab055624–5.

Okimoto, R., H. M. Chamberlin, J. L. MacFarlane, and R. Wolstenholme. 1991. Repeated sequence sets in mitochondrial DNA molecules of root knot nematodes (Meloidogyne): Nucleotide sequences, genome location and potential for host race identification. Nucleic Acids Research 19:1619–1626.

Okusu, A. 2002. Embryogenesis and development of *Epimenia babai* (Molusca, Neomeniomorpha). Biological Bulletin 203:87–103.

Page, L. R. 1997. Larval shell muscles in the Abalone *Haliotis kamtschatkana*. Biological Bulletin 193:30–46.

Page, R. D. M. 2000. Comparative analysis of secondary structure of insect mitochondrial small subunit ribosomal RNA using maximum weighted matching. Nucleic Acids Research 28:3839–3845.

Parsch, J., J. M. Braverman, and W. Stephan. 2000. Comparative sequence analysis and patterns of covarion in RNA secondary structures. Genetics 154:909–921.

Passamonti, M., and V. Scali. 2001. Gender-associated mitochondrial DNA heteroplasmy in the venerid clam *Tapes philippinarum* (Mollusca Bivalvia). Current Genetics 39:117–124.

Patel, N. 1994. Developmental evolution: Insights from studies of insect segmentation. Science 266:581–590.

Peterson, K. J., and D. J. Eernisse. 2001. Animal phylogeny and the ancestry of bilaterians: Inferences from morphology and 18S rDNA gene sequences. Evolution and Development 3:1–35.

Rawlings, T. A., T. M. Collins, and R. Bieler. 2001. A major mitochondrial gene rearrangement among closely related species. Molecular Biology and Evolution 18:1604–1609.

Reindl, S., W. Salvenmorser, and G. Haszprunar. 1995. Fine structural and immunocytochemical investigations of the caeca of *Argyrotheca cordata* and *Argyrotheca cuneata* (Brachiopoda, Terebratulida, Terebratellacea). Journal of Submicroscopic Cytology and Pathology 27:543–556.

Reindl, S., W. Salvenmorser, and G. Haszprunar. 1997. Fine structural and immunocytochemical studies on the eyeless aesthetes of *Leptochiton algesirensis,* with comparison to *Leptochiton cancellatus* (Mollusca, Polyplacophora). Journal of Submicroscopic Cytology and Pathology 29:135–151.

Rokas, A., and P. W. H. Holland. 2000. Rare genomic changes as a tool for phylogenetics. Trends in Ecology and Evolution 14:454–459.

Rosenberg, G., G. S. Kuncio, G. M. Davis, and M. G. Harasewych. 1994. Preliminary ribosomal RNA phylogeny of Gastropod and Unionoidean Bivalve mollusks. The Nautilus, supp. 2:111–121.

Rosenberg, G., S. Tillier, A. Tillier, G. S. Kuncio, R. T. Hanlon, M. Masselot, and C. J. Williams. 1997. Ribosomal RNA phylogeny of selected major clades in the Mollusca. Journal of Molluscan Studies 63:301–309.

Ruiz-Trillo, I., M. Riutort, D. T. J. Littlewood, E. A. Herniou, and J. Baguña. 1999. Acoel flatworms: Earliest extant bilaterian metazoans, not members of Platyhelminthes. Science 283:1919–1923.

Salvini-Plawen, L. von. 1991. Origin, phylogeny, and classification of the phylum Mollusca. Iberus 9:1–33.

Salvini-Plawen, L. von, and G. Steiner. 1996. Synapomorphies and plesiomorphies in higher classification of Mollusca. Pp. 29–51 in Origin and Evolutionary Radiation of the Mollusca (J. Taylor, ed.). Oxford University Press, Oxford.

Scheltema, A. H. 1993. Aplacophora as progenetic aculiferans and the coelomatic origin of mollusks as the sister taxon of Sipuncula. Biological Bulletin 184:57–78.

Scheltema, A. H. 1996. Phylogenetic position of Sipuncula, Mollusca and the progenetic Aplacophora. Pp. 53–58 in Origin and Evolutionary Radiation of the Mollusca (J. Taylor, ed.). Oxford University Press, Oxford.

Schram, F. R. 1991. Cladistic analysis of metazoan phyla and the placement of fossil problematica. Pp. 35–46 in The Early Evolution of Metazoa and the Significance of Problematic Taxa (A. Simonetta and S. Conway Morris, eds.). Cambridge University Press, Cambridge.

Shen, X., A. M. Belcher, P. K. Hansma, G. D. Stucky, and D. E. Morse. 1997. Molecular cloning and characterization of Lustrin A, a matrix protein from shell and pearl nacre of *Haliotis rufescens*. Journal of Biological Chemistry 272:32472–32481.

Sibinski, D. O. F., C. Gallagher, and C. M. Beynon. 1994. Mitochondrial DNA inheritance. Nature 368:817–818.

Smith, F. G. W. 1935. The development of *Patella vulgata*. Philosophical Transactions of the Royal Society of London B 225:95–125.

Sudo, S., T. Fujikawa, T. Nagakura, T. Ohkubo, K. Sakaguchi, M. Tanaka, and K. Nakashima. 1997. Structures of mollusc shell framework proteins. Nature 387:563–564.

Suzuki, T., T. Ban, and T. Furukohri. 1997a. Evolution of phosphagen kinase V. cDNA-derived amino acid sequences of two molluscan arginine kinases from the chiton *Liophura japonica* and the turbanshell *Battilus cornutus*. Biochimica et Biophysica Acta 1340:1–6.

Suzuki, T., H. Fukuta, H. Nagato, and M. Umekawa. 2000a. Arginine kinases from *Nautilus pompilius*, a living fossil. Journal of Biological Chemistry 275:23884–23890.

Suzuki, T., N. Inoue, T. Higashi, R. Mizobuchi, N. Sugimura, K. Yokouchi, and T. Furukohri. 2000b. Gastropod arginine kinases from *Cellana grata* and *Aplysia kurodai*. Isolation and cDNA-derived amino acid sequences. Comparative Biochemistry and Physiology Part B 127:505–512.

Suzuki, T., Y. Kawasaki, T. Furukohri, and W. R. Ellington. 1997b. Evolution of phosphagen kinase. VI. Isolation, characterization and cDNA-derived amino acid sequence of lombricine kinase from the earthworm *Eisenia foetida*, and identification of a possible candidate for the guanidine substrate recognition site. Biochimica et Biophysica Acta 1348:152–159.

Suzuki, T., Y. Kawasaki, Y. Unemi, Y. Nishimura, T. Soga, M. Kamidochi, Y. Yazawa, and T. Furukohri. 1998. Gene duplication and fusion have occurred frequently in the evolution of phosphagen kinases: A two-domain arginine kinase from the clam *Pseudocardium sachalinensis*. Biochimica et Biophysica Acta 1388:253–259.

Taylor, J. 1996. Origin and Evolutionary Radiation of the Mollusca. Oxford University Press, Oxford.

Terret, J. A., S. Miles, and R. H. Thomas. 1996. Complete DNA sequence of the mitochondrial genome of *Cepaea nemoralis* (Gastropoda: Pulmonata). Journal of Molecular Evolution 42:160–168.

Thollesson, M. 2000. Increasing fidelity in parsimony analysis of dorid nudibranchs by differential weighting, or a tale of two genes. Molecular Phylogenetics and Evolution 16:161–172.

van den Biggelaar, J. A. M. 1977. Development of dorsoventral polarity and mesentoblast determination in *Patella vulgata*. Journal of Morphology 154:157–186.

van den Biggelaar, J. A. M. 1996. The significance of the early cleavage pattern for the reconstruction of gastropod phylogeny. Pp. 155–160 in Origins and Evolutionary Radiation of the Mollusca (J. Taylor, ed.). Oxford University Press, Oxford.

van den Biggelaar, J. A. M., J. A. G. Dictus, and F. Serras. 1994. Molluscs. Pp. 77–91 in Embryos: Color Atlas of Development (J. L. Bard, ed.). Wolfe Publishing, Singapore.

van den Biggelaar, J. A. M., and G. Haszprunar. 1996. Cleavage patterns and mesentoblast formation in the Gastropoda: An evolutionary perspective. Evolution 50:1520–1540.

Van de Peer, Y., P. De Rijk, J. Wuyts, T. Winkelmans, and R. de Wachter. 2000. The European small subunit ribosomal RNA database. Nucleic Acids Research 28:175–176.

van Holde, K. E., and K. I. Miller. 1995. Hemocyanins. Advances in Protein Chemistry 47:1–81.

Verdonk, N. H., J. A. M. van de Biggelaar, and A. S. Tompa. 1983. Development. Academic Press, New York.

Wanninger, A. B., and G. Haszprunar. 2001. The expression of an engrailed protein during embryonic shell formation of the tusk-shell, *Antalis entalis* (Mollusca, Scaphopoda). Evolution and Development 3:312–321.

Wanninger, A., B. Ruthensteiner, W. J. A. G. Dictus, and G. Haszprunar. 1999a. The development of the musculature in the limpet *Patella* with implications on its role in the process of ontogenetic torsion. Invertebrate Reproduction and Development 36:211–215.

Wanninger, A., B. Ruthensteiner, S. Lobenwein, W. Salvenmoser, W. J. A. G. Dictus, and G. Haszprunar. 1999b. Development of the musculature in the limpet *Patella* (Mollusca, Patellogastropoda). Development, Genes and Evolution 209:226–238.

Westenholme, D. R. 1992. Animal mitochondrial DNA: Structure and evolution. International Review of Cytology 141:173–216.

Wierzejski, A. 1905. Embryologie von *Physa fontinalis*. Zeitschrift für Wissenschaftliche Zoologie 83:502–706.

Wilding, C. S., P. J. Mill, and J. Grahame. 1999. Partial sequence of the mitochondrial genome of *Littorina saxatilis:* Relevance to gastropod phylogenetics. Journal of Molecular Evolution 48:348–359.

Winchell, C. J., J. Sullivan, C. B. Cameron, B. J. Swalla, and J. Mallat. 2002. Evaluating hypothesis of deuterostome phylogeny of chordate evolution with new LSU and SSU ribosomal DNA data. Molecular Biology and Evolution 19:762–776.

Winnepenninckx, B., T. Backeljau, and R. De Wachter. 1994. Small ribosomal subunit RNA and the phylogeny of Mollusca. The Nautilus, supp. 2:98–110.

Winnepenninckx, B., T. Backeljau, and R. De Wachter. 1995. Phylogeny of protostome worms derived from 18S rRNA sequences. Molecular Biology and Evolution 12:641–649.

Winnepenninckx, B., T. Backeljau, and R. De Wachter. 1996. Investigation of molluscan phylogeny on the basis of 18S rRNA sequences. Molecular Biology and Evolution 13:1306–1317.

Wollscheid-Lengeling, E., J. L. Boore, W. M. Brown, and H. Wägele. 2001. The phylogeny of Nudibranchia (Opisthobranchia, Gastropoda, Mollusca) reconstructed by three molecular markers. Organisms, Diversity and Evolution 1:241–256.

Wray, C. G., D. K. Jacobs, R. Krostriken, A. P. Vogler, R. Baker, and R. DeSalle. 1995. Homologues of the engrailed gene from five molluscan classes. FEBS Letters 365:71–74.

Wuyts, S., P. De Rijk, Y. Van de Peer, T. Winkelmans, and R. De Wachter. 2001. The European large subunit ribosomal RNA database. Nucleic Acids Research 29:175-177.

Yalazaki, N., R. Ueshima, J. Terret, S. I. Yokobori, M. Kaifu, R. Segawa, T. Kobayashi, K. I. Numachi, T. Ueda, E. Nishikawa, R. Watanabe, and R. Thomas. 1997. Evolution of pulmonate gastropod mitochondrial genomes: Comparisons of gene organizations of *Euhadra, Cepaea* and *Albinaria* and implications of unusual tRNA secondary structures. Genetics 145:749–758.

Zouros, E., A. D. Ball, C. Saavedra, and K. R. Freeman. 1994. An unusual type of mitochondrial DNA inheritance in the blue mussel *Mytilus*. Proceedings of the National Academy of Sciences U.S.A. 91:7463–7467.

Zrzavy, J., S. Mihulka, P. Kepka, A. Bezdek, and D. Tietz. 1998. Phylogeny of the Metazoa based on morphological and 18S ribosomal DNA evidence. Cladistics 14:249–285.

GONZALO GIRIBET AND DANIEL L. DISTEL

3

BIVALVE PHYLOGENY AND MOLECULAR DATA

Imagine a bizarre group of animals whose common ancestor lost its head and all its related organs, including the brain and buccal apparatus. Imagine too that certain members of this lineage later expanded and redesigned their respiratory organs to act as a functional filter-feeding mechanism to attract and direct food to the newly formed "mouth." Others completely transformed their already modified gills, creating special chambers that could be used as a sucking pump to capture prey with the aid of already muscular siphons. Still others loaded their colorful mantles with symbiotic zooxanthellae and exposed them to the sky to capture energy from the sun. These and many other strange events contribute to the remarkable evolutionary history of the class Bivalvia, the second-largest group of extant mollusks.

Bivalves include aquatic (primarily marine), typically bilaterally symmetrical mollusks characterized by a laterally compressed body with an external bivalved shell. The valves are hinged dorsally, connected by a partly calcified elastic ligament, and held together by adductor muscles (one or two) that attach to their inner surfaces. The valves are opened by the ligament and closed by contraction of the adductor muscles. As mentioned, anatomy diverges considerably from the basic mollusk plan in relation to the head. There is no buccal or radular apparatus and the mantle lobes are either joined or free ventrally. At least three lineages have developed posterior fusion of the mantle margin to form inhalant and exhalant siphons, which are hypertrophied in several lineages. The spacious mantle cavity is lateroventral, with left and right cavities extending upward on each side of the visceral mass. Most bivalves have a pair of ctenidia suspended laterally in the mantle cavity, which may be enlarged, lamellate, and plicate. The mouth and anus are located at opposite ends of the body and the gut is typically convoluted. A pair of ciliated labial palps connects the ctenidia to the mouth and directs food particles into it. The foot is extensible and either elongated or laterally compressed.

Ecologically, bivalves are mainly restricted to benthic environments, although their sedentary lifestyle shows several interesting variations, such as attachment by byssal threads or cementation to hard substrates, or boring and excavating lifestyles with some interesting adaptations for boring in coral, rock, or wood. Some pectinid and limid species have developed jet-propulsory systems to avoid prey, and a few galeommatoids have a crawling foot that functions much like that of gastropods.

The fact that bivalves present so many modifications from the putative plesiomorphic mollusk condition has made comparisons with other molluscan classes difficult. The result is a set of controversial hypotheses on bivalve sister-group relationships (see some of the most influential hypotheses in Figure 3.1). Problems arise in the difficulty in homologizing (and polarizing) certain structures useful for their taxonomy, but not present in the other molluscan classes. In addition, paleontologists and neontologists have disputed the monophyly and phylogenetic position of many groups, such as Anomalodesmata, Protobranchia, and Palaeoheterodonta. Those who previously used molecular sequence data have also openly questioned the monophyly of the Protobranchia or Heterodonta, as well as the monophyly of several bivalve orders including Nuculoida, Veneroida, and Myoida.

After several decades of detailed study, the bivalve evolutionary tree is becoming more stable, thanks in part to several symposia and meetings of the bivalve research community. The first bivalve symposium, "Evolutionary Systematics of Bivalve Molluscs," organized by Charles M. Yonge and Thomas E. Thompson, was held in 1978 (papers published in the Philosophical Transactions of the Royal Society of London B, vol. 284, pp. 199–436). This symposium was followed by a memorial symposium to honor Sir Charles M. Yonge held in Edinburgh in 1986 (Morton 1990). Molecular biologists joined students of bivalve evolution at the International Symposium on the Paleobiology and Evolution of the Bivalvia held in Drumheller in 1995 (Johnston and Haggart 1998). Four years later, in September 1999, bivalve workers met again, this time in Cambridge, United Kingdom, at "The Biology and Evolution of the Bivalvia" meeting organized on behalf of the Malacological Society of London (Harper et al. 2000b). In these volumes we can find many sources of morphological characters and, more recently, sequence data. Since the appearance of the first molecular study on bivalve systematics, molecular characters have gained importance in estimating phylogenetic events within the bivalves. In this chapter,

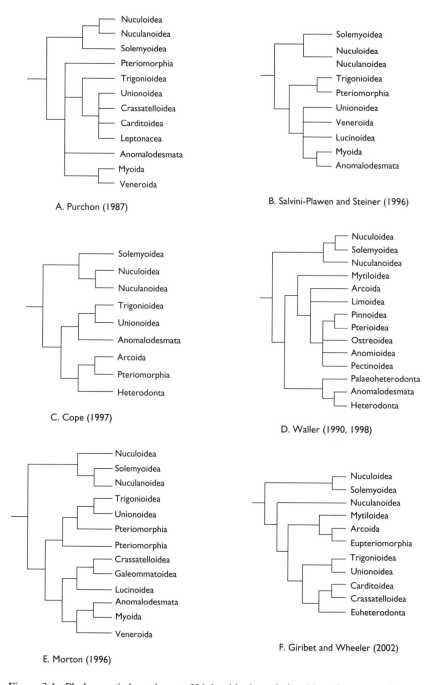

Figure 3.1. Phylogenetic hypotheses of higher bivalve relationships. "Leptonacea" from Purchon (1987) comprises Galeommatoidea and Cyamioidea.

we follow this line of inquiry, one that makes possible the "rapid" gathering of phylogenetically informative data.

Although innumerable phylogenetic hypotheses for bivalve phylogenetics have been (and continue to be) proposed, consensus seems to have been reached regarding several important questions, such as the monophyly of autobranchs, pteriomorphians, heterodonts (including Anomalodesmata) and the nonmonophyly of the classical orders "Veneroida" and "Myoida" (Adamkewicz et al. 1997; Campbell 2000; Steiner and Hammer 2000). Other issues, such as the paraphyly of "Protobranchia," the monophyly of Palaeoheterodonta, or the sister-group relationship of the Crassatelloidea + Carditoidea to the remaining heterodonts, have recently been suggested as well (Giribet and Wheeler 2002). Knowledge of the relationships among the protobranchiate families is still in its infancy. Nonmonophyly of the "Nuculoida" is, however, consistently obtained, with the Nuculoidea probably being closer to Solemyoidea than to Nuculanoidea (Waller 1998; Carter et al. 2000; Giribet and Wheeler 2002). Relationships of the monophyletic Pteriomorphia have recently received special attention (Carter 1990; Waller 1998; Steiner and Hammer 2000; Giribet and Wheeler 2002). Monophyly of Palaeoheterodonta has recently been settled based on sperm morphology and molecular and morphological data analyses (Healy 1989; Hoeh et al. 1998; Giribet and Wheeler 2002). However, the internal relationships of the heterodont families are still mostly unsettled, and the few modern studies evaluating heterodont relationships with explicit data matrices (Campbell 2000; Park and Ó Foighil 2000; Steiner and Hammer 2000; Giribet and Wheeler 2002) lack a large part of the diversity within the group. Nonetheless, some results are consistent among these studies. For example, when represented, Astartidae and Carditidae constitute the sister group to the remaining heterodonts, which also include a monophyletic anomalodesmatan clade and multiple lineages of myoid bivalves.

Morphological cladistic analyses of bivalve relationships are scarce (Salvini-Plawen and Steiner 1996; Waller 1998; Carter et al. 2000; Giribet and Wheeler 2002) and disagreement is still present among the four independent studies. Carter et al. (2000) are the only investigators to have compiled a data matrix of extant and extinct taxa analyzed in a parsimony framework. However, our lack of knowledge of bivalve fossils makes it difficult to incorporate those observations in this analysis, which we therefore restrict to living taxa. Giribet and Wheeler (2002) are the only researchers to combine morphological and molecular information in a total evidence framework, suggesting a pattern of bivalve phylogeny that agrees generally with the study of Waller (1998) or even with the classification systems of Newell (1969; see also Beesley et al. 1998). In this chapter, we follow a similar scheme, but expand the previous molecular matrix presented by Giribet and Wheeler (2002) to include a much larger familial and generic diversity and a more thorough molecular sampling that includes sequence data from the nuclear protein-encoding locus histone H3. Because of time constraints, we have not attempted the enormous task of developing a new morphological matrix for the bivalves. Giribet and Wheeler (2002) showed that molecular data are highly congruent with morphological characters. However, levels of resolution of both sets of characters used are slightly different: morphology resolved deeper nodes better than did the molecules used (nuclear ribosomal genes and mitochondrial cytochrome *c* oxidase subunit I, COI), whereas molecules in general resolved more recent splits better than morphology. This is not, in principle, a property inherent of either set of characters, but most certainly a bias in the investigators' choice of taxa, characters, and genes.

Given this background, the aim of this analysis is to attempt to settle some of the most conflicting current issues in bivalve evolution, such as the status of the protobranchiate bivalves and relationships among the heterodont families. With this aim, we have added new sequence data of 14 families not represented in the previous analysis of Giribet and Wheeler (2002): Malleidae, Isognomidae, Gryphaeidae, and Plicatulidae for the Pteriomorphia; Margaritiferidae for the Palaeoheterodonta; and Fimbriidae, Thyasiridae, Ungulinidae, Sportellidae, Donacidae, Glossidae, Petricolidae, Pholadidae, and Xylophagidae for the Heterodonta. In addition to members of the new families, we have added more genera and species for many bivalve families, totaling 133 bivalve taxa plus 9 outgroups. All species are represented by 18S rRNA sequence data, 77 species are represented by 28S rRNA sequence data, and 54 species are represented by COI sequence data. In addition, we have added a new molecular marker, histone H3, for 26 species representing all outgroup classes and all bivalve subclasses.

TAXON SAMPLING

This analysis includes representatives of the mollusk classes Polyplacophora, Gastropoda, Cephalopoda, and Scaphopoda but not of the Monoplacophora because we have not been able to obtain material suitable for DNA work. Most bivalve superfamilies (Beesley et al. 1998), with the exception of Dimyoidea, Muteloidea, Pholadomyoidea, Thracioidea, Clavagelloidea, Verticordioidea, and Poromyoidea, are represented (see Table 3.1). We sampled a total of 55 of approximately 94 families currently recognized (58% of familial diversity), with major gaps located within the protobranchiate bivalves, the Unionoida, and the anomalodesmatan families. We sampled 108 bivalve genera. Current taxonomic changes in the bivalve literature were taken into consideration to a certain degree. For example, the evidence for the derived status of the genera Tridacna and Hippopus within the Cardiidae (Schneider 1992, 1998; Schneider and

Table 3.1

Taxa sampled and genes sequenced for each taxon[a]

Taxon	GenBank Accession Numbers			
	18S rRNA	28S rRNA	COI	H3
Class Polyplacophora				
Lepidopleurus cajetanus	**AF120502**	**AF120565**	**AF120626**	**AY070142**
Acanthochitona crinita	**AF120503**	**AF120566**	**AF120627**	**AY070143**
Class Cephalopoda				
Nautilus pompilius	**AF207641**	**AF411688**	**AF120628**	AF033704
Loligo pealei	**AF120505**	**AF120568**	**AF120629**	
Sepia elegans	**AF120506-7**	**AF120569**		
Class Gastropoda				
Haliotis tuberculata	**AF120511**	**AF120570**		AY070145
Sinezona confusa	**AF120512**	**AF120571**	**AF120631**	
Diodora graeca	**AF120513**	**AF120572**	**AF120632**	
Class Scaphopoda				
Antalis pilsbryi	**AF120522**	**AF120579**	**AF120639**	
Rhabdus rectius	**AF120523**	**AF120580**	**AF120640**	**AY070144**
Class Bivalvia				
Superfamily Solemyoidea				
Family Solemyidae				
Solemya velum	**AF120524**	**AF120581**	U56852	**AY070146**
Solemya reidi	**AF117737**			
Superfamily Nuculoidea				
Family Nuculidae				
Nucula sulcata (**Med.**)	**AF120525**	**AF120582**		**AY070147**
Nucula sulcata (**NA**)	**AF207642**	**AF207649**	**AF207654**	
Nucula proxima	**AF120526**	**AF120583**	**AF120641**	
Acila castrensis	**AF120527**	**AF120584**		
Superfamily Nuculanoidea				
Family Yoldiidae				
Yoldia limatula	**AF120528**	**AF120585**	**AF120642**	**AY070149**
Yoldia myalis	**AF207643**	**AF207650**	**AF207655**	
Yoldiella nana	AJ389659			
Family Nuculanidae				
Nuculana minuta	**AF120529**	**AF120586**	**AF120643**	
Nuculana pernula	**AF207644**	**AF207651**		
Nuculana pella	**AY070111**	**AY070124**	**AY070138**	**AY070148**
Family Neilonellidae				
Neilonella subovata	**AF207645**	**AF207652**	**AF207656**	
Superfamily Mytiloidea				
Family Mytilidae				
Modiolus modiolus	**AF124210**			
Lithophaga lithophaga	**AF120530**	**AF120588**	**AF120644**	
Geukensia demissa	L33450		U56844	
Brachidontes domingensis	**AF117736**			
Mytilus edulis	L33448	**AF120587**	U68773	**AY070150**
Musculus discors	**AF124206**			
Brachidontes variabilis	AJ389643			
Septifer bilocularis	AF229622			

Table 3.1 continued

Taxon	GenBank Accession Numbers			
	18S rRNA	28S rRNA	COI	H3
Superfamily Arcoidea				
Family Arcidae				
Arca noae	X90960			
Barbatia barbata	**AF207646**	**AF120589**	**AF120645**	
Acar plicata	AJ389630			
Family Noetiidae				
Striarca lactea	**AF120531**	**AF120590**	**AF120646**	
Family Glycymerididae				
Glycymeris insubrica	**AF207647**	**AF120591**		
Glycymeris pectunculus	AJ389631			
Superfamily Pinnoidea				
Family Pinnidae				
Atrina fragilis	X90961	**AF120593**	**AF120648**	
Pinna muricata	AJ389636			
Superfamily Pterioidea				
Family Pteriidae				
Pteria hirundo	**AF120532**	**AF120592**	**AF120647**	
Pteria macroptera	AJ389637			
Pinctada margaritifera	AJ389638			
Electroma alacorvi	AJ389641			
Family Malleidae				
Malvifundus regulatus	AJ389640			
Vulsella sp.	AJ389642			
Family Isognomidae				
Isognomon legumen	AJ389639			
Isognomon isognomum	AF229621			
Superfamily Ostreoidea				
Family Ostreidae				
Ostrea edulis	L49052	AF137047	**AF120651**	**AY070151**
Lopha cristagalli	AJ389635	AF137038		
Crassostrea virginica	X60315	AF137050	AF152566	
Saccostrea cucullata	AJ389634	Z29553	AY038076	
Family Gryphaeidae				
Hyotissa hyotis	AJ389632	AF137036		
Parahyotissa numisma	AJ389633	AF137035		
Superfamily Limoidea				
Family Limidae				
Lima lima	**AF120533**	**AF120594**	**AF120649**	
Limaria hians	**AF120534**	**AF120595**	**AF120650**	**AY070152**
Ctenoides annulatus	AJ389653			
Superfamily Anomioidea				
Family Anomiidae				
Anomia ephippium	**AF120535**	**AF120598**		
Pododesmus caelata	AJ389650			
Superfamily Plicatuloidea				
Family Plicatulidae				
Plicatula plicata	AJ389651			
Plicatula australis	AF229626			

Continued on next page

Table 3.1 continued

Taxon	GenBank Accession Numbers			
	18S rRNA	28S rRNA	COI	H3
Superfamily Pectinoidea				
Family Pectinidae				
Pecten maximus	L49053		X92688	
Pecten jacobaeus	**AY070112**	**AY070125**		**AY070153**
Flexopecten glaber	AJ389662			
Argopecten gibbus	AF074389			
Argopecten irradians	L11265			
Adamussium colbecki	AJ242534			
Excellichlamys spectabilis	AJ389648			
Mimachlamys varia	L49051	**AF120597**		**AY070154**
Chlamys hastata	L49049			
Chlamys islandica	L11232			
Pedum spondyloideum	AJ389649			
Placopecten magellanicus	X53899	AF342798		
Crassadoma gigantea	L49050			
Family Spondylidae				
Spondylus sinensis	AF229629		AB033683	
Spondylus hystrix	AJ389647			
Spondylus crassisquamatus	AJ389646			
Superfamily Trigonioidea				
Family Trigoniidae				
Neotrigonia bednalli		**AF120538**		
Neotrigonia margaritacea	**AF411690**	**AF411689**	U56850	**AY070155**
Superfamily Unionoidea				
Family Margaritiferidae				
Margaritifera margaritifera	AF229612		U56847	
Family Unionidae				
Psilunio littoralis	**AF120536**	**AF120599**	**AF120652**	
Lampsilis cardium	**AF120537**	**AF120600**	**AF120653**	
Elliptio complanata	**AF117738**			
Superfamily Carditoidea				
Family Carditidae				
Cardita calyculata	**AF120549**	**AF120610**	**AF120660**	**AY070156**
Cardites antiquata	**AF120550**	**AF120611**	**AF120661**	
Carditamera floridana	AF229617			
Superfamily Crassatelloidea				
Family Astartidae				
Astarte castanea	**AF120551**	AF131001	**AF120662**	
Superfamily Lucinoidea				
Family Lucinidae				
Codakia orbiculata	**AF120546**	**AF120607**	**AF120657**	
Cardiolucina semperiana	AJ389655			
Ctena divergens	AJ389656			
Myrtea spinifera	**AY070115**		**AY070139**	**AY070157**
Family Ungulinidae				
Diplodonta cfr. *subrotundata*	AJ389654			
Family Fimbriidae				
Fimbria fimbriata	**AY070116**	**AY070128**		

Table 3.1 continued

Taxon	GenBank Accession Numbers			
	18S rRNA	28S rRNA	COI	H3
Family Thyasiridae				
Thyasira sarsi	**AY070117**	**AY070129**		
Superfamily Pandoroidea				
Family Pandoridae				
Pandora arenosa	**AF120539**	**AF120601**		
Pandora pinna	**AY070113-4**	**AY070127**		
Family Lyonsiidae				
Lyonsia floridana	**AF120540**	**AF120602**	**AF120654**	
Superfamily Cuspidarioidea				
Family Cuspidariidae				
Cuspidaria cuspidata	**AF120541-2**	**AF120603**	**AF120655**	
Myonera **sp.**	**AF120544**	**AF120605**		
Tropidomya abbreviata	AJ389657			
Superfamily Galeommatoidea				
Family Galeommatidae				
Galeomma turtoni	**AF120547**	**AF120608**		
Family Lasaeidae				
Lasaea **sp.**	**AF120548**	**AF120609**	**AF120659**	
Superfamily Cyamioidea				
Family Sportellidae				
Basterotia elliptica	AF229616			
Superfamily Solenoidea				
Family Pharidae				
Ensis ensis	**AF120555**	**AF120616**		**AY070159**
Ensiculus cultellus	AF229614			
Superfamily Tellinoidea				
Family Semelidae				
Abra **cfr.** *prismatica*	**AF120554**			**AY070160**
Family Donacidae				
Capsella variegata	**AY070118**	**AY070126**		**AY070158**
Donax trunculus	AJ309018			
Superfamily Chamoidea				
Family Chamidae				
Chama gryphoides	**AF120545**	**AF120606**	**AF120656**	
Superfamily Cardioidea				
Family Cardiidae				
Parvicardium exiguum	**AF120553**	**AF120614**	**AF120664**	
Fragum unedo	D84664			
Fragum fragum	D84663			
Fulvia mutica	D88911			
Vasticardium flavum	D88910			
Corculum cardissa	D88909			
Hippopus hippopus	D84660			
Tridacna gigas	D84189			
Tridacna maxima	D84659			
Superfamily Dreissenoidea				
Family Dreissenidae				
Dreissena polymorpha	**AF120552**	**AF120613**	**AF120663**	**AY070165**

Continued on next page

Table 3.1 continued

Taxon	GenBank Accession Numbers			
	18S rRNA	28S rRNA	COI	H3
Superfamily Mactroidea				
Family Mactridae				
Spisula subtruncata	L11271	**AF120615**	**AF207657**	M17876
Tresus nuttallii	L11269			
Mactromeris polynyma	L11230			
Mulinia lateralis	L11268	AF131003		
Superfamily Arcticoidea				
Family Arcticidae				
Arctica islandica	U93555			
Family Vesicomyidae				
Calyptogena magnifica	**AF120556**	**AF120617**	**AF120665**	
Superfamily Glossoidea				
Family Glossidae				
Glossus humanus	**AY070119**	**AF458072**		**AY070163**
Superfamily Corbiculoidea				
Family Corbiculidae				
Corbicula fluminea	**AF120557**	AF131009	**AF120666**	**AY070161**
Superfamily Sphaerioidea				
Family Sphaeriidae				
Sphaerium striatinum	**AF120558**	AF131013	**AF120667**	**AY070162**
Superfamily Veneroidea				
Family Veneridae				
Mercenaria mercenaria	**AF120559**	AF131019	**AF120668**	
Callista chione	AJ007613			
Venus verrucosa	AJ007613			
Family Petricolidae				
Petricola pholadiformis	**AY070120**	**AY070130**		
Superfamily Myoidea				
Family Myidae				
Mya arenaria	**AF120560**	**AF120621**	**AY070140**	**AY070164**
Family Corbulidae				
Varicorbula disparilis	**AF120561**	**AF120622**	**AF120669**	
Superfamily Gastrochaeonoidea				
Family Gastrochaenidae				
Gastrochaena dubia	**AF120562**	**AF120623**	**AF120670**	
Gastrochaena stimpsonii	AF229615			
Superfamily Hiatelloidea				
Family Hiatellidae				
Hiatella arctica	**AF120563**	**AF120624**	**AY070166**	
Hiatella **sp.**	**AY070121**			
Superfamily Pholadoidea				
Family Pholadidae				
Pholas dactylus	**AY070122**	**AY070131**	**AY070141**	
Xylophaga atlantica	**AY070123**	**AY070132**		
Family Teredinidae				
Bankia carinata	**AF120564**	**AF120625**		

[a] Taxa and GenBank accession numbers in bold font indicate species sequenced in the authors' laboratories.

Ó Foighil 1999) allow us to disregard the classical superfamily Tridacnoidea and family Tridacnidae used in many systematic and taxonomic lists. Evidence for independent origins of the families Sphaeriidae and Corbiculidae is over-whelming (Park and Ó Foighil 2000; Giribet and Wheeler 2002), rejecting the superfamily Corbiculoidea *sensu stricto* (in the strict sense) and thus assigning the Sphaeridae to its own superfamily, Sphaerioidea. Arcticoidea is also non-monophyletic (Giribet and Wheeler 2002), a taxonomic consideration already implemented elsewhere, in which the families Kelliellidae and Vesicomyidae were transferred to the superfamily Glossoidea (Coan et al. 2000). The rela-tionships of Arcticidae, Vesicomyidae, and Glossidae are evaluated in the ab-sence of sequence data for Kelliellidae and Trapeziidae.

Other taxonomic considerations for the future include the status of the Ar-coida. Traditionally, the Arcoida groups the superfamilies Arcoidea (with Arcidae, Cucullaeidae, and Noetiidae) and Limopsoidea (with Limopsidae, Glycymerididae, and Philobryidae) (Beesley et al. 1998), but alternative classi-fications have been proposed, with the Limopsoidea split into three superfam-ilies (Coan et al. 2000). Molecular data, however, suggests that the Glycymeri-didae and the Noetiidae nest within the Arcidae (Steiner and Hammer 2000; Giribet and Wheeler 2002). This suggests the invalidation of perhaps several families and superfamilies within the Arcoida. These issues cannot be addressed with the present taxon sampling because they do not include representatives of the families Cucullaeidae, Limopsidae, and Philobryidae. More representatives of the Glycymerididae and Noetiidae would also be needed.

DNA SEQUENCE DATA

DNA sequence data have been obtained for 80 of the 142 sampled terminals. Some of the sequences used here have been published in our previous studies on bivalve phylogeny and evolution (Distel 2000; Giribet and Wheeler 2002). Readers interested in detailed protocols for DNA isolation, amplification, and sequencing should refer to these publications. Other fundamental contributions from which sequences have been obtained are the studies of Campbell (2000) and Steiner and Hammer (2000).

Tissues were collected and preserved in different ways, but DNA was mainly extracted from fresh, frozen, or ethanol-preserved tissues (ethanol concentra-tions of 70 to 99%). Results for tissues preserved in 70% ethanol are not as sat-isfactory as at higher concentrations, but in general, this method of preserving collection material (without the use of formalin) works well for amplifying ri-bosomal genes.

DNA Isolation

Genomic DNA samples were obtained by digesting tissues in a solution of guanidinium thiocyanate homogenization buffer following a modified protocol for RNA extraction (Chirgwin et al. 1979). The tissues were homogenized in 400 μl of 4 M guanidinium thiocyanate and 0.1 M b-mercaptoethanol for 1 hour, followed by a standard protocol of phenol purification and 3 M sodium acetate precipitation.

DNA Amplification

The 18S rDNA loci were PCR-amplified in two or three overlapping fragments of about 950, 900, and 850 bp each, using primer pairs 1F-5R, 3F-18Sbi, and 5F-9R, respectively, although some taxa (e.g., Anomalodesmata) present large insertions in many of these positions (Giribet and Wheeler 2001). Primers used in amplification and sequencing are described elsewhere (Giribet et al. 1996; Whiting et al. 1997). The 28S rDNA D3 fragment was amplified and sequenced using primers 28Sa (5′–GAC CCG TCT TGA AAC ACG GA–3′) and 28Sb (5′–TCG GAA GGA ACC AGC TAC–3′) (Whiting et al. 1997). The larger 28S rRNA fragment was amplified using primers D1F (5′–GGG ACT ACC CCC TGA ATT TAA GCA T–3′) and D6R (5′–CCA GCT ATC CTG AGG GAA ACT TCG–3′) (Park and Ó Foighil 2000). The COI fragment was amplified and sequenced using primers LCO1490 (5′–GGT CAA CAA ATC ATA AAG ATA TTG G–3′) and HCO2198 (5′–TAA ACT TCA GGG TGA CCA AAA AAT CA–3′) (Folmer et al. 1994). Histone H3 was amplified and sequenced using primers H3a F (5′–ATG GCT CGT ACC AAG CAG AC(ACG) GC–3′) and H3a R (5′–ATA TCC TT(AG)GGC AT(AG) AT(AG) GTG AC–3′) (Colgan et al. 1998).

Amplification was carried out in a 50-μl volume reaction, with 1.25 units of AmpliTaq DNA Polymerase (Perkin Elmer), 200 μM of dNTPs and 1 μM of each primer. The PCR program consisted of an initial denaturing step at 94°C for 60 seconds, 35 amplification cycles (94°C for 15 seconds, 49°C for 15 seconds, 72°C for 15 seconds), and a final step at 72°C for 6 minutes in a GeneAmp PCR System 9700 (Perkin Elmer). The annealing temperature to amplify the protein-encoding genes was 46°C.

DNA Sequencing

PCR-amplified samples were purified with the GENECLEAN III kit (BIO 101 Inc.), and directly sequenced using an automated ABI Prism 377 DNA se-

quencer or an ABI Prism 3700 DNA Analyzer. Cycle-sequencing with Ampli-
Taq DNA Polymerase, FS (Perkin-Elmer) using dye-labeled terminators (ABI
PRISM BigDye Terminator Cycle Sequencing Ready Reaction Kit) was performed in a GeneAmp PCR System 9700 (Perkin Elmer). The sequencing reaction was carried out in a 10-µl volume reaction: 4 µl of Terminator Ready Reaction Mix, 10 to 30 ng/ml of PCR product, 5 pmoles of primer, and dH$_2$0 to 10
µl. The cycle-sequencing program consisted of an initial step at 94°C for 3 minutes, 25 sequencing cycles (94°C for 10 seconds, 50°C for 5 seconds, 60°C for
4 minutes), and a rapid thermal ramp to 4°C and hold. The BigDye-labeled PCR
products were cleaned with AGTC Gel Filtration Cartridges (Edge BioSystems).

DNA Editing

Chromatograms obtained from the automated sequencer were read and con-
tigs made using the sequence editing software Sequencher 4.0. Complete sequences were then edited in GDE (Genetic Data Environment; Smith et al.
1994), in which they were split according to primer-delimited regions and secondary structure features (Giribet and Wheeler 2001). The external primers 1F
and 9R (for the 18S rRNA loci), 28Sa and 28Sb (for the 28S fragment),
LCO1490 and HCO2198 (for the COI fragment), and H3aF and H3aR (for histone H3) were excluded from the analyses. All of the new sequences have been
deposited in GenBank (see accession codes in Table 3.1).
    The following molecular loci were used in this study.

18S rRNA   The complete sequence of the small nuclear ribosomal subunit has
proven useful in previous studies on bivalve phylogeny (Campbell et al. 1998;
Steiner 1999; Distel 2000; Giribet and Wheeler 2002) and was chosen as the
"skeleton" of the tree. This locus has been sampled for 142 terminals; the total
length (excluding primer pairs 1F and 9R) ranges from 1,760 to 2,485 bp. The
18S rRNA sequences were divided into 37 fragments according to primer regions and secondary structure features (Giribet 2001; Giribet and Wheeler
2001). Five of these regions showed large sequence length heterogeneity (fragments biv6, biv10, biv15, biv28, biv36) and were excluded from the analyses.

28S rRNA   The D1–D3 fragments of the large nuclear ribosomal subunit have
also been used in previous analyses of bivalve phylogeny (Park and Ó Foighil
2000; Giribet and Wheeler 2002). Here we used sequences of the D3 fragment
for 77 specimens, a total length (excluding primers 28Sa and 28Sb) of 290 to
600 bp. The fragment was divided into five pieces, one of which (biv28s7) was
excluded from the analyses for showing too much length variation. In addition,

19 autobranchs were represented by a larger fragment (ca. 1.1 kb), comprising the D1–D3 regions (Park and Ó Foighil 2000).

COI   A fragment of 660 to 675 bp of the mitochondrial protein coding gene cytochrome *c* oxidase subunit I (COI) was sequenced for 54 species. This fragment has been used in previous phylogenetic studies of bivalve phylogeny (Hoeh et al. 1998) and in combination with ribosomal genes (Giribet and Wheeler 2002). Because of its unequal length and the ambiguous amino acid–based alignments, the fragment was divided into eight regions and analyzed without constraints about insertion/deletion events. New analyses have shown that two of the COI sequences obtained by Giribet and Wheeler (2002), *Bankia carinata* (AF120671) and *Galeomma turtoni* (AF120658), are probably of bacterial origin, and were avoided in this study.

Histone H3   Histone H3 (a fragment of 327 bp) has previously been used in phylogenetic studies of arthropods (Colgan et al. 1998; Giribet et al. 2001) and annelids (Brown et al. 1999). Histone H3 is highly conserved at the amino acid level and is easily amplifiable. It was sequenced for 26 species and analyzed as a single prealigned fragment (command "–prealigned").

In total, we included 1.8 to 4 kb of sequence data per taxon, as well as information about all of the molecular markers that have been used to estimate higher-level relationships within the Bivalvia. The new sequences have been deposited in GenBank under accession codes AY070111–AY070132, AY070138–AY070166, and AF458072 (see Table 3.1).

PHYLOGENETIC ANALYSES

We analyzed the data in three independent sets (ribosomal genes, COI, and histone H3) and in combination using the direct optimization method (Wheeler 1996) as implemented in the computer program POY (Wheeler and Gladstein 2000). We chose this method because it accommodates unequal-length sequences in a way that other programs cannot, without requiring previously aligned matrices. The method therefore allows for exploring data in a sensitivity analysis framework (Wheeler 1995) without requiring primary homology statements ("alignments") obtained using different criteria than the ones used to reconstruct the phylogeny. This one-step process is applicable to any other method based on an optimality criterion, such as maximum likelihood, if a likelihood function has been incorporated into the existing software.

Unequal sequence length is common among ribosomal genes, especially when comparing sequences of taxa that diverged many million years ago. Protein-encoding genes are generally less prone to length variation because of the necessity of preserving a reading frame, as in the case of histone H3. The mitochondrial COI is generally conserved in length across animal phyla, including most mollusks; however, bivalves present a great deal of length variation, especially within some of the autobranch taxa.

Splitting sequences into regions according to primer sequences or secondary structure features has previously been proposed as a useful technique to postulate accurate primary homology statements (Giribet et al. 2000; Giribet and Wheeler 2001). Giribet (2001) empirically tested the effect of splitting the sequences into regions; results show that tree calculations are obtained faster and the tree length is comparable to that obtained when using the sequences as a single fragment, at least in cases where length variation is not substantial.

Tree searches were conducted in parallel using a cluster of 128 dual-processor nodes of 1 GHz per processor, using pvm language. Multiple Wagner trees were generated through random addition sequences and submitted to a combination of subtree pruning and regrafting (SPR) and tree bisection and reconnection (TBR) branch swapping (the default in POY). Although this may appear redundant in other software, the way it is programmed in POY greatly increases search efficiency by the use of SPR to narrow the searches before TBR. This implementation also allows the use of different "maxtrees" during each step, avoiding unnecessary swapping on multiple suboptimal trees if small numbers of "–sprmaxtrees" and "–tbrmaxtrees" are specified. TBR branch swapping was followed by a combination of treefusing and treedrifting (Goloboff 1999) to optimize tree searches. The commands "–slop" and "–checkslop" were used to reduce error derived from the heuristic operations (Wheeler 2001). Each search was performed for each partition independently, as well as for the combined analysis of all data.

Each analysis was conducted multiple times, for six combinations of parameters considering insertion/deletion events and transversion/transition ratios. This allowed us to test for stability of clades—those groupings that are not affected by the model/parameter choice.

Character congruence (incongruence length difference, ILD) among partitions was measured by the ILD metrics (Mickevich and Farris 1981; Farris et al. 1995; see Table 3.2). This value is calculated by dividing the difference between the overall tree length and the sum of the lengths of the individual datasets by the combined tree length:

ILD = (LengthCombined – Sum LengthIndividualSets) ÷ LengthCombined.

Character congruence is used as an optimality criterion to choose the "best" tree; the tree that minimizes character conflict among all of the data. This is un-

Table 3.2

Tree lengths for individual and combined datasets at different parameter set values, and character congruence (ILD) for the combined analyses of all data

Parameter Set	Tree Length				
	RIB	H3	COI	MOL	ILD
111	6313	620	7067	14528	0.0363
121	9473	901	11094	22189	0.0325
141	15536	1439	18733	36933	0.0332
211	*7235*	*620*	*7238*	*15397*	*0.0197*
221	11191	901	11363	23956	0.0209
241	18929	1439	19290	40581	0.0227

Note: Values for the parameter set that minimizes incongruence are italicized.

RIB = ribosomal, 18S rDNA + 28S rDNA; H3 = histone H3; COI = cytochrome c oxydase I; MOL = molecular (ribosomal + 28S + COI + H3).

derstood as an extension of parsimony (or any other minimizing criteria); in the same sense that parsimony tries to minimize the number of overall steps in a tree, the "character congruence analysis" attempts to find the "model" that maximizes overall congruence for all of the data sources. Obviously, trying to generalize an evolutionary model (viewed as an inferential model we use to make sense of observations) for all taxa, all loci, and all regions may be too simplistic, especially when evaluating divergences ranging from the Cambrian to the Miocene. However, evaluating for two general parameters ("gap/change ratio" and "transversion/transition ratio") will be a starting point in evaluating many other parameters when faster computers become available for exploring hypotheses in phylogenetic analysis.

All of the input files, including specific step matrices used in the analyses, parameter files, command files (.bat), report files (.err), and output files containing trees (.out), are available online at http://www.mcz.harvard.edu/Departments/InvertZoo/giribet_data.htm.

PHYLOGENETIC HYPOTHESES

Tree lengths and congruence analyses for all of the partitions and combined analyses are presented in Table 3.2, where each partition (ribosomal, COI, H3, and molecular [ribosomal + COI + H3]) shows the minimum tree length found. The ILD analysis shows that the parameter set that minimizes overall character incongruence is the one in which gaps receive twice the weight of any other

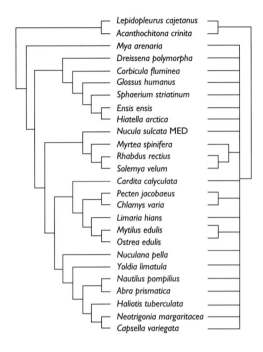

Figure 3.2. Phylogenetic relationships of bivalve and outgroup taxa proposed on the basis of histone H3 sequence data. The tree on the left represents the single shortest tree at 620 weighted steps for the parameter set that minimizes overall incongruence (parameter set 211). The tree on the right represents the strict consensus of all trees obtained for all examined parameter sets, and shows nodes that are stable to parameter variation.

transformations, and where transversions are equally weighted to transitions (parameter set 211—see the Web page mentioned above for details).

Results are presented for each partition, showing the tree (or consensus tree) obtained for the parameter set that minimizes incongruence (211), as well as the strict consensus of all of the trees obtained under all analyzed parameter sets. The first hypothesis represents our favored tree of the trees obtained. The second hypothesis is a more conservative hypothesis aiming to show agreement and conflict for the different parameter sets analyzed. Results from the independent partitions are not to be considered better explanations than the more inclusive combined analysis of all partitions, which is our final hypothesis. Representing the partitioned analyses is intended to show only at what levels those partitions may be influencing the bivalve phylogenetic estimation exercise. The combined tree is also represented by the "preferred" tree, the one obtained under parameter scheme 211, and by the most conservative hypothesis, the one obtained by the consensus of the results obtained under all of the parameter sets explored.

Independent analyses of histone H3 (Figure 3.2) and COI (Figure 3.3) generate topologies that combine generally low-level relationships (families and superfamilies), which agree with morphological analyses. On the other hand, these genes suggest other groupings that involve more distantly related taxa

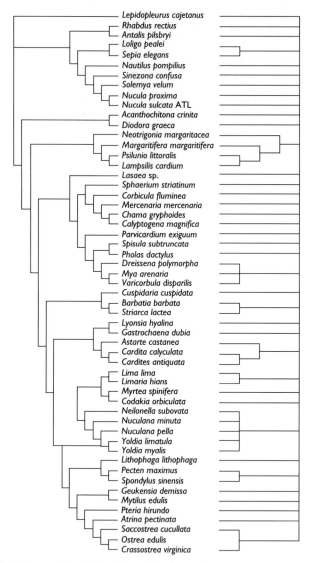

Figure 3.3. Phylogenetic relationships of bivalve and outgroup taxa proposed on the basis of cytochrome *c* oxidase subunit I sequence data. The tree on the left represents the strict consensus of the four shortest trees at 7,238 weighted steps for the parameter set that minimizes overall incongruence (parameter set 211). The tree on the right represents the strict consensus of all trees obtained for all examined parameter sets, and shows nodes that are stable to parameter variation.

(from superfamilies, orders, subclasses, or even classes) that are basically at odds with morphology. This is typical of genes that resolve recent cladogenetic events, but not distant ones. For example, among the relationships resolved by all of the parameter sets for histone H3, a relationship between *Pecten* and *Chlamys* is obtained, but a relationship between *Myrtea, Rhabdus,* and *Solemya* is also obtained (Figure 3.2). The first group, two pectinids, would be expected, but the second group mentioned involves a heterodont bivalve, a scaphopod, and a solemyoid bivalve. The H3 tree for parameter set 211 (Figure 3.2) obtains monophyly of certain higher groupings, such as Pteriomorphia, as well as certain heterodont clades obtained in the combined analysis of all genes.

COI (Figure 3.3) data fails in obtaining monophyly for Polyplacophora, Gastropoda, or Cephalopoda under most parameter sets. Among the groupings found throughout the parameter space explored, we found the Coleoidea, Nuculanoidea, Arcidae, Pectinoidea, Ostreidae, Limidae, Palaeoheterodonta, Unionoidea, Unionidae, Astartidae + Carditidae, Carditidae, and a clade composed of Dreissenidae + Myoidea. Some of these results contrast with the trees obtained from the combined analysis of the ribosomal genes, which resolve the classes Polyplacophora, Gastropoda, and Scaphopoda, and in general resolve most families and superfamilies of bivalves. The COI tree for parameter set 211 contains structure, and the resolution agrees with morphology at low taxonomic units, but not in the deeper nodes. These results are to be expected from a mitochondrial gene that evolves rapidly.

The analysis of up to 2.8 kb of ribosomal genes agrees well with hypotheses based on morphology, but again the deepest nodes conflict with morphology or are unstable to parameter set variation (Figure 3.4). All parameters agree on the monophyly of the following groups: Solemyidae, Nuculidae, Nuculanoidea, Mytilidae, Glycymeridae, (Arcoidea + Glycymeridae), Pinnidae, Limidae, Anomiidae, Plicatulidae, (Anomiidae + Plicatulidae), Ostreidae, Ostreoidea, Isognomidae, Pterioidea, (Ostreoidea + Pterioidea), Spondylidae, Pectinidae, Pectinoidea, Trigoniidae, Unionidae, (Astartidae + Carditidae), Gastrochaenidae, Pharidae, Hiatellidae, Donacidae, Tellinoidea, Galeommatoidea, (Sportellidae + Galeommatoidea), Veneridae, Mactridae, (Fimbriidae + Lucinidae), (Dreissena + Pholadoidea + Myoidea), Cardiidae, and (Chamidae + Cardiidae). More importantly, none of the groupings supported by ribosomal data for all parameter sets conflicts with current taxonomy, with the exception of a few families or genera that do not appear resolved or that appear paraphyletic.

The ribosomal tree for parameter set 211 (Figure 3.4A) fails to recover monophyly of Bivalvia because some outgroups are nested within the bivalves, forming a clade with the Solemyidae, Anomalodesmata, and Astartidae + Carditidae. Palaeoheterodonta is a monophyletic group that appears as the sister

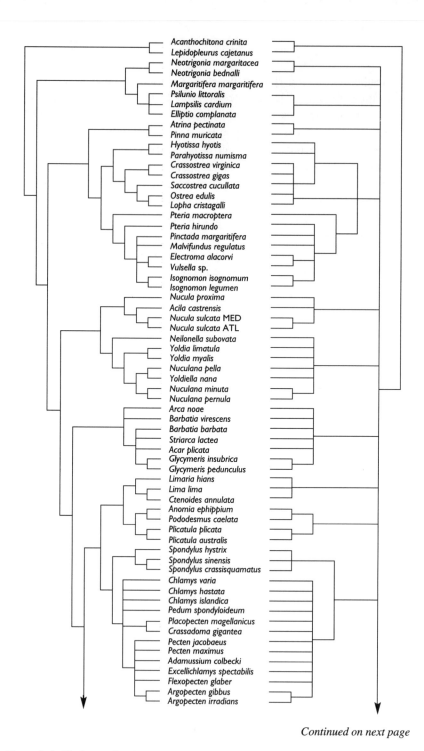

Acanthochitona crinita
Lepidopleurus cajetanus
Neotrigonia margaritacea
Neotrigonia bednalli
Margaritifera margaritifera
Psilunio littoralis
Lampsilis cardium
Elliptio complanata
Atrina pectinata
Pinna muricata
Hyotissa hyotis
Parahyotissa numisma
Crassostrea virginica
Crassostrea gigas
Saccostrea cucullata
Ostrea edulis
Lopha cristagalli
Pteria macroptera
Pteria hirundo
Pinctada margaritifera
Malvifundus regulatus
Electroma alacorvi
Vulsella sp.
Isognomon isognomum
Isognomon legumen
Nucula proxima
Acila castrensis
Nucula sulcata MED
Nucula sulcata ATL
Neilonella subovata
Yoldia limatula
Yoldia myalis
Nuculana pella
Yoldiella nana
Nuculana minuta
Nuculana pernula
Arca noae
Barbatia virescens
Barbatia barbata
Striarca lactea
Acar plicata
Glycymeris insubrica
Glycymeris pedunculus
Limaria hians
Lima lima
Ctenoides annulata
Anomia ephippium
Pododesmus caelata
Plicatula plicata
Plicatula australis
Spondylus hystrix
Spondylus sinensis
Spondylus crassisquamatus
Chlamys varia
Chlamys hastata
Chlamys islandica
Pedum spondyloideum
Placopecten magellanicus
Crassadoma gigantea
Pecten jacobaeus
Pecten maximus
Adamussium colbecki
Excellichlamys spectabilis
Flexopecten glaber
Argopecten gibbus
Argopecten irradians

Continued on next page

Figure 3.4. Phylogenetic relationships of bivalve and outgroup taxa proposed on the basis of 18S rRNA and 28S rRNA sequence data. The tree on the left represents the strict consensus of 22 cladograms at 7,235 weighted steps for the parameter set that

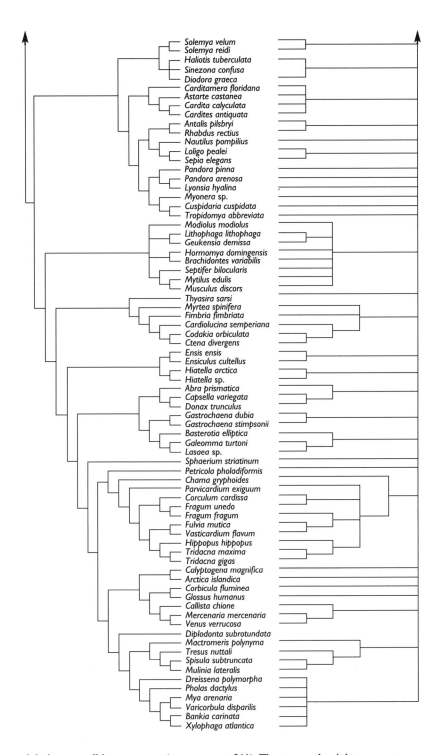

minimizes overall incongruence (parameter set 211). The tree on the right represents
the strict consensus of all trees obtained for all examined parameter sets, and shows
nodes that are stable to parameter variation.

group to the remaining conchiferans, followed by a clade containing Nuculoidea + Nuculanoidea, and then a paraphyletic assemblage of pteriomorphian super-families, as well as the remaining bivalves. Heterodonta are monophyletic, with the exception of Astartidae, Carditidae, and Anomalodesmata, which appear within a clade that contains the conchiferan outgroups, as mentioned above. However, resolution of superfamilies and families is highly congruent with mor-phology, and these results are stable to parameter set variation, as shown in the strict consensus of all parameter sets.

None of the partitions explored—histone H3, COI, or the ribosomal genes—adds a stable signal to higher-level (deeper) relationships of bivalves. Interest-ingly, the combined analysis of all partitions adds structure to the higher-level relationships of bivalves by recognizing monophyly of the Palaeoheterodonta and of certain higher Heterodonta (Figure 3.5). The combined analysis of all markers still fails to recover monophyly of Bivalvia, because the Nuculidae and the Solemyidae form a clade with the conchiferan outgroups. Polyplacophora, Gastropoda, Scaphopoda, and Cephalopoda are each monophyletic. Scapho-poda appear to be related to Cephalopoda, both being sister groups to Gas-tropoda. The nonnuculid, nonsolemyid bivalves are monophyletic, showing the following structure: (Palaeoheterodonta (Pteriomorphia (Nuculanoidea–Heterodonta))), a somewhat unorthodox topology for not recognizing the mono-phyly of Autolamellibranchiata. This topology resembles the topologies of both the ribosomal and the COI partitions, which place some of the protobranchiate bivalves with the conchiferan outgroups, and place the Palaeoheterodonta branching earlier than the split between Heterodonta and Pteriomorphia. How-ever, these results are not obtained under all parameter sets and do not withstand combination with morphological data (Giribet and Wheeler 2002).

Internal phylogeny of Palaeoheterodonta is well resolved, recognizing the Trigoniidae, Unionoidea, and Unionidae for all analytical conditions. Mono-phyly of Heterodonta is, however, not found in all analyses. The preferred tree shows a clade of Astartidae–Carditidae as sister group to Nuculanoidea, being the clade sister group to Euheterodonta. (Thyasiridae (Fimbriidae + Lucinidae)) are sister to the remaining euheterodonts, including Anomalodesmata (see Fig-ures 3.5 and 3.6). Some euheterodont groups found under all analytical condi-tions are (*Ctena* + *Codakia*), Galeommatoidea, Sportellidae + Galeommatoidea, Gastrochaenidae, Cuspidariidae, Pharidae, Hiatellidae, Donacidae, Tellinoidea, and an unnamed clade containing the families Sphaeriidae, Chamidae, Cardi-idae, Petricolidae, Glossidae, Veneridae, Corbiculidae, Vesicomyidae, Arctici-dae, Ungulinidae, Mactridae, Pholadidae, Corbulidae, Myidae, Dreissenidae, Teredinidae, and Xylophagidae. Stable structure within this clade is mainly found for Myidae + Corbulidae, Mactridae, and Chamidae + Cardiidae.

Other relationships found under more restricted conditions, including parameter set 211, are Anomalodesmata, Arcticoidea, and Pholadoidea (Figures 3.5 and 3.6). Neither Veneroida nor Myoida are monophyletic groups under any parameter set explored; Myoidea and Pholadoidea often appear to be related to Dreissenidae and Mactridae, but not to Gastrochaenidae or Hiatellidae.

A relationship between Nuculanoidea and Pteriomorphia is suggested by some parameter sets. All of the pteriomorphian superfamilies, with the exception of Arcoidea and Pectinoidea, are recognized as monophyletic under all parameter schemes. Arcoida, and Anomioidea + Plicatuloidea constitute the only supra-superfamilial groupings within Pteriomorphia obtained in all analyses.

This analysis represents the largest molecular analysis, in terms of taxon and character sampling, for any higher group of bivalves published to date. The only comparable studies are two 18S rRNA analyses for bivalves (Campbell 2000; Steiner and Hammer 2000) and a combined analysis of morphology and three loci (Giribet and Wheeler 2002). This analysis, built to improve the molecular sampling (both taxon and character-based) of the previous study of Giribet and Wheeler (2002), presents certain limitations in the resolution of the deeper nodes of the bivalve phylogenetic tree. "Failure" in resolving some of the earliest relationships is judged on the basis of lack of stability in the deeper nodes, whereas they are highly stable in the combined analysis of morphology and molecules of Giribet and Wheeler (2002). Perhaps the addition of genes more conserved than the ones used here will add stability to the deeper nodes, but no such candidate genes have yet been identified.

Molecular data do not suggest a monophyletic origin of bivalves, or a clear sister-group relationship of bivalves to a particular class of conchiferans. Instead, the data suggest that bivalves are polyphyletic, with certain protobranchiate taxa (Nuculidae and Solemyidae) being sister group to the remaining conchiferans, and the remaining bivalves forming a monophyletic group. This contrasts with the more orthodox hypotheses of bivalve monophyly and a sister-group relationship with Scaphopoda (Götting 1980; Lauterbach 1983; Runnegar 1996; Salvini-Plawen and Steiner 1996), although the monophyly of the three clades of conchiferan outgroups, Gastropoda, Scaphopoda, and Cephalopoda, is supported in recent studies of morphological and molecular data (Waller 1998; Haszprunar 2000; Giribet and Wheeler 2002). A sister-group relationship of Bivalvia to Polyplacophora and Aplacophora has also been suggested (Scarlato and Starobogatov 1978), but this hypothesis has generally been ignored in the subsequent literature.

Since the most important contribution of the molecular data presented here is at low taxonomic rank, we next discuss bivalve superfamilies and their status and resolution based on the data. Figure 3.6 shows a summary tree of the relationships obtained for the best parameter set when all of the data are combined.

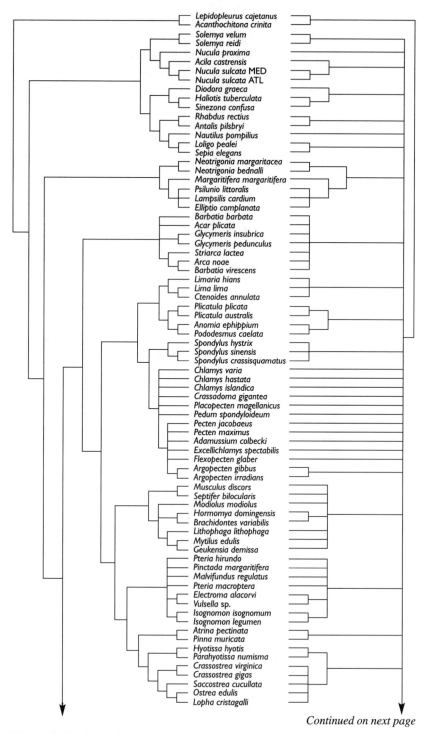

Continued on next page

Figure 3.5. Phylogenetic relationships of bivalve and outgroup taxa proposed on the basis of combined molecular sequence data (18S rRNA and 28S rRNA, COI, and histone H3). The tree on the left represents the strict consensus of 17 cladograms at

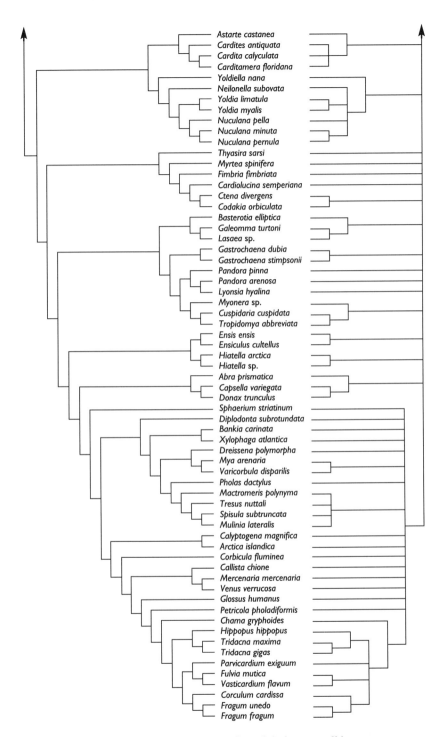

Astarte castanea
Cardites antiquata
Cardita calyculata
Carditamera floridana
Yoldiella nana
Neilonella subovata
Yoldia limatula
Yoldia myalis
Nuculana pella
Nuculana minuta
Nuculana pernula
Thyasira sarsi
Myrtea spinifera
Fimbria fimbriata
Cardiolucina semperiana
Ctena divergens
Codakia orbiculata
Basterotia elliptica
Galeomma turtoni
Lasaea sp.
Gastrochaena dubia
Gastrochaena stimpsonii
Pandora pinna
Pandora arenosa
Lyonsia hyalina
Myonera sp.
Cuspidaria cuspidata
Tropidomya abbreviata
Ensis ensis
Ensiculus cultellus
Hiatella arctica
Hiatella sp.
Abra prismatica
Capsella variegata
Donax trunculus
Sphaerium striatinum
Diplodonta subrotundata
Bankia carinata
Xylophaga atlantica
Dreissena polymorpha
Mya arenaria
Varicorbula disparilis
Pholas dactylus
Mactromeris polynyma
Tresus nuttali
Spisula subtruncata
Mulinia lateralis
Calyptogena magnifica
Arctica islandica
Corbicula fluminea
Callista chione
Mercenaria mercenaria
Venus verrucosa
Glossus humanus
Petricola pholadiformis
Chama gryphoides
Hippopus hippopus
Tridacna maxima
Tridacna gigas
Parvicardium exiguum
Fulvia mutica
Vasticardium flavum
Corculum cardissa
Fragum unedo
Fragum fragum

15,397 weighted steps for the parameter set that minimizes overall incongruence (parameter set 211). The tree on the right represents the strict consensus of all trees obtained for all examined parameter sets, and shows nodes that are stable to parameter variation.

BIVALVE SUPERFAMILIES

Solemyoidea

The two members of the genus *Solemya* appear in a clade containing the conchiferan outgroups and the nuculids in most combined analyses. A relationship with Scaphopoda is also suggested by the analyses based on histone H3, and a relationship with the conchiferan outgroups and the nuculids is suggested by COI and ribosomal gene sequence data. Previous molecular analyses using sequence data of *Solemya* agree with our findings (Hoeh et al. 1998; Giribet and Wheeler 2002). The 18S rRNA study of Steiner and Hammer (2000) placed *Solemya togata* within a clade containing the protobranchiate bivalves and the Pteriomorphia. We have not included the 18S rRNA sequence data of *S. togata* because the sequence deposited in GenBank presents anomalous deletions, whereas the rest of the molecule is largely similar to the other two species of the genus we analyzed.

The position of the Solemyoidea differs from that of other studies in which they are considered sister group to the remaining protobranchs (Salvini-Plawen and Steiner 1996), or sister group to the Nuculoidea (Waller 1998; Giribet and Wheeler 2002), forming part of a monophyletic class Bivalvia, for which a great deal of morphological evidence exists.

Nuculoidea

The members of the family Nuculidae represented in our study (the other family of nuculoideans, Pristiglomidae, was not available) form a monophyletic group under all parameter sets we explored, and appear to be related to the Solemyoidea and to the conchiferan outgroups, not forming a clade with the remaining bivalves. This topology seems to be driven by COI data, as previously shown by other COI studies (Hoeh et al. 1998; Giribet and Wheeler 2002), because under certain parameter sets, ribosomal data place Nuculoidea with Nuculanoidea (Figure 3.4). Certain morphological analyses have previously proposed a relationship between Nuculoidea and Solemyoidea (Waller 1998), or have shown irresolution between the protobranchiate superfamilies (Giribet and Wheeler 2002).

Nuculanoidea

The nuculanoideans are represented by members of three families, Yoldiidae, Nuculanidae, and Neilonellidae. The members of these three families constitute a monophyletic lineage for all analyses and all partitions, with the exception

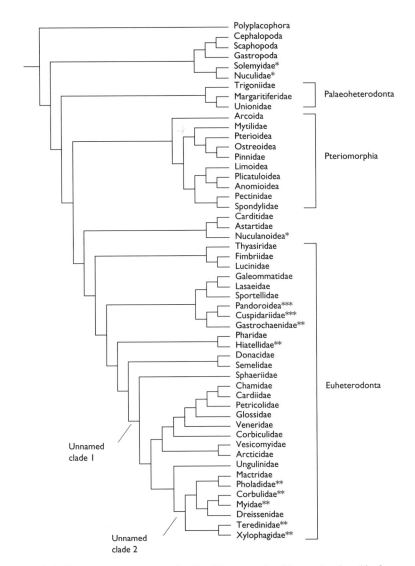

Figure 3.6. Summary tree representing families, superfamilies, and orders (the lowest
rank monophyletic category) for the bivalve taxa when all data are analyzed in combination for the parameter set that minimizes overall incongruence (parameter set
211). * Protobranchiate bivalves; ** "Myoida"; and *** anomalodesmatans are indicated. Unnamed clades 1 and 2 correspond to clades discussed in the text. Euheterodonta corresponds to "unnamed clade 3" of Giribet and Wheeler (2002), all heterodonts except for carditoids and crassatelloids.

of some H3 trees. These results are consistent with all previous molecular analyses of bivalve relationships. However, the position of Nuculanoidea relative to the other bivalve superfamilies is somewhat ambiguous. Previous analyses have suggested a relationship to Nuculoidea (forming the Nuculoida; Salvini-Plawen and Steiner 1996), a sister group to the remaining protobranchiate bivalves (Waller 1998), or a sister-group relationship to Autolamellibranchiata (Giribet and Wheeler 2002). However, molecular data alone is ambiguous and may suggest a relationship to Nuculoidea based on ribosomal data alone (Figure 3.4) (Campbell 2000), a relationship to Solemyoidea (Giribet and Wheeler 2002), or monophyly of Protobranchia (Steiner and Hammer 2000). COI sequence data, however, may suggest a relationship of Nuculanoidea to certain pteriomorphians and lucinids (Figure 3.3).

Most of the parameter sets for the combined analysis suggest a relationship of Nuculanoidea to the Pteriomorphia and/or the clade containing Astartidae and Carditidae. This topology clearly contrasts with the topologies suggested by morphological analyses and combined analyses of morphology and molecules. A relationship of Nuculanoidea to Astartidae + Carditidae (Figure 3.5) is however not supported by all parameter sets tested. The internal phylogeny of the Nuculanoidea needs to be better studied by including representatives of all of the families (Allen and Hannah 1986; Maxwell 1988).

Trigonioidea and Unionoidea

Trigonioidea, represented by two species of the single genus *Neotrigonia,* and Unionoidea, represented by members of the families Unionidae and Margaritiferidae, both constitute monophyletic superfamilies in all of the analyses performed so far. The higher grouping Palaeoheterodonta is also found in all analyses for all partitions, as suggested in previous molecular and morphology-based analyses (Newell 1965; Healy 1989, 1996; Lynn 1994; Cope 1997; Hoeh et al. 1998; Waller 1998; Graf and Ó Foighil 2000; Giribet and Wheeler 2002; for more details, see Chapter 4, this volume).

Monophyly of Palaeoheterodonta has been disputed (Purchon 1987; Salvini-Plawen and Steiner 1996; Cope 2000). Cope (2000) explicitly suggested separating the Trigonioida from the Unionida, leaving the latter within the Heteroconchia, but considering the Trigonioida as an independent autolamellibranchiate group. This is, however, not supported by molecular data, which consistently suggests palaeoheterodont monophyly (Figures 3.3 to 3.5). What is interesting, however, is that molecular data generally places Palaeoheterodonta before the split between Pteriomorphia–Nuculanoidea and Heterodonta. The indefinite position of Palaeoheterodonta based on sequence data alone, however, contrasts with the stability of the monophyly of Heteroconchia when molecular data are analyzed in combination with morphology, even when none of the independent sets of characters unambiguously support such a position (Giribet and Wheeler 2002).

Arcoidea and Glycymeridoidea

Relationships among the families of Arcoida are not well understood. Most current taxonomic studies divide the Arcoida into two superfamilies, Arcoidea (with the families Arcidae, Cucullaeidae, and Noetiidae), and Limopsoidea (with the families Limopsidae, Glycymerididae, and Philobryidae; Beesley et al. 1998). Other authors, however, have given to each limopsoidean family a superfamilial status (Coan et al. 2000). In the absence of the families Cucullaeidae, Limopsidae, and Philobryidae, molecular data suggest monophyly of the Arcoida (= Neotaxodonta *sensu* Cope 1996, 1997), but they do not support monophyly of Arcoidea, which constitute a paraphyletic assemblage that includes the Glycymerididae (Figures 3.4 and 3.5). This is consistent with all previous molecular analyses (Campbell 2000; Steiner and Hammer 2000; Giribet and Wheeler 2002) and suggests the need for serious taxonomic rearrangements within the Arcoida, a point shared by a study on the noetiid ligaments (Thomas et al. 2000).

The relationships of Arcoida to the other pteriomorphian families are mostly unsettled. Previous analyses based on simultaneous analyses of morphological and molecular data suggest that Arcoida are the sister group to the nonmytiloid pteriomorphians (Steiner and Hammer 2000; Giribet and Wheeler 2002), a hypothesis obtained under the immediate suboptimal parameter set (241). Similarly, Waller (1998) placed Arcoidea and Limopsoidea as forming an unresolved trichotomy with the nonmytiloid pteriomorphians, although he did not recognize the monophyly of Arcoida. The optimal parameter set for all combined data, however, proposes a scheme similar to Cope's (1996, 1997) idea of Arcoidea (= Neotaxodonta) being an independent lineage of the remaining Pteriomorphia.

Mytiloidea

Mytiloideans, represented by eight genera of the single family Mytilidae, are recognized as a monophyletic clade in all analyses when all data are combined and in most partitioned analyses, agreeing with previous molecular studies (Distel 2000; Distel et al. 2000; Steiner and Hammer 2000; Giribet and Wheeler 2002). The internal relationships of the mytilid genera mostly agree with those of Distel (2000) and Steiner and Hammer (2000) and suggest that subfamilial relationships among mytilids (Newell 1969) need to be revisited.

The relationships of mytilids to other pteriomorphians are equally in need of revision. Although molecular data alone have suggested a relationship of mytilids to Ostreoidea, Pterioidea, and Pinnoidea (Figures 3.5 and 3.6; Steiner and Hammer 2000), the combination of molecular data with morphology suggests a sister-group relationship of mytilids to the remaining pteriomorphians (Steiner and Hammer 2000; Giribet and Wheeler 2002), as suggested by other authors based on morphological data alone (Waller 1998; Carter et al. 2000).

Pinnoidea

The two members of the single family Pinnidae are monophyletic in all of the analyses performed, a result consistent with a previous molecular analysis (Steiner and Hammer 2000). The phylogenetic position of the Pinnoidea has shown conflict in previous analyses of pteriomorphian phylogeny, being either related to Pterioidea (Starobogatov 1992; Cope 1996; Morton 1996; Waller 1998), to Ostreoidea + Plicatuloidea (Carter 1990), or to Pterioidea + Ostreoidea (Steiner and Hammer 2000; Giribet and Wheeler 2002). A relationship of Pinnoidea with Ostreoidea and Pterioidea is also supported by most parameters in our study (Figures 3.3 to 3.5), although the exact relative position of the three families is ambiguous.

Ostreoidea

Ostreoidea, including the families Ostreidae and Gryphaeidae, is monophyletic in all analyses under all parameter sets. In current classifications, Ostreoidea is considered one of the five superfamilies of the order Ostreoida, the other families being Plicatuloidea, Dimyoidea, Pectinoidea, and Anomioidea (Beesley et al. 1998; Coan et al. 2000). This classification, however, disagrees with most molecular datasets, which place the Ostreoidea as more closely related to Pinnoidea and Pterioidea than to the other "Ostreoida" (Figures 3.3 to 3.6; Campbell 2000; Steiner and Hammer 2000; Giribet and Wheeler 2002; see a discussion in Campbell et al. 1998).

Pterioidea

Pterioidea includes the families Pteriidae, Malleidae, Isognomonidae, and Pulvinitidae (the latter not included in this study), and constitutes a monophyletic group in all of the analyses performed (Figures 3.4 and 3.5). Superfamilial relationships of Pterioidea are with Ostreoidea and Pinnoidea, but monophyly of the order Pteroida is generally not supported (Steiner and Hammer 2000; Giribet and Wheeler 2002). Monophyly of Pteriidae or Malleidae is not supported by the molecular data.

Anomioidea and Plicatuloidea

The two superfamilies, represented by two species each, constitute the single stable relationship of pteriomorphian superfamilies, appearing in all analyses and under all parameter sets. This finding has been proposed in previous molecular studies (Campbell 2000; Steiner and Hammer 2000), despite conflicting with morphology-based hypotheses (Carter 1990; Starobogatov 1992; Cope 1996; Morton 1996). Because the molecular results were stable, we suggest that these two superfamilies constitute a monophyletic group. Steiner and Hammer (2000) proposed right-valve cementation as a putative synapomorphy for the clade, but the cementation is different, with a mineralized byssus in the cementing Anomioidea and shell cementation in Plicatuloidea (David Campbell, University of Alabama, personal communication). Furthermore, it has been claimed that *Plicatula* is ambidextrous in cementing (Lal 1996), whereas right-valve cementation is also common in Pectinoidea.

Limoidea

This monofamilial taxon is represented in our study by three genera that are monophyletic in all analyses (Figures 3.3 to 3.5). The position of Limidae is variable, and relationships have been proposed with Anomioidea and Pectinoidea (Carter 1990), with Pectinoidea and Plicatuloidea (Starobogatov 1992), with Ostreoidea (Morton 1996), with Pectinoidea (Steiner and Hammer 2000), or with Anomioidea and Pectinoidea (Giribet and Wheeler 2002). The combined analysis under the best parameter set suggests a sister-group relationship to ((Anomioidea + Plicatuloidea) Pectinoidea), and most analyses seem to agree with both Steiner and Hammer (2000) and Giribet and Wheeler (2002) in constituting a monophyletic clade that includes Limoidea, Anomioidea, Plicatuloidea, and Pectinoidea, as suggested by Starobogatov (1992).

Pectinoidea

Only two of the four pectinoidean families are represented in our study, Pectinidae and Spondylidae. These families constitute a monophyletic group in most analyses and for most parameter sets (Figures 3.2 to 3.5), as shown in previous molecular work (Campbell 2000; Steiner and Hammer 2000; Giribet and Wheeler 2002). In current classifications (Beesley et al. 1998; Coan et al. 2000),

Pectinoidea belong, together with Ostreoidea, Plicatuloidea, Dimyoidea, and Anomioidea, to the order "Ostreoida," which results in a polyphyletic group, since Pectinoidea appears to form a clade with Anomioidea, Plicatuloidea, and Limoidea under most parameter sets, whereas the Ostreoidea appears to be more closely related to the pterioidean families.

Carditoidea and Crassatelloidea

In our study, Crassatelloidea is represented by a single species of the family Astartidae (*Astarte castanea*), whereas the family Crassatellidae is not represented. Carditoidea is represented by three genera of the family Carditidae (*Cardita, Cardites,* and *Carditamera*), whereas the family Condylocardiidae is not represented. The monophyly of Carditidae + Astartidae is obtained throughout the parameter space for all partitions and combined analyses (Figures 3.3 to 3.5), as was found in previous analyses of heterodont relationships (Giribet and Wheeler 2002). This relationship is supported by the presence of similar hinge structures and by sperm morphology (Healy 1995; Giribet and Wheeler 2002), but analyses that include members of the families Condylocardiidae and Crassatellidae are still needed.

A sister-group relationship of this clade to the remaining heterodonts is found under several parameter sets we analyzed (but not in our preferred tree), as it was found in the combined analysis of Giribet and Wheeler (2002). Also, all previous molecular analyses that included any of the members of these families have consistently found them to be the sister group to the remaining heterodonts (Campbell 2000; Park and Ó Foighil 2000). These results are consistent with a previous hypothesis that considers Crassatelloidea and Carditoidea the most primitive (i.e., sister group to the remaining heterodonts) of modern eulamellibranchs (Yonge 1969). A similar result was also proposed by Purchon (1987), who removed Crassatelloidea, Carditoidea, and Leptonoidea (= Galeommatoidea) from the Veneroida and placed them together with the Unionoidea in the suborder Unionoida. The basal position of Crassatelloidea and Carditoidea contrasts with the notion that Lucinoidea are the earliest diverging heterodonts (Carter et al. 2000). Other researchers have considered Lucinoidea, Crassatelloidea, and Galeommatoidea to form a clade of the most basal extant heterodont bivalves, but regarded Carditoidea as derived heterodonts (Morton 1996). However, our data strongly suggest that (Carditoidea + Crassatelloidea) are monophyletic and constitute either the sister group to the remaining heterodonts or the sister group to the Nuculanoidea (Figure 3.5), although they are never nested within the remaining heterodonts (Euheterodonta).

Lucinoidea

Lucinoidea includes several families of heterogeneous bivalves, the Lucinidae, Fimbriidae, Thyasiridae, Ungulinidae, and Mactromyidae (Beesley et al. 1998), of which the Mactromyidae are not represented here. The Lucinoidea have received separate status from other heterodonts (Salvini-Plawen and Steiner 1996), and have been proposed to be the sister group to the remaining heterodonts (Carter et al. 2000), or to constitute a clade with Crassatelloidea and Galeommatoidea to form the sister group to the remaining heterodont bivalves (Morton 1996). Molecular data on Lucinoidea are scarce, with just a few sequences published for the Lucinidae and Ungulinidae (Steiner and Hammer 2000; Giribet and Wheeler 2002). The addition of new sequences of Lucinidae, Fimbriidae, and Thyasiridae in our study suggests that these three families are monophyletic (under most parameter sets), but do not form a group with Ungulinidae, as found in previous 18S rRNA analyses (Steiner and Hammer 2000). The lucinoidean families form a monophyletic clade with the noncarditoid, noncrassatelloid heterodonts (unnamed node 3 of Giribet and Wheeler 2002). The nonungulinid lucinoids are monophyletic under most parameter sets and for all partitions (Figures 3.3 to 3.5), and they generally constitute the sister group to the remaining euheterodonts (Figures 3.4 and 3.5), or form a clade together with Anomalodesmata. Another topology shown by some parameter sets is ((Carditoidea + Crassatelloidea) (Lucinoidea (Anomalodesmata + remaining euheterodonts))). However, the relationships between Lucinoidea and Anomalodesmata still need a great deal of new information. The exact position of *Diplodonta* (Ungulinidae) apart from the remaining lucinoids is highly unstable, but it seems clear that this taxon forms part of a much more derived clade of euheterodont bivalves.

Anomalodesmata

The bivalve "class" Anomalodesmata is one of the most interesting heterodont groups, traditionally composed of the superfamilies Pholadomyoidea, Thracioidea, Clavagelloidea, Pandoroidea, Verticordioidea, Poromyoidea, and Cuspidarioidea. The phylogeny of the group has recently been settled in a cladistic analysis of morphological data (Harper et al. 2000a). Molecular data are still lacking for most families, but complete 18S rRNA sequences are available for Pandoridae, Lyonsiidae (Pandoroidea), and Cuspidariidae (Cuspidarioidea) (Campbell 2000; Steiner and Hammer 2000; Giribet and Wheeler 2002), and for a short fragment of the 18S rRNA locus of Periplomatidae (Thracioidea) (Adamkewicz et al. 1997). In this study, we are limited to six anomalodesmatans (two Pandoridae, one Lyonsiidae, and three Cuspidariidae) represented by ribosomal sequence data, two of which have been sequenced for COI. This taxon sampling is obviously not optimal, but allows for testing the monophyly of the group (orders Pholadomyoidea and Septibranchia), which has been questioned by some authors. However, the inclusion of the Verticordioidea seems especially important in this respect.

Our results suggest monophyly of Anomalodesmata (not with the rank of a subclass) for most parameter sets, as well as the monophyly of Pandoroidea and Cuspidariidae. Monophyly of Anomalodesmata, and its position within the Euheterodonta, has previously been supported by molecular data (Campbell 2000; Giribet and Wheeler 2002). In fact, the heterodont condition of anomalodesmatans has been proposed by earlier authors that considered Anomalodesmata as sister group to Myoida (Morton 1996; Salvini-Plawen and Steiner 1996). The currently well-corroborated position of Anomalodesmata makes the "classical" Heterodonta (Cope 1997, 2000; Waller 1998) a paraphyletic group, and the use of the term *Heterodonta* in this article follows that of Giribet and Wheeler (2002).

Galeommatoidea and Cyamioidea

Galeommatoidea includes the families Galeommatidae and Lasaeidae, containing many symbiotic and commensal forms, usually associated with large infaunal invertebrates. The superfamily has been considered a cul-de-sac by some authors and is therefore suspected to be polyphyletic (Coan et al. 2000). Cyamioidea also groups three families of small bivalves, Cyamiidae, Neoleptonidae, and Sportellidae, but again, the characters defining these families are in need of revision, and the group may also be polyphyletic, as suggested by recent work proposing to place the Neoleptonidae within the Veneroidea (Salas and Gofas 1998).

Galeommatoidean and cyamioidean sequences have seldom been used in phylogenetic studies of bivalves (Adamkewicz et al. 1997; Campbell 2000; Steiner and Hammer 2000; Giribet and Wheeler 2002). Monophyly of Galeommatoidea (genera *Lasaea* and *Galeomma*) and of Galeommatoidea + Sportellidae is found in most analyses (Figures 3.4 to 3.6), as it is found in previous molecular analyses based on ribosomal data (Campbell 2000). A previously proposed relationship of *Galeomma* to the Teredinid *Bankia* (Giribet and Wheeler 2002) is now suspected to be incorrect, the error is probably due to the inadvertent amplification of a bacterial COI sequence from a *Bankia* symbiont and from some sort of bacterial sequence in *Galeomma*. The results from the ribosomal sequence data seem to agree with morphology; thus, it is here proposed that the represented families of Galeommatoidea and Cyamioidea constitute a monophyletic group. Obviously, the position of many other genera of these three families, as well as the members of the Cyamiidae and Neoleptonidae, need to be investigated on the basis of molecular sequence data.

Tellinoidea

The Tellinoidea is the largest bivalve superfamily and includes five families of laterally compressed forms that live infaunally buried in the sediment up to several times their length: the Tellinidae, Semelidae, Psammobiidae, Solecurtidae, and Donacidae. Despite being locally common in many areas, we studied only molecular data on Semelidae and Donacidae. The results consistently support monophyly of Tellinoidea (Figures 3.4 to 3.6), as it does previous work (Adamkewicz et al. 1997; Campbell 2000; Park and Ó Foighil 2000; Steiner and Hammer 2000; Canapa et al. 2001). Relationships of this important bivalve superfamily to other euheterodont groups are not very stable; candidate groups include Arcticoidea (Morton 1996), Dreissenoidea (Park and Ó Foighil 2000), and Chamoidea + Cardioidea (Giribet and Wheeler 2002).

The position of Tellinoidea is highly unstable in our analyses, but a relationship to Gastrochaenoidea + Galeommatoidea is supported by ribosomal sequence data (Figure 3.4), and a relationship to *Lasaea* is supported by some parameter sets for the combined molecular data. In general, the Tellinoidea appear near to galeommatoideans, hiatellids, and pharids, but every parameter set seems to suggest an alternative position for the group, which is totally unsatisfactory if stability in hypotheses is sought.

Gastrochaenoidea

Monophyly of the Gastrochaenidae is found in all of the analyses we performed, although the position of this enigmatic family of endolithic or infaunal tube-building marine bivalves remains a mystery. Classical taxonomy places the Gastrochaenidae within the order Myoida, but molecular data consistently fails to recognize the monophyly of Myoida and places Gastrochaenidae in different positions, although a relationship with Galeommatoidea and Cyamioidea (Figure 3.4) (Campbell 2000; Giribet and Wheeler 2002) or with Anomalodesmata (Figure 3.5) seems to be somewhat supported.

Hiatelloidea

Monophyly of the two members of the genus *Hiatella* is supported by all of the analyses (Figures 3.4 to 3.5). The position of the family relative to other euheterodonts is unstable, as suggested in previous analyses (Giribet and Wheeler 2002), but many parameter sets suggest a relationship with the Solenoidea. It is clear that Hiatellidae does not group with any of the pholadoidean or myoidean families under any parameter set for any partition, thus supporting the hypothesis of Myoida polyphyly.

Solenoidea

From the two families of razor-clams, Solenidae and Pharidae, only the latter has been represented in molecular analyses (Adamkewicz et al. 1997; Campbell 2000; Steiner and Hammer 2000; Canapa et al. 2001; Giribet and Wheeler 2002); they seem to be sister group to most of the Euheterodonta (Campbell 2000; Canapa et al. 2001). However, this relationship is ambiguous in our analyses and, under several parameter sets, Solenoidea forms a monophyletic clade with Hiatelloidea (Figures 3.4 and 3.5).

Arcticoidea and Glossoidea

The families Arcticidae and Trapezidae belong to the Arcticoidea, and Glossidae belongs to the Glossoidea, whereas the Vesicomyidae and Kelliellidae have been disputed between these superfamilies (Newell 1965; Boss and Turner 1980; Beesley et al. 1998; Coan et al. 2000). Arcticoidea + Glossoidea constitute a monophyletic group under parameter set 111, but they generally do not form a monophyletic clade. They do, however, tend to group with members of the Veneridae and Corbiculidae, as shown in previous analyses that did not include Glossidae (Giribet and Wheeler 2002). Morton (1996) previously recognized the relationship of Glossoidea and Veneroidea.

The status of Arcticoidea is contentious and certain authors have suggested it to be polyphyletic (Giribet and Wheeler 2002), a taxonomic consideration also implemented elsewhere (Coan et al. 2000). In the combined tree of all data for parameter set 211, Arcticoidea constitutes a monophyletic group, sister to a clade containing Corbiculidae, Veneridae, Glossidae, Petricolidae, and Cardiidae. Arcticoidea is also monophyletic for parameter set 211 for the ribosomal data analysis, suggesting that the monophyly of Arcticoidea may depend on taxonomic sampling, since it was not obtained in previous analyses with less heterodont taxa.

Corbiculoidea and Sphaerioidea

The Corbiculidae and Sphaeriidae are two "corbiculoidean" families recognized in current taxonomy; however, recent studies using molecular data have proposed a polyphyletic status of the group (Park and Ó Foighil 2000; Giribet and

Wheeler 2002), a result consistent with the our study (Figures 3.3 to 3.5). Corbiculidae and Sphaeriidae form part of a large clade of euheterodonts (unnamed clade 1 in Figure 3.6), which is obtained under all examined parameter sets for the combined data. Corbiculidae tends to group with Arcticoidea, Veneridae, Petricolidae, Glossidae, Chamidae, and Cardiidae. Sphaeridae tends to appear as the sister group to the remaining families constituting unnamed clade 1 (Figure 3.6), but several other topologies appear with different parameter sets.

Veneroidea

The Veneroidea comprises the families Veneridae and Petricolidae, although other minor families have been assigned to the group, such as Glauconomidae (Boss 1982; Beesley et al. 1998) and Turtoniidae (Boss 1982; Coan et al. 2000); however, molecular data are only available for the two main families. Monophyly of Veneroidea is not found under any parameter set we explored or for any partition, Veneridae appeared to be generally related to Corbiculidae, Arcticidae, and Glossidae, whereas the position of Petricolidae is highly unstable to parameter set variation. Both families are constituents of unnamed clade 1.

Cardioidea and Chamoidea

Cardioidea (*sensu* Schneider) includes representatives of the old families Cardiidae and Tridacnidae (Schneider 1992), and the Hemidonacidae (the latter, represented by the single genus *Hemidonax,* is not represented in any molecular studies). Molecular data clearly support monophyly of Cardiidae, including the members of the Tridacninae (Figures 3.4 and 3.5), as proposed in earlier phylogenetic studies (Schneider 1998; Campbell 2000; Steiner and Hammer 2000; Canapa et al. 2001; Giribet and Wheeler 2002). The phylogenetic position of Cardioidea within the Euheterodonta is unstable to parameter sets, although ribosomal data strongly support a sister-group relationship with Chamoidea (Figures 3.4 to 3.6), as do all parameter sets for the combined analyses, including the optimal one (Figure 3.5). A relationship of Chamoidea and Cardioidea was proposed by Giribet and Wheeler (2002), although it disagrees with other hypotheses of heterodont relationships (Morton 1996), and with the idea that juvenile chamids resemble venerids before becoming cemented.

Mactroidea

This superfamily comprises four families, Mactridae, Mesodesmatidae, Anatinellidae, and Cardiliidae, from which only Mactridae has been used in molecular analyses. Monophyly of Mactridae is evident from all morphological analyses performed (Figures 3.4 and 3.5), as has been found in previous molecular analyses (Campbell 2000; Park and Ó Foighil 2000; Steiner and Hammer 2000; Canapa et al. 2001; Giribet and Wheeler 2002). However, the relative position of Mactroidea among the Euheterodonta is not definitively solved. A relationship of Mactroidea to Solenoidea (Morton 1996) or to Dreissenoidea + Myoidea (Giribet and Wheeler 2002) has been proposed. Again, the definitive position of the Mactroidea is not completely clear, but under most parameter sets, data indicate that Mactroidea forms part of a clade that also contains Dreissenoidea, Pholadoidea, and Myoidea, and the ungulinid *Diplodonta rotundata* may be the sister group to this clade (Figure 3.6). This hypothesis contrasts with the previous hypothesis of Giribet and Wheeler (2002), but the addition of new pholadoidean taxa and the realization of the putative artifactual grouping of *Galeomma* with *Bankia* led us to propose this hypothesis.

Pholadoidea

Members of the Pholadoidea, Pholadidae, and Teredinidae are represented by *Pholas dactylus, Xylophaga atlantica,* and *Bankia carinata.* An unidentified teredinid and a partial 18S rRNA sequence of a pholadid have previously been reported (Adamkewicz et al. 1997; Campbell 2000). A representative of the Pholadoidea used in previous molecular analyses was suggested to be the sister group to the Galeommatidae *Galeomma turtoni* (Giribet and Wheeler 2002). This result was stable to parameter variation, even when combined with morphological data. However, subsequent analyses indicate that both sequences are of bacterial origin. The *B. carinata* sequence likely originated from bacterial symbionts known to be associated with teredinid bivalves. The origin of the *Galeomma* sequence is unknown, but may have arisen from a symbiont, pathogen, saprophyte, or casual contaminant of the *G. turtoni* specimen. Hence the sister-group association indicated by COI data but not supported by other loci is likely an artifact. The inclusion of new sequence data of other pholadoid species of the genera *Pholas* and *Xylophaga* strongly suggests a monophyletic clade composed of the Pholadoidea, Myoidea, and Dreissenoidea, this result being found in all parameter sets explored for the ribosomal sequence data (Figure 3.4) and for most combined analyses of all data (see also Campbell 2000). Therefore, the Pholadoidea and Pholadidae are not monophyletic, and *Bankia* appears to be sister group to *Xylophaga* in most analyses. The members of the polyphyletic Pholadoidea constitute a monophyletic clade with the Myoidea and the freshwater Dreissenoidea (Figure 3.6). To our knowledge, this clade has previously not been proposed.

Myoidea

This superfamily consists of the families Myidae and Corbulidae, here represented by the genera *Mya* and *Varicorbula*. The Myoidea are found to be monophyletic under most analytical conditions, in terms of parameter sets and data partitions. Interestingly, the myoidean species appear to be related to the Dreissenoidea, Pholadoidea, and Mactroidea for most parameter sets for the COI, ribosomal, and all data partitions. Morphological support for this clade is unclear, although many of its members show fusion of mantle margins of type C (Yonge 1957, 1982b), but clearly not in *Dreissena*. Ligamental structure in Mactroidea and Myoidea also shows similarities that were assumed to have been achieved independently (Yonge 1982a), although common ancestry of these traits could be postulated, given the proposed scenario.

Dreissenoidea

This superfamily is represented in our analyses by a single species. The analyses reveal a relationship of Dreissenoidea to Myoidea, Pholadoidea, and Mactroidea, similar to our previous study (Giribet and Wheeler 2002), although 28S rRNA sequence data have shown alternative relationships for *Dreissena* (Park and Ó Foighil 2000).

The protobranchiate and most pteriomorphian superfamilies are well-established groups morphologically, the molecular data largely agreeing with such a classification. However, certain orders (Nuculoida, Pterioida, and Ostreoida) fail to be monophyletic based on molecular data. This is not surprising, since many of the relationships among the superfamilies constituting those orders have been disputed (Carter 1990; Starobogatov 1992; Cope 1996; Morton 1996; Waller 1998; Steiner and Hammer 2000). Superfamilial and ordinal classifications within the Heterodonta, and especially within the Euheterodonta (unnamed node 3 of Giribet and Wheeler 2002), have received much less attention than those of Pteriomorphia, and will need special treatment in future bivalve phylogenetic studies. Some superfamilies (Lucinoidea, Veneroidea, and Corbiculoidea) appear to be ill-defined in the current taxonomical treatises or, alternatively, poorly resolved by molecular data. Ordinal taxonomy of the heterodonts is in need of revisionary work since none of the recognized orders (e.g., Veneroida and Myoida) seem to support the data, and the Anomalodesmata (order Pholadomyoida) falls within the euheterodonts. Within the Euheterodonta, certain unorthodox relationships appear highly stable to data selection and parameter sensitivity, such as the sister-group relationship of Chamidae and Cardiidae, or the unnamed clades 1 and 2 in Figure 3.6.

The phylogenetic scheme suggested by the molecular data alone (Figure 3.6) bears a large degree of resemblance to the phylogenetic hypothesis proposed by Cope (1997; Figure 3.1C), in which the Palaeoheterodonta diverge before the split between Pteriomorphia and Heterodonta, and where the Pteriomorphia are divided between the Arcoida and the remaining groups (Neotaxodonta and Pteriomorphia *sensu* Cope 1997). Obviously, the positions of the Anomalodesmata and of the Nuculanoidea differ in the two schemes, as does that of Nuculoidea + Solenomyoidea, which clusters with the outgroup in our analysis. However, this strongly conflicts with other analyses based on morphological data (Waller 1998; Carter et al. 2000; Giribet and Wheeler 2002).

FINAL REMARKS

Interest in the fascinating field of bivalve phylogeny and evolution has increased remarkably in recent times, especially since the incorporation of molecular data into the toolkit of evolutionary biologists. Bivalves have emerged as fascinating models for testing the phylogenetic power of molecular data (Steiner and Müller 1996; Giribet and Carranza 1999; Steiner 1999), multiple colonization of freshwater environments (Park and Ó Foighil 2000), colonization of deep-sea environments (Distel et al. 2000), larval evolution (Gustafson and Reid 1986; Gustafson and Lutz 1992; Zardus and Morse 1998), the study of chemoautotrophic endosymbioses (Cavanaugh 1983; Distel 1998), and symbioses with zooxanthellae (Yonge 1980; Ohno et al. 1995).

Although bivalves are for the most part inconspicuous creatures, they have proven to be far more fascinating, and their history more perplexing, than their appearance would suggest. To better understand these enigmatic creatures and the interesting biological questions that surround them, we must continue to reveal the patterns of their evolutionary tree. Some aspects of bivalve evolution that seem most urgent and important include the following:

- Relationships of the protobranchiate families and their position in relation to other bivalves
- Taxonomic status of the Arcoida, with emphasis on the position of the Glycymerididae and other "Limopsoidea"
- Taxonomic revision of the Mytilidae at the subfamilial level
- The position of the families Condylocardiidae and Crassatellidae relative to Astartidae and Carditidae. (These four families could well constitute the sister group to the remaining heterodonts.)
- Internal relationships of the Heterodonta, with emphasis on the relationships and taxonomic status of the heterodont superfamilies

- Origins and relationships of Anomalodesmata, and origins of carnivorism in septibranchs
- Colonization of freshwater environments
- Origins of rock- and wood-boring habits
- Origins of the bacterial and algal symbioses
- Colonization of the deep sea.

Initiatives such as the recent bivalve meeting in Cambridge (Harper et al. 2000b) seem a proper way to promote research and to pursue and share knowledge among bivalve evolutionary biologists.

REFERENCES

Adamkewicz, S. L., M. G. Harasewych, J. Blake, D. Saudek, and C. J. Bult. 1997. A molecular phylogeny of the bivalve mollusks. Molecular Biology and Evolution 14:619–629.

Allen, J. A., and F. J. Hannah. 1986. A reclassification of the recent genera of the subclass Protobranchia (Mollusca: Bivalvia). Journal of Conchology 32:225–249.

Beesley, P. L., G. J. B. Ross, and A. Wells. 1998. Mollusca: The Southern Synthesis. CSIRO Publishing, Melbourne, Australia.

Boss, K. J. 1982. Mollusca. Pp. 1092–1166 in Synopsis and Classification of Living Organisms (S. P. Parker, ed.). McGraw Hill, New York.

Boss, K. J., and R. D. Turner. 1980. The giant white clam from the Galapagos Rift, *Calyptogena magnifica* species novum. Malacologia 20:161–194.

Brown, S., G. W. Rouse, P. Hutchings, and D. J. Colgan. 1999. Assessing the usefulness of histone H3, U2 snRNA and 28S rDNA in analyses of polychaete relationships. Australian Journal of Zoology 47:499–516.

Campbell, D. C. 2000. Molecular evidence on the evolution of the Bivalvia. Pp. 31–46 in The Evolutionary Biology of the Bivalvia (E. M. Harper, J. D. Taylor, and J. A. Crame, eds.). The Geological Society of London, London.

Campbell, D. C., K. J. Hoekstra, and J. G. Carter. 1998. 18S ribosomal DNA and evolutionary relationships within the Bivalvia. Pp. 75–85 in Bivalves: An Eon of Evolution (P. A. Johnston and J. W. Haggart, eds.). University of Calgary Press, Calgary, Alberta.

Canapa, A., M. Barucca, A. Marinelli, and E. Olmo. 2001. A molecular phylogeny of Heterodonta (Bivalvia) based on small ribosomal subunit RNA sequences. Molecular Phylogenetics and Evolution 21:156–161.

Carter, J. G. 1990. Evolutionary significance of shell microstructure in the Palaeotaxodonta, Pteriomorphia and Isofilibranchia (Bivalvia: Mollusca). Pp. 135–296 in Skeletal Biomineralization: Patterns, Processes and Evolutionary Trends (J. G. Carter, ed.). Vol. 1. Van Nostrand Reinhold, New York.

Carter, J. G., D. C. Campbell, and M. R. Campbell. 2000. Cladistic perspectives on early bivalve evolution. Pp. 47–79 in The Evolutionary Biology of the Bivalvia (E. M. Harper, J. D. Taylor, and J. A. Crame, eds.). The Geological Society of London, London.

Cavanaugh, C. M. 1983. Symbiotic chemoautotrophic bacteria in marine invertebrates from sulphide-rich habitats. Nature 302:58–61.

Chirgwin, J. M., A. E. Przybyla, R. J. MacDonald, and W. J. Rutter. 1979. Isolation of biologically active ribonucleic acid from sources enriched in ribonuclease. Biochemistry 18:5294–5299.

Coan, E. V., P. V. Scott, and F. R. Bernard. 2000. Bivalve seashells of Western North America. Marine bivalve mollusks from Arctic Alaska to Baja California. Santa Barbara Museum of Natural History, Santa Barbara, CA.

Colgan, D. J., A. McLauchlan, G. D. F. Wilson, S. P. Livingston, G. D. Edgecombe, J. Macaranas, G. Cassis, and M. R. Gray. 1998. Histone H3 and U2 snRNA DNA sequences and arthropod molecular evolution. Australian Journal of Zoology 46:419–437.

Cope, J. C. W. 1996. The early evolution of the Bivalvia. Pp. 361–370 in Origin and Evolutionary Radiation of the Mollusca (J. D. Taylor, ed.). Oxford University Press, Oxford.

Cope, J. C. W. 1997. The early phylogeny of the class Bivalvia. Palaeontology 40:713–746.

Cope, J. C. W. 2000. A new look at early bivalve phylogeny. Pp. 81–95 in The Evolutionary Biology of the Bivalvia (E. M. Harper, J. D. Taylor, and J. A. Crame, eds.). The Geological Society of London, London.

Distel, D. L. 1998. Evolution of chemoautotrophic endosymbioses in bivalves. Bivalve-bacteria chemosymbioses are phylogenetically diverse but morphologically similar. BioScience 48:277–286.

Distel, D. L. 2000. Phylogenetic relationships among Mytilidae (Bivalvia): 18S rRNA data suggest convergence in mytilid body plans. Molecular Phylogenetics and Evolution 15:25–33.

Distel, D. L., A. R. Baco, E. Chuang, W. Morrill, C. Cavanaugh, and C. R. Smith. 2000. Do mussels take wooden steps to deep-sea vents? Nature 403:725–726.

Farris, J. S., M. Källersjö, A. G. Kluge, and C. Bult. 1995. Testing significance of incongruence. Cladistics 10:315–319.

Folmer O., M. Black, W. Hoeh, R. Lutz, and R. C. Vrijenhoek. 1994. DNA primers for amplification of mitochondrial cytochrome c oxidase subunit I from diverse metazoan invertebrates. Molecular Marine Biology and Biotechnology 3:294–299.

Giribet, G. 2001. Exploring the behavior of POY, a program for direct optimization of molecular data. Cladistics 17:S60–S70.

Giribet, G., and S. Carranza. 1999. What can 18S rDNA do for bivalve phylogeny? Journal of Molecular Evolution 48:256–261.

Giribet, G., S. Carranza, J. Baguñà, M. Riutort, and C. Ribera. 1996. First molecular evidence for the existence of a Tardigrada + Arthropoda clade. Molecular Biology and Evolution 13:76–84.

Giribet, G., D. L. Distel, M. Polz, W. Sterrer, and W. C. Wheeler. 2000. Triploblastic relationships with emphasis on the acoelomates and the position of Gnathostomulida, Cycliophora, Plathelminthes, and Chaetognatha: A combined approach of 18S rDNA sequences and morphology. Systematic Biology 49:539–562.

Giribet, G., G. D. Edgecombe, and W. C. Wheeler. 2001. Arthropod phylogeny based on eight molecular loci and morphology. Nature 413:157–161.

Giribet, G., and W. C. Wheeler. 2001. Some unusual small-subunit ribosomal RNA sequences of metazoans. American Museum Novitates 3337:1–14.

Giribet, G., and W. C. Wheeler. 2002. On bivalve phylogeny: A high-level analysis of the Bivalvia (Mollusca) based on combined morphology and DNA sequence data. Invertebrate Biology 121:271–324.

Goloboff, P. A. 1999. Analyzing large data sets in reasonable times: Solutions for composite optima. Cladistics 15:415–428.

Götting, K-J. 1980. Origins and relationships of the Mollusca. Zeitschrift für Zoologische Systematik und Evolutionsforschung 18:24–27.

Graf, D. L., and D. Ó Foighil. 2000. The evolution of brooding characters among the freshwater pearly mussels (Bivalvia: Unionoidea) of North America. Journal of Molluscan Studies 66:157–170.

Gustafson, R. G., and R. A. Lutz. 1992. Larval and early post-larval development of the protobranch bivalve *Solemya velum* (Mollusca: Bivalvia). Journal of the Marine Biological Association of the U.K. 72:383–402.

Gustafson, R. G., and R. G. B. Reid. 1986. Development of the pericalymma larva of *Solemya reidi* (Bivalvia: Cryptodonta: Solemyidae) as revealed by light and electron microscopy. Marine Biology 93:411–427.

Harper, E. M., E. A. Hide, and B. Morton. 2000a. Relationships between the extant Anomalodesmata: A cladistic test. Pp. 129–143 in The Evolutionary Biology of the Bivalvia (E. M. Harper, J. D. Taylor, and J. A. Crame, eds.). The Geological Society of London, London.

Harper, E. M., J. D. Taylor, and J. A. Crame. 2000b. The Evolutionary Biology of the Bivalvia. The Geological Society of London, London.

Haszprunar, G. 2000. Is the Aplacophora monophyletic? A cladistic point of view. American Malacological Bulletin 15:115–130.

Healy, J. M. 1989. Spermiogenesis and spermatozoa in the relict bivalve genus *Neotrigonia:* Relevance to trigonioid relationships, particularly with Unionoidea. Marine Biology 103:75–85.

Healy, J. M. 1995. Sperm ultrastructure in the marine bivalve families Carditidae and Crassatellidae and its bearing on unification of the Crassatelloidea with the Carditoidea. Zoologica Scripta 24:21–28.

Healy, J. M. 1996. Spermatozoan ultrastructure in the trigonioid bivalve *Neotrigonia margaritacea* Lamarck (Mollusca): Comparison with other bivalves, especially Trigonioida and Unionoida. Helgoländer Meeresuntersuchungen 50:259–264.

Hoeh, W. R., M. B. Black, R. G. Gustafson, A. E. Bogan, R. A. Lutz, and R. C. Vrijenhoek. 1998. Testing alternative hypotheses of *Neotrigonia* (Bivalvia: Trigonioida) phylogenetic relationships using cytochrome *c* oxidase subunit I DNA sequences. Malacologia 40:267–278.

Johnston, P. A., and J. W. Haggart. 1998. Bivalves: An Eon of Evolution. University of Calgary Press, Calgary, Alberta.

Lal, S. 1996. Thoughts on *Plicatula* Lamarck (Pectinacea: Bivalvia). Journal of the Paleontological Society of India 41:139–143.

Lauterbach, K-E. 1983. Erörterungen zur Stammesgeschichte der Mollusca, insbesondere der Conchifera. Zeitschrift für zoologische Systematik und Evolutionsforschung 21:201–216.

Lynn, J. W. 1994. The ultrastructure of the sperm and motile spermatozeugmata released from the freshwater mussel *Anodonta grandis* (Mollusca, Bivalvia, Unionidae). Canadian Journal of Zoology 72:1452–1461.

Maxwell, P. A. 1988. Comments on "A reclassification of the Protobranchia (Mollusca: Bivalvia)" by J. A. Allen and F. J. Hannah (1986). Journal of Conchology 33:85–96.

Mickevich, M. F., and J. S. Farris. 1981. The implications of congruence in *Menidia*. Systematic Zoology 27:143–158.

Morton, B. 1990. The Bivalvia: Proceedings of a Memorial Symposium in Honour of Sir Charles Maurice Yonge, Edinburgh, 1986. Hong Kong University Press, Hong Kong, 355 pp.

Morton, B. 1996. The evolutionary history of the Bivalvia. Pp. 337–359 in Origin and Evolutionary Radiation of the Mollusca (J. D. Taylor, ed.). Oxford University Press, Oxford.

Newell, N. D. 1965. Classification of the Bivalvia. American Museum Novitates 2206:1–25.

Newell, N. D. 1969. Classification of Bivalvia. Pp. N205-N244 in Treatise on Invertebrate Paleontology, Part N, Mollusca 6. Vol. 1. Bivalvia. (R. Moore, ed.). Geological Society of America and University of Kansas, Boulder-Lawrence, KS.

Ohno, T., T. Katoh, and T. Yamasu. 1995. The origin of algal-bivalve photo-symbiosis. Palaeontology 38:1–21.

Park, J. K., and D. Ó Foighil. 2000. Sphaeriid and corbiculid clams represent separate heterodont bivalve radiations into freshwater environments. Molecular Phylogenetics and Evolution 14:75–88.

Purchon, R. D. 1987. Classification and evolution of the Bivalvia: An analytical study. Philosophical Transactions of the Royal Society of London B 316:277–302.

Runnegar, B. 1996. Early evolution of the Mollusca: The fossil record. Pp. 77–87 in Origin and Evolutionary Radiation of the Mollusca (J. D. Taylor, ed.). Oxford University Press, Oxford.

Salas, C., and S. Gofas. 1998. Description of four new species of *Neolepton* Monterosato, 1875 (Mollusca: Bivalvia: Neoleptonidae), with comments on the genus and on its affinity with the Veneracea. Ophelia 48:35–70.

Salvini-Plawen, L., and G. Steiner. 1996. Synapomorphies and plesiomorphies in higher classification of Mollusca. Pp. 29–51 in Origin and Evolutionary Radiation of the Mollusca (J. D. Taylor, ed.). Oxford University Press, Oxford.

Scarlato, O. A., and Y. I. Starobogatov. 1978. Phylogenetic relations and the early evolution of the class Bivalvia. Philosophical Transactions of the Royal Society of London B 284:217–224.

Schneider, J. A. 1992. Preliminary cladistic analysis of the bivalve family Cardiidae. American Malacological Bulletin 9:145–155.

Schneider, J. A. 1998. Phylogeny of the Cardiidae (Bivalvia): Phylogenetic relationships and morphological evolution within the subfamilies Clinocardiinae, Lymnocardiinae, Fraginae and Tridacninae. Malacologia 40:321–373.

Schneider, J. A., and D. Ó Foighil. 1999. Phylogeny of giant clams (Cardiidae: Tridacninae) based on partial mitochondrial 16S rDNA gene sequences. Molecular Phylogenetics and Evolution 13:59–66.

Smith, S. W., R. Overbeek, C. R. Woese, W. Gilbert, P. M. Gillevet. 1994. The genetic data environment: An expandable GUI for multiple sequence analysis. Computer Applications in the Biosciences 10:671–675.

Starobogatov, Y. I. 1992. Morphological basis for phylogeny and classification of Bivalvia. Ruthenica 2:1–25.

Steiner, G. 1999. What can 18S rDNA do for bivalve phylogeny? Response. Journal of Molecular Evolution 48:258–261.

Steiner G., and S. Hammer, S. 2000. Molecular phylogeny of the Bivalvia inferred from 18S rDNA sequences with particular reference to the Pteriomorphia. Pp. 11–29 in The Evolutionary Biology of the Bivalvia (E. M. Harper, J. D. Taylor, and J. A. Crame, eds.). The Geological Society of London, London.

Steiner, G., and M. Müller. 1996. What can 18S rDNA do for bivalve phylogeny? Journal of Molecular Evolution 43:58–70.

Thomas, R. D. K., A. Madzvamuse, P. K. Maini, and A. J. Wathen. 2000. Growth patterns of noetiid ligaments: Implications of developmental models for the origin of an evolutionary novelty among arcoid bivalves. Pp. 279–289 in The Evolutionary Biology of the Bivalvia (E. M. Harper, J. D. Taylor, and J. A. Crame, eds.). The Geological Society of London, London.

Waller, T. R. 1990. The evolution of ligament systems in the Bivalvia. Pp. 49–71 in The Bivalvia (B. Morton, ed.). Hong Kong University Press, Hong Kong.

Waller, T. R. 1998. Origin of the molluscan class Bivalvia and a phylogeny of major groups. Pp. 1–45 in Bivalves: An Eon of Evolution (P. A. Johnston and J. W. Haggart, eds.). University of Calgary Press, Calgary, Alberta.

Wheeler, W. C. 1995. Sequence alignment, parameter sensitivity, and the phylogenetic analysis of molecular data. Systematic Biology 44:321–331.

Wheeler, W. C. 1996. Optimization alignment: The end of multiple sequence alignment in phylogenetics? Cladistics 12:1–9.

Wheeler, W. C. 2001. Optimization alignment: Down, up, error, and improvements. Pp. in Techniques in Molecular Systematics and Evolution (R. DeSalle, G. Giribet, and W. C. Wheeler, eds.). Birkhäuser, Basel, Switzerland.

Wheeler, W. C., and D. Gladstein. 2000. POY: The optimization of alignment characters. Program and documentation available online at ftp.amnh.org/pub/molecular.

Whiting, M. F., J. M. Carpenter, Q. D. Wheeler, and W. C. Wheeler. 1997. The Strepsiptera problem: Phylogeny of the holometabolous insect orders inferred from 18S and 28S ribosomal DNA sequences and morphology. Systematic Biology 46:1–68.

Yonge, C. M. 1957. Mantle fusion in the Lamellibranchia. Pubblicazioni della Stazione zoologica di Napoli 29:151–171.

Yonge, C. M. 1969. Functional morphology and evolution within the Carditacea (Bivalvia). Proceedings of the Malacological Society of London 38:493–527.

Yonge, C. M. 1980. Functional morphology and evolution in the Tridacnidae (Mollusca: Bivalvia: Cardiacea). Records of the Australian Museum 33:735–777.

Yonge, C. M. 1982a. Ligamental structure in Mactracea and Myacea (Mollusca: Bivalvia). Journal of the Marine Biological Association of the U.K. 62:171–186.

Yonge, C. M. 1982b. Mantle margins with a revision of siphonal types in Bivalvia. Journal of Molluscan Studies 48:102–103.

Zardus, J. D., and M. P. Morse. 1998. Embryogenesis, morphology and ultrastructure of the pericalymma larva *Acila castrensis* (Bivalvia: Protobranchia: Nuculoida). Invertebrate Biology 117:221–244.

KEVIN J. ROE AND WALTER R. HOEH

4

SYSTEMATICS OF FRESHWATER
MUSSELS (BIVALVIA: UNIONOIDA)

Bivalves of the order Unionoida (also called freshwater mussels or naiades) are a diverse group of freshwater organisms (about 175 genera) with a broad distribution that currently includes all continents except Antarctica (Simpson 1896, 1900, 1914; Haas 1969a; Starobogatov 1970). The group has a fossil record extending back to at least the Triassic (e.g., Henderson 1935; Haas 1969b; Waller 1990, 1998). Habitat destruction and other anthropogenic perturbations have led to rapid population declines in North America and elsewhere such that freshwater mussels may represent the most endangered group of animals (e.g., Bogan 1993, 1998; Williams et al. 1993; Neves et al. 1997). Particular genera (e.g., *Epioblasma*) have been so severely affected that most of the species contained within them are either endangered or presumed extinct (Johnson 1978; Turgeon et al. 1998; Buhay et al. 2002).

Despite the substantial amount of research conducted on them, unionoids are still poorly understood from a systematic perspective. Systematic insights can contribute markedly to unionoid evolutionary and conservation biology research initiatives (e.g., see Moritz 1996; Lydeard and Roe 1998; Roe and Lydeard 1998a; Roe et al. 2001). This chapter provides an historical review of the systematics of unionoid mussels. The review is in two sections: lower-level (genera and below) and higher-level relationships. The lower-level section is almost entirely devoted to North American taxa because virtually all published molecular systematic studies at the lower-level have focused on North American taxa. Many more lower-level studies are needed from other regions of the world. The next section presents a reexamination of higher-level unionoid relationships using an alternative coding of the morphological data (Hoeh et al. 2001). This review will provide a background on unionoid systematics and encourage further study of the group.

LOWER-LEVEL SYSTEMATICS

Unionoid bivalves are a diverse group of mollusks with worldwide distribution. The last decade has seen an increase in interest in all aspects of unionoid mussel biology, particularly in the United States. This increased interest has arisen primarily because of concerns about the continued survival of many species. Some estimates indicate that roughly 70% of all species of unionids in North America are threatened or endangered (Williams et al. 1993; Neves et al. 1997; Master et al. 1998). Before the advent of allied molecular techniques (allozymes, restriction fragment analysis, DNA sequencing), the history of lower-level systematics of unionoids often consisted of the description and redescription of species, largely on the basis of conchological features. The most prolific of all workers was undoubtedly Isaac Lea, who described 850 species of unionoid mussels worldwide (not including fossil taxa).

North America represents a region with a very high diversity of unionoid bivalves (Burch 1975; Lydeard and Mayden 1995; Neves et al. 1997; Turgeon et al. 1998). Despite the increase in interest in unionoids on other continents (for a recent review, see Nagel and Badino 2001), most studies involving generic or species-level relationships have been produced in the United States. Recent studies have indicated that the Unionidae is a paraphyletic group, therefore, we use the term "unionoid" when referring to freshwater mussel taxa as a whole. In this section, we review studies concerning the systematics of unionoid mussels at the generic level and below and examine the utility of various types of character data for investigating evolutionary relationships among them.

Characters

Morphology   Many researchers believed that accurately determining the number of unionoid taxa was hindered by having to rely too heavily on conchological differences (e.g., Davis 1984). Although much of the current classification is based on conchological characters, the high degree of conchological variation exhibited by unionoids can make it impossible to clearly delineate the boundaries between some species based on conchology alone (Davis et al. 1981; Kat 1983a; Davis 1983; Williams and Mulvey 1994). Species delineation can be further complicated by the many conchological convergences and parallelisms in shell shape and external morphology that some think are driven by environmental variables such as substrate composition or water velocity (e.g.,

Johnson 1970; Watters 1994). Typical conchological characters used in the systematics of unionoids include the presence or absence and position or shape of hinge teeth, position of adductor muscle scars, presence, shape, and position of external shell sculpture (pustules, knobs, ridges, sulci, spines), shell thickness, and overall shape. Although most conchological characters used in unionoid systematics are derived from the interior and exterior valve surface, Kat (1983b) examined the taxonomic utility of the microstructure of conchiolin layers deposited within the shells of freshwater mussels. Although the external "anatomy" of unionoid bivalves has provided the bulk of the morphological characters used in systematic studies, the internal or soft anatomy has remained underexploited. Kat (1983a) examined several anatomical characters such as stomach anatomy, mantle edge characteristics, and number of papillae on the exhalant and inhalant siphons. Smith (1986) also examined stomach morphology in comparing members of the Margaritiferidae. In an earlier publication, Smith (1980) compared the mantle and neural anatomy of two margaritiferids and provided several characters that could be included in explicit tests of the relationships within the genus *Margaritifera*. In addition to adult morphology, the utility of larval morphology has also been explored, albeit on a limited basis. In a survey of the larval characters of unionid mussels, Hoggarth (1988, 1999) provided a wealth of morphological characters that have yet to be fully utilized for lower-level systematic analyses. Hoeh (1990) used some of these larval characters and additional characters from Kat (1983c) to develop a morphology-based phylogenetic hypothesis for eastern *Anodonta*.

With few exceptions, the systematic relationships presented in most of the morphological studies mentioned above were arrived at by intuition and not by explicitly analyzing the characters. Within the context of modern character-based phylogenetic analyses, only a few studies have explored the utility of morphological characters to elucidate the relationships between species and genera. Hoeh's (1990) morphology-only tree was based on 10 characters and, not surprisingly, was poorly resolved. A combined analysis of the morphological characters with 24 presumptive allozyme loci, however, did produce a well-resolved topology. Roe and Lydeard (1998b), in examining the molecular systematics of the genus *Potamilus,* tested the hypothesis that the genus was diagnosed by ax-head–shaped glochidia and that the presence of hooklike teeth on the valve edges was indicative of sister relationships of species within *Potamilus*. The results of their analyses did not support the monophyly of *Potamilus,* but indicated that it was paraphyletic with respect to its presumed sister genus *Leptodea,* and that the presence of hooklike teeth on the valve edges was a homoplastic character and not always indicative of phylogenetic relatedness within *Potamilus*. The confidence in the homology of some adult shell characters used to diagnose genera has also been called into question. The genus

Quincuncina is diagnosed primarily by the presence of chevron-shaped ridges on the disk and posterior slope. Lydeard et al. (2000) provided evidence that the chevron-shaped ridges that diagnose *Quincuncina* are not homologous and that *Quincuncina,* as currently recognized, is not a natural group.

The two studies just presented call into question the utility of morphological characters for diagnosing lower-level relationships and seem to support the contention of Davis (1984) and others that morphological characters are often too variable to be useful in determining relationships among species of unionids. It is worth reiterating that the synapomorphic status of most of the morphological characters used by malacologists to diagnose genera or species of unionoid mussels (e.g., Smith 2000) have never been tested within the framework of a modern phylogenetic analysis. Of the few characters that have undergone explicit testing in a phylogenetic context, some (e.g., larval teeth), although homoplastic in some instances, were diagnostic at some level. This indicates that not all morphological characters are too variable for use in determining species-level relationships among unionids. Wiens (2000, 2001) provides invaluable guidelines and methods for including morphological characters in phylogenetic analyses. The methods outlined by Wiens (2001) for analyzing continuous characters and Zelditch et al.'s (2000) discovery of phylogenetic characters in morphometric data seem particularly appropriate for analyzing unionoids.

Allozymes   Davis et al. (1981) and Davis (1984) examined relationships between and within genera of freshwater mussels. In particular, their intrageneric examinations focused on members of the genus *Elliptio*. Davis et al. (1981) determined that sympatric populations of mussels in the *Elliptio complanata* species-group that exhibited distinct phenotypes displayed extremely low levels of genetic divergence. Despite these results, Davis et al. (1981) consistently rejected the concept that the various phenotypes observed represented a single polymorphic species. Instead they interpreted the low genetic diversity as evidence of a recent origin of these species. In a similar study, Davis (1984) found that measures of genetic polymorphism and heterozygosity indicated that the number of species of *Elliptio* has been "considerably underestimated" and that lanceolate-shaped shells appeared to have arisen at least three times in the genus *Elliptio*. Berg et al. (1998) examined intraspecific variation in *Quadrula quadrula* and also found significant differences at 3 of the 10 allozyme loci surveyed. Genetic differences in this study were small, but were positively correlated with geographic distance.

Kat (1983d) examined the amount of genetic variation among species of *Lampsilis* from the Atlantic slope and found that allozyme variation was consistent with the morphological differentiation observed in stomach anatomy. Kat and Davis (1984) tested hypotheses of dispersal by comparing the number of alleles present, allele frequencies, and heterozygosity of peripheral populations of freshwater mussels in Nova Scotia. The results of this study revealed two groups of mussels: one was characterized by low levels of heterozygosity, whereas the other group exhibited moderate levels typical of conspecifics from the northeastern United States. The authors attributed the observed differences in heterozygosity to differences in the dispersal ability of the various species, which was related to whether they used anadromous or saltwater-tolerant host species. This example illustrates how aspects of the life history of an organism can affect the genetic variation in a species. Other aspects of the complex life histories of unionoids have been found to have some effect on the genetic structure of species as well. Hoeh et al. (1998a) examined the allozymic variation among populations of the genus *Utterbackia* and found that hermaphroditic populations of *U. imbecillis* and another undescribed species of *Utterbackia* displayed low levels of within-population variation relative to the dioecious species *U. peggyae* and *U. peninsularis*. These results suggested a high degree of self-fertilization in the hermaphroditic species, and concomitantly high levels of among–population-level variation, which the authors attributed to a combination of self-fertilization and founder events.

Allozyme data have been used to develop explicit hypotheses for the systematic relationships of freshwater mussels, as they have for other organisms. Most allozyme studies of animals have focused on intrageneric relationships (e.g., Gutierrez et al. 1983; Hafner and Nadler 1988), often because at higher levels of divergence, taxa share few electromorphs. Analysis of allozyme data has changed over time. Buth (1984) provides a useful review of the application of allozyme data to systematic questions. Studies on unionoid phylogenetic relationships have used a variety of coding strategies for allozyme data. For example, Hoeh (1990) examined relationships of eastern *Anodonta* and treated the allele as the character, whereas Hoeh et al. (1995) sought to elucidate phylogenetic relationships and the evolution of simultaneous hermaphroditism in the genus *Utterbackia* and coded the locus as the character.

There has also been some debate over inferring evolutionary trees by using the allele frequencies rather than coding the presence/absence of alleles. Opponents of retaining frequency information have cited cases where ancestral taxa are reconstructed as having impossible allele frequencies. Swofford and Berlocher (1987) advocate retaining allele frequency data and have often criticized reducing frequency data to simple presence/absence as ignoring valuable phylogenetic information. Swofford and Berlocher (1987) proposed a parsimony-based method that would allow retention of allele frequency information without the undesirable result of unrealistic estimations of ancestral frequencies.

Early allozyme-based examinations of unionoids frequently used values of genetic similarity, often derived from studies of unrelated organisms as indicators of whether the operational taxonomic units included represented distinct species, subspecies, or members of the same population. Many earlier studies that used allozyme data for examining relationships among freshwater mussels did not use character-based analyses and did not generate explicit phylogenetic hypotheses. This makes it difficult to evaluate the strength of the phylogenetic signal contained within the data. Later studies, such as Hoeh et al. (1995), reported high levels of homoplasy at some loci included in their dataset, although in this case the allozyme-based phylogeny was corroborated by an independent mitochondrial DNA (mtDNA) dataset (Knazek et al. 2001). A review of the studies presented in this section supports Hoeh's (1990) contention that phylogenetic analysis of allozyme data is informative in studies of specific and generic-level relationships.

RFLPs and RAPDs   Restriction fragment length polymorphisms (RFLPs) have been widely used as molecular markers (see Dowling et al. 1996 for a review). Despite their popularity, few RFLP-based studies have been conducted on unionids (White et al. 1994, 1996). Kandl et al. (2001) used a variety of data types including RFLPs, allozyme electrophoresis, and mitochondrial cytochrome *c* oxidase subunit I (COI) sequences to assess the genetic distinctiveness of populations of several species of *Pleurobema* along the eastern Gulf Coast. The authors also wished to clarify the species status of *P. pyriforme* and putative species *P. bulbosum,* and *P. reclusum.* Whereas all data types in the study were able to discriminate between currently recognized species of *Pleurobema,* specimens referable to *P. bulbosum* and *P. reclusum* were not found to be distinct from *P. pyriforme* based on allozyme and RFLP data. However, *Pleurobema reclusum* was found to be genetically distinguishable from *P. pyriforme* using the COI sequence data.

Randomly amplified polymorphic DNA (RAPD; Welsh and McClelland 1990) involves screening genomic DNA for interpretable polymorphisms using a variety of short primers of arbitrary sequence to amplify at random using polymerase chain reaction (PCR). Lieberman (2000), in the only study to date to use RAPDs on unionids, examined the population structure of *Amblema plicata* and analyzed the results for similarities to patterns produced by freshwater fishes (Mayden 1988). The difficulty of assessing the homology of RAPD markers has generally restricted their application to population-level studies (Avise 1994; Hillis 1994), where the potential for comparing nonorthologous bands is minimized. Lieberman's (2000) results reveal some geographic structure in the data and some similarity to Mayden's (1988) results, but as the author states, "such congruence is quite incomplete." One possible explanation for the lack of congruence observed between these two area cladograms is that, unlike the fishes examined by Mayden (1988), which are generally endemic to habitats marked by cool, clear, high-gradient streams, *Amblema plicata* is a widespread species that inhabits large rivers such as the Mississippi, as well as its many tributaries. The cosmopolitan distribution of *Amblema plicata,* combined with the widespread distribution of its 13 known or suspected host fishes, would suggest a panmictic population and little or no geographic structure. That the results did show some local geographic correlation would tend to support the utility of RAPDs for more restricted studies.

DNA Sequences   The advent of the polymerase chain reaction (PCR) (Saiki et al. 1985) has revolutionized both the manner and scope of systematic studies. As early as 1991, DNA sequence data accounted for approximately 25% of all systematic studies published that year (Sanderson et al. 1993). That percentage has undoubtedly increased over the past decade and, although the limitations of DNA sequence data have been realized (see below), sequence data will assuredly play a major role in systematics for many years to come.

The phenomenon of doubly uniparental inheritance (DUI) of mitochondria has been documented in unionids and some other bivalve taxa such as *Mytilus* (Hoeh et al. 1996; Chapter 2, this volume). In most animals, mtDNA is inherited along a matrilineal line, but in taxa that exhibit DUI, males are heteroplasmic, with female mitotypes in the somatic and gonadal tissues and a unique male mitotype in the gonads. Thus, to ensure that orthologous genes are being compared for a phylogenetic study, care should be taken to ensure that all female or all male mitotypes are being examined (see Quesada et al. 1996 for associated problems).

Many molecular systematists working on unionoids have used the primers developed by either Folmer et al. (1994) or Lydeard et al. (1996) that amplified portions of the first subunit of the cytochrome *c* oxidase gene (COI) and the 16S rDNA (16S) genes, respectively. Both of these genes are somewhat conserved, but the protein coding the COI gene often provides ample variation at the third codon position for resolving relationships within and among genera (e.g., Roe and Lydeard 1998b).

One of the earliest studies to include DNA sequences to examine lower-level relationships in unionoids was Mulvey et al. (1997). In this study, the authors used both DNA sequences of the 16S gene and allozyme data in an attempt to identify diagnosably distinct evolutionary entities within *Amblema* and *Megalonaias.* Their results supported the recognition of three species of *Amblema* (*A.*

ellioti, A. neislerii, and *A. plicata*), but only one species of *Megalonaias*. The conservation implications of these findings were significant for species in both of these commercially important genera, because two of the three species of *Amblema—A. neislerii* and *A. ellioti—*occupy restricted ranges. The results did not support the recognition of *Megalonaias boykiniana*, which had previously been considered for federal protection by the U.S. Fish and Wildlife Service (Butler 1993).

Roe and Lydeard (1998b) attempted to resolve the phylogenetic relationships within the genus *Potamilus* and to address the question of relationships between *Potamilus* and its putative sister taxon *Leptodea*, using a portion of the COI gene. The results of the analysis supported retaining most of the species of *Potamilus* as a natural group; however, the placement of *P. capax* rendered *Potamilus* paraphyletic. The authors also examined the degree of genetic variation between two populations of the federally threatened Alabama heelsplitter, *Potamilus inflatus*. The range of this species has been greatly reduced and known reproducing populations were restricted to the Amite River in Louisiana and the Black Warrior River in Alabama. The analyses of Roe and Lydeard (1998b) identified both populations as genetically distinct evolutionary entities and the degree of genetic divergence between these populations exceeded that observed between other species of *Potamilus*.

King et al. (1999) was the first and, to date, the only to investigate the evolutionary relationships within a unionoid genus that included DNA sequence data from a nuclear locus. The authors used sequence data from the first internal transcribed region (ITS-1), which lies between the 5.8S and 18S rDNAs in the nuclear genome. They also used mitochondrial COI gene sequences to investigate the phylogeography of the green floater, *Lasmigona subvirdis*. The authors found significant genetic differentiation between northern and southern populations of *L. subvirdis* and recommended that these populations be managed as separate conservation units. In the course of examining genetic variation within *L. subvirdis*, King et al. also presented a phylogeny of *Lasmigona* based on COI sequences that indicated the genus *Lasmigona* was not a monophyletic group.

Turner et al. (2000) compared two different approaches for determining the level of interspecific variation in *Lampsilis hydiana:* nested cladistic analysis (Templeton et al. 1995) and analysis of molecular variation (AMOVA) on a portion of the 16S gene. Turner et al. (2000) examined individual variation through single-stranded conformation polymorphism (SSCP). This method detects point mutations and insertion–deletion events that can affect the folding of the single-stranded fragments and, by doing so, alter their mobility during electrophoresis. Turner et al. (2000) also sequenced two representatives of each unique haplotype identified using the SSCP method. In general, the results indicated two "fragmentation events," one separating populations in the Arkansas River from the Saline and Ouachita rivers to the southwest and a second more recent event separating the upper Saline populations from those in the lower Saline and the Ouachita rivers. The results of the methodological comparison revealed that under some conditions, nested cladistic analysis provided less insight into the processes shaping population differentiation than traditional AMOVA.

Using sequences from both the COI and the 16S genes, Roe et al. (2001) examined the relationships among four threatened and endangered species of the genus *Lampsilis*. These four species are the only freshwater mussels known to produce superconglutinate lures that are endemic to the Gulf Coast drainages of the United States. Additional goals of the study were to compare the zoogeographic patterns of these taxa to patterns produced by other organisms. The results of the study supported the recognition of these four species as a natural group, although the monophyly of three of the four species was not supported. The authors attributed the lack of observed monophyly of the species to a combination of the recent origin of these species lineages and the conservative nature of the COI and 16S sequences. As did King et al. (1999), Roe et al. (2001) also urged that any plans for augmenting existing population through captive rearing of freshwater mussels should maintain the genetic identity of the respective populations.

As can be seen from these examples, examinations of relationships between species of unionids are often driven, at least in part, by questions relating to the conservation of freshwater mussel species. The importance of phylogenetic systematics to the conservation of freshwater mussels was outlined by Lydeard and Roe (1998). Knowledge of the genetic diversity both within and between populations is crucial to maximizing the benefits of captive breeding and reintroduction of endangered mussel species. In addition, systematic studies can identify previously unrecognized diversity (e.g., cryptic species) in need of protection.

Protecting areas of endemism is also crucial for preserving mussel diversity. Biodiversity hotspots, such as the rivers of the southeastern United States, are also in need of protection. Systematic studies that include these areas can improve estimates of biotic diversity. For example, Lydeard et al. (2000) used COI and 16S sequences in an examination of relationships of mussels from the Gulf Coast drainages. Their analyses identified several genera as polyphyletic and not representative of natural groups. Specifically, *Obovaria rotulata* was found to be more closely related to *Fusconaia ebena* than it was to its putative congeners *O. unicolor* and *O. olivaria*. Furthermore, *F. succisa* was found to be sister to *Quincuncina infucata* and *Q. burkei* sister to *F. escambia*.

Undoubtedly, DNA sequence data will continue to be a valuable source of data for improving our estimates of species boundaries and the systematic relationships of unionoid mussels. What is required, however, is the development of additional markers better suited to the range of systematic questions that unionoids present. These new markers should include nuclear as well as mitochondrial loci and ideally will encompass a wide range of evolutionary rates.

HIGHER-LEVEL SYSTEMATICS

Most recent classifications of the Unionoida (e.g., Boss 1982; Vaught 1989; Bogan and Woodward 1992; Bonetto 1997; but see Starobogatov 1970 for an alternative classification) have recognized two superfamilies: the Unionoidea, containing three families: Hyriidae, Margaritiferidae, and Unionidae, and the Etherioidea (formerly Muteloidea, see Kabat 1997), containing three families: Etheriidae, Iridinidae, and Mycetopodidae. The current geographic ranges of these families are restricted with respect to continental landmasses (e.g., Simpson 1896; Parodiz and Bonetto 1963). Within the Unionoidea, representatives of the Unionidae are found in North America, Eurasia, and Africa. Those of the Hyriidae are restricted to Australasia and South America, and the Margaritiferidae inhabit North America, northwest Africa, and Eurasia. Within the Etherioidea, representatives of the Iridinidae (formerly Mutelidae; see Kabat 1997) are restricted to Africa, whereas those of the Mycetopodidae are found only in Central and South America. The Etheriidae, typically understood to represent the freshwater oysters, is distributed in Africa, India, and South America. However, a recent phylogenetic analysis indicated that the freshwater oysters are a polyphyletic assemblage (Bogan and Hoeh 2000).

The Unionoida contains an array of distinctive morphological characteristics, many of which are associated with reproduction. Two highly differentiated larval morphologies exist within the Unionoida. Species typically included in the Unionoidea possess a bivalved larva called a glochidium, whereas members of the Etherioidea have a strikingly distinct univalved lasidium/haustorium larva (e.g., see Bonetto 1951, 1997; Parodiz and Bonetto 1963; Wachtler et al. 2001). The morphological differences between these larval types are so extraordinary that Parodiz and Bonetto (1963) hypothesized that unionoidean and etherioidean bivalves represented two independent invasions of freshwater. The recent corroboration of unionoid bivalve monophyly (Hoeh et al. 1998b; Graf and Ó Foighil 2000b) suggests that an evaluation of the polarity of the glochidium–lasidium/haustorium transition(s) would indeed be appropriate. Another aspect of reproduction that varies among unionoid higher taxa is the location of larval brooding in modified gills called marsupia. Etherioidean and hyriid bivalves use the two inner demibranchs only (endobranchy), whereas margaritiferid bivalves use all four demibranchs (tetrageny) for brooding. Unionids use either all four or only the two outer demibranchs (ectobranchy) for brooding (e.g., Heard and Guckert 1970). Elucidating the polarity and order of evolutionary transitions among these reproductive character states depends on the availability of robust estimates of phylogeny for the Unionoida.

The particular classification presented above does not imply that a consensus has been reached regarding unionoid evolutionary relationships. Quite the contrary, unionoid bivalves have a long history of protean classification schemes. Early classifications were artificial and subjective, owing to different a priori character selections and weightings, which produced little consensus (see reviews in Heard and Guckert 1970; Davis and Fuller 1981; Parmalee and Bogan 1998). Lamarck (1805, 1812) established the family Nayades (subsequently changed to Naiades [Lamarck 1830]) for the freshwater mussels and placed all species in two genera, *Anodonta* and *Unio*. Rafinesque (1820, 1831, 1832) departed from the typical custom of placing each species in one of a very small number of genera by erecting 37 new freshwater bivalve genera and more than 100 new species. He is also credited with the first use of *Unio* as the root of a freshwater mussel higher taxon (Unioninae) and thus the current ordinal name, Unionoida, dates from Rafinesque 1820 and not Fleming 1828 as has been credited for many years (Bowden and Heppell 1968).

Isaac Lea, in his four editions of the *Synopsis* (1836, 1838, 1852, 1870), was one of the first workers to attempt an in-depth global view for the Unionoida. As was the case for most early unionoid taxonomists, Lea's classifications relied largely on conchological characteristics and placed all species in one of two large genera, *Margaron* and *Platiris*. Specifically, Lea used adult shell sculpture, shell form, and the presence or absence of dorsal shell wings, while at the same time realizing that a better understanding of anatomy would be required for a more permanent classification.

Subsequently, other workers made use of different suites of unionoid characteristics to enable their classifications. For example, Swainson (1840) used shell characteristics and a quinary system of contiguous circles (each representing a group of related species) to divide the Unionidae into five subfamilies. Gray (1840, 1847) established a classification for the freshwater mussels, based on anatomical characteristics, which contained three families: Mutelidae, Mycetopodidae, and Unionidae. Troschel (1847) used anatomical characteristics and von Ihering (1893) used larval characteristics to inform their classifications of the Unionoida. The latter author divided the freshwater mussels into two families: Mutelidae (possessing lasidial larvae) and Unionidae (possessing glochidial larvae). Troschel first noted that "*Anodonta*" from South America (now in the genus *Anodontites*) was actually more closely related to certain African forms than to *Anodonta* of the northern hemisphere.

Charles T. Simpson made significant contributions to global unionoid systematics by producing classifications based on conchological, larval/marsupial characteristics, and the presence/absence of sexual dimorphism (Simpson 1896, 1900, 1914). Simpson (1900, 1914), following von Ihering (1893), divided the freshwater mussels into two families, the Mutelidae and Unionidae, based largely on the presence of glochidial larvae in the latter and lasidial larvae in the former (Figure 4.1A). His 1914 volume is still the most comprehensive compendium on global unionoid alpha-level taxonomy.

Arnold E. Ortmann (e.g., 1910, 1911, 1912, 1916, 1921) greatly extended the work of Simpson. Ortmann integrated all available morphological characteristics, especially adult soft part anatomy, into his classification schemes. Ortmann (1911) was the first to officially recognize the extreme anatomical distinctness of the margaritiferids from other unionoids by raising that group to familial status (e.g., Smith 2001). His later freshwater mussel classifications (1912, 1921) contained three families: Margaritanidae (now Margaritiferidae, Kennard et al. 1925; Haas 1940), Mutelidae, and Unionidae. Ortmann (1912) noted that the Margaritiferidae was, undoubtedly, the most ancient of these three families. Notwithstanding his familial elevation of the margaritiferids, Ortmann's classification (1912, 1921) is substantively distinct from the seemingly similar versions of von Ihering (1893) and Simpson (1900, 1914) in that Ortmann placed the hyriines in his Mutelidae (Figure 4.1B). The other two authors placed them in the Unionidae. The placement of the hyriines in the Unionidae was largely based on the presumed importance of a single, potentially plesiomorphic characteristic, the shared larval type (glochidium), whereas its placement in the Mutelidae was based on multiple shared apomorphic adult anatomical characteristics. Most subsequent unionoid classification schemes have included the hyriines in the Unionidae or Unionoidea (e.g., Modell 1942, 1949, 1964; Parodiz and Bonetto 1963; Haas 1969a, 1969b; Heard and Guckert 1970; Starobogatov 1970; Morrison 1973; Boss 1982; Vaught 1989; Bogan and Woodward 1992; Bonetto 1997; Walker et al. 2001), whereas relatively few have supported Ortmann's position (e.g., Hannibal 1912; Thiele 1935; McMichael and Hiscock 1958; Heard and Guckert 1970: Fig. 1).

The relatively recent integration of explicit tree-building methodologies with biochemical/molecular genetics (e.g., Hillis et al. 1996b) has contributed significantly to our current understanding of unionoid systematics because of their ability to explicitly test hypotheses of evolutionary relationships. Davis and Fuller (1981) made the initial attempt to investigate higher-level unionoid relationships using these new techniques. This investigation was prompted by a

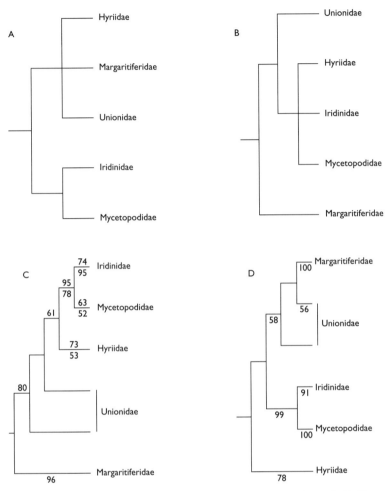

Figure 4.1. Unionoid familial relationships based on the classifications of (A) Simpson (1900) and (B) Ortmann (1912). (C) Phylogenetic analyses of morphological characters from Graf (2000) and Hoeh et al. (2001). (D) Phylogenetic analyses of morphological plus molecular characters after Hoeh et al. (2001). In C and D, bootstrap percentages above branches are from Graf (2000) and those below branches are from Hoeh et al. (2001). Unionid paraphyly is indicated in both C and D.

unionoidean classification, based largely on the duration and location of larval brooding, proposed by Heard and Guckert (1970). Davis and Fuller used immunoelectrophoretic techniques to build distance matrices for 52 species representing 27 genera of the North American Unionoidea (i.e., excluding the hyriids). The matrices were analyzed using ordination and minimum spanning tree techniques to produce two-dimensional plots of relationships. Their results (e.g., Davis and Fuller 1981: Fig. 2) were largely consistent with the unionid classification of Ortmann (1910) in that margaritiferines, anodontines, and unionines + lampsilines were portrayed as three very distinct groups. This led Davis and Fuller (1981: Fig. 4) to propose subfamilial status, within the Unionidae, for those three groups: Margaritiferinae, Anodontinae, and Ambleminae. Additional systematic work, using allozyme data (Davis et al. 1981; Davis 1984), was largely congruent with these determinations. A principal inference from these studies was that unionid reproductive character (e.g., marsupial placement and length of the brooding season) evolution has often been homoplasious. This finding rejects the underlying basis for the unionid classification proposed by Heard and Guckert (1970).

A significant follow-up to the landmark Davis and Fuller study was that of Lydeard et al. (1996). The latter examined 29 species representing 23 genera of North American Unionoidea by comparing 16S ribosomal DNA sequences alone and in conjunction with a morphological dataset. This was the first use of DNA sequences to specifically examine freshwater mussel evolutionary relationships. Parsimony and neighbor-joining techniques were used for their phylogenetic analyses. The resulting phylogenies gave strong support for (1) unionid monophyly and (2) sister taxa status for anodontines and amblemines, thus offering general corroboration of the phylogenetic and reproductive character evolution hypotheses presented in Davis and Fuller (1981). As with the latter hypotheses, those of Lydeard et al. (1996) offer little support for the views of Heard and Guckert (1970) regarding unionid relationships.

Although both Davis and Fuller (1981) and Lydeard et al. (1996) have contributed significantly to our understanding of North American unionid evolutionary relationships, these studies displayed the same significant flaw. As with other regional studies (e.g., Nagel and Badino 2001; Walker et al. 2001), they evaluated unionoid phylogeny and character state transitions. These evaluations were based on estimates of phylogeny for a paraphyletic group (i.e., North American unionoids) given the absence of representative hyriids, iridinids, and mycetopodids from their trees and the paucity of earlier support for Unionidae + Margaritiferidae monophyly. The analyses of nonmonophyletic groups to estimate phylogeny and character state transitions can produce misleading inferences (e.g., see Eldredge and Cracraft 1980; Brooks and McLennan 1991). Thus, to obtain robust estimates of phylogeny and character state transitions for unionoids, multiple representatives from each of the suprageneric taxa within a recognized monophyletic assemblage (i.e., the Unionoida) should be included in the phylogenetic analyses (e.g., see Swofford et al. 1996). Subsequent unionoid systematic work has begun to address this problem.

Two recent parsimony analyses of nonmolecular character matrices have largely supported Ortmann's view (1912, 1921) of higher-level unionoid relationships. Both Graf (2000) and Hoeh et al. (2001) performed unweighted parsimony analyses on qualitative unionoid characteristics that were coded in a multistate fashion. Although these two matrices had only partial overlap in the characters used, the resulting tree topologies were quite similar (Figure 4.1C). Graf's (2000) analysis, based on 38 characters, found a single most parsimonious tree (Graf 2000: Fig. 1), which indicated a "basal" Margaritiferidae, paraphyletic Unionidae and Unionoidea, and a monophyletic group that contained etheriids, hyriids, mycetopodids, and iridinids. This tree's placement of hyriids as the sister taxon to etheriids + mycetopodids + iridinids is inconsistent with the classification of Simpson (1900, 1914), but consonant with that of Ortmann (1912, 1921). The Hoeh et al. (2001) morphological analysis, based on 28 morphological characters, found 240 equally parsimonious trees, where the strict consensus tree (Hoeh et al. 2001: Fig. 14.2) indicated the same major features as Graf's (2000) analysis. However, evidential support for most nodes in these two studies was meager. The general topology supported in both studies is consistent with tetragenous brooding of glochidial larvae as the ancestral unionoid condition.

Recent phylogenetic analyses of partial COI sequences confirmed the monophyly of the Paleoheterodonta (Unionoida + *Neotrigonia*) and the sister taxa status for *Neotrigonia* and the Unionoida (Hoeh et al. 1998b; Graf and Ó Foighil 2000b). Graf and Ó Foighil's (2000b: Fig. 2B) analysis of partial COI sequences did not corroborate the nonmolecular analyses of Ortmann (1912), Graf (2000), and Hoeh et al. (2001). However, the Graf and Ó Foighil study did not include a representative of the Mycetopodidae and lacked replicate taxon sampling for the iridinids and hyriids. Nevertheless, after the application of successive weighting techniques (Farris 1969), this study produced relatively high nodal support values (jackknife percentages) for the Paleoheterodonta (99), Unionoida (96), Unionoidea (86), Margaritiferidae (99), Anodontinae (76), Ambleminae (94), and Lampsilini (72).

In addition to the morphology-based analysis mentioned above, a total evidence analysis (partial COI sequences plus morphology) of unionoid phylogeny was presented in Hoeh et al. (2001: Fig. 1D). This topology suggests that both the Unionidae and Unionoidea (Hyriidae + Margaritiferidae + Unionidae), as currently conceived, are not monophyletic groups. This topology is distinct from those presented above, given the placement of the Hyriidae as a product of the earliest cladogenic event within the ancestral unionoid lineage. However, the results of this analysis and that of Graf and Ó Foighil (2000b) are consistent with the hypothesis that endobranchy is the brooding state of the ancestral

unionoid. It is worthwhile to emphasize that the family-level evolutionary relationships indicated by the non-molecule–based phylogenetic analyses (Graf
2000; Hoeh et al. 2001) have not been corroborated (or strongly rejected because of relatively weak support at crucial nodes) by these additional analyses
using many more characters (partial COI DNA sequences, Graf and Ó Foighil
2000b; partial COI DNA sequences together with morphology, Hoeh et al. 2001).

Attempts at rigorous phylogenetic analyses of unionoid nonmolecular traits
are hampered by a relatively small number of coded characters that can contain
a significant level of homoplasy (e.g., Graf 2000; Hoeh et al. 2001). These situa-
tions can yield multiple, equally parsimonious trees with low nodal confidence
scores. Although additional traits are currently being added to the matrices mentioned above, further exploration of alternative character coding and analytical techniques is warranted at this time. The appropriateness of multistate versus presence/absence coding for qualitative characteristics has recently been
debated at length (e.g., Pimentel and Riggins 1987; Meier 1994; Pleijel 1995;
Wilkinson 1995; Lee and Bryant 1999; Strong and Lipscomb 1999; Seitz et al.
2000), with no clear consensus emerging. The same could be said for the use of
"successive weighting" in parsimony analyses (e.g., Farris 1969; Carpenter
1988; Carpenter et al. 1993; Carpenter 1994; Swofford et al. 1996; Cunningham 1997b). Presence/absence coding has yet to explored for use in unionoid
phylogenetic studies, but there is precedent for the use of successive weighting techniques (Lydeard et al. 1996; Graf and Ó Foighil 2000b). Nevertheless,
further comparisons of the results using alternatively coded qualitative characteristics and *a posteriori* weighting techniques would be a useful undertaking
given the disparate and relatively unresolved estimates of higher-level unionoid
evolutionary history we currently have.

To explore the usefulness of alternative coding and analytical techniques for
morphology-based unionoid phylogenetics, the 28-character, multistate-coded
data matrix from Hoeh et al. (2001) was recoded into a 57-character, pres-
ence/absence matrix (Table 4.1) from 31 bivalve taxa (Table 4.2), and analyzed
with the parsimony algorithm in PAUP* v4.0b8 (Swofford 2002) using successive weighting based on the rescaled consistency index. Ultimately, six
equally parsimonious trees were found and the strict consensus tree is presented
in Figure 4.2. The sister taxa relationship for the Hyriidae and Etherioidea, indicated by the non-molecule–based analyses of Graf (2000) and Hoeh et al.
(2001), is supported by this tree; however, unlike the previous two analyses,
Figure 4.2 indicates that the Unionidae and Margaritiferidae are sister taxa, albeit with little evidential support. Generally, the nodal support percentages,
based on 10,000 replicates, are higher in this analysis than in those of the previously mentioned two studies. Nevertheless, the evolutionary propinquity of

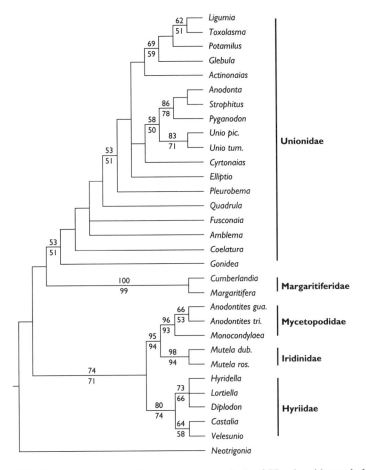

Figure 4.2. Strict consensus tree from parsimony analysis of 57 unionoid morphological traits using presence/absence character coding and successive weighting. Numbers represent bootstrap (above branch) and jackknife (below branch) percentages from 10,000 replicates using fast-heuristic searches. Only percentages ≥50% are shown.

the Hyriidae and Etherioidea was indicated when either equally weighted analyses of multistate or successively weighted analysis of presence/absence coded qualitative nonmolecular characters was used to estimate unionoid phylogeny. The amount of resolution and nodal support levels were generally higher in the latter analysis.

An examination of both nonmolecular and molecular datasets in a total evidence approach (e.g., Miyamoto 1985; Kluge 1989, 1998), using presence/absence coding of the nonmolecular characters and successive weighting tech-

Table 4.1

Presence/absence morphology data matrix for the bivalve taxa listed in Table 4.2

	Character Number					
Taxon/Node	123456789	1111111111 0123456789	2222222222 0123456789	3333333333 0123456789	4444444444 0123456789	55555555 01234567
Neotrigonia	010000100	01??011000	1000000000	0000000001	0000000100	11001000
Actinonaias	110010000	1010010100	1001011001	0010000010	0100100010	00010100
Amblema	110010000	1010011000	1010011001	0010000100	0100101000	00010100
Anodonta	110010000	1010011000	1001011001	1010011000	0000000010	00010100
Anodontites gua.	010010000	1001010110	0100010101	0101000000	0000010000	00101000
Anodontites tri.	010010000	1001010110	0100010101	0101000000	0000010000	00101000
Castalia	001010000	1010010110	0100011010	0010011001	0100100100	01000110
Coelatura	110010000	1010011000	1010011010	0010000100	0100100001	000101??
Cumberlandia	010000001	0010101000	1010000000	0010000100	1001000010	10010100
Cyrtonaias	110010000	1010011000	1001111001	001000???0	0100101000	00010100
Diplodon	001010000	0110010110	0100011010	0010011000	1001000100	00010110
Elliptio	110010000	1010011000	1001011001	0010000010	0100100010	00010100
Fusconaia	110001000	1010011000	1010011001	0010000100	0100100010	00010100
Glebula	11001000?	??10010100	1001111001	0010000010	0100100000	00010101
Gonidea	11000100?	??10011000	1010010100	0010000100	0100100010	00010100
Hyridella	00101000?	????010101	0100011010	0010011000	1001000001	01000110
Ligumia	110010010	1010010100	1001111001	0010000010	0100101000	00100101
Lortiella	???????0?	????01????	?100011010	001001???0	1001000000	0???????
Margaritifera	010010001	0010101000	1010000000	0010000100	1001000010	10010100
Monocondylaea	001010000	0101010110	01000101??	0101000000	0000110100	00101000
Mutela dub.	00010000?	????010110	0100010101	0000100000	0010010000	001001??
Mutela ros.	00010000?	????010110	0100010101	0000100000	0010010000	001001??
Pleurobema	110001000	1010011000	1001011001	0010000010	0100100010	00010100
Potamilus	110010000	1010010100	1001111001	0010000000	0100100010	01000101
Pyganodon	110010000	1010011000	1001011001	1010011000	0000001000	00010100
Quadrula	110000100	1010011000	1010011001	0010000010	0100100010	00010100
Strophitus	110010000	1010011000	1001011001	1010011000	0000000010	00010100
Toxolasma	11001001?	??10011000	1001111001	0010000010	0100100010	01000101
Unio pic.	110010000	1010011000	1001011010	0010011000	0100101000	00010100
Unio tum.	110010000	1010011000	1001011010	0010011000	0100101000	00010100
Velesunio	0???1000?	????010101	0100011010	0010011000	0100100000	01000110

Morphological characters and character states based on literature and specimen examination:

0 = absent, 1 = present in each:

1. Supra-excurrent opening.

2. Posterior end of mantle sheets unfused, with simple incurrent and excurrent openings.

3. Posterior end of mantle sheets with a simple incurrent opening, but fused to provide a short, muscular excurrent siphon.

4. Posterior end of mantle sheets fused to provide short, muscular incurrent and excurrent siphons.

5. Simple incurrent papillae.

6. Branched incurrent papillae.

7. Arborescent incurrent papillae.

8. Mantle margin of females with specialized structures (e.g., flaps, caruncles, etc.) anteroventral to the incurrent opening.

Table 4.1 continued

9. Position of anus on the posterior adductor muscle at the dorsal aspect of posterior adductor muscle.

10. Position of anus on the posterior adductor muscle at the posterior aspect of posterior adductor muscle.

11. Position of anus on the posterior adductor muscle at the posteroventral aspect of posterior adductor muscle.

12. Intestine morphology simple and undifferentiated.

13. Intestine morphology complex, with three compartments.

14. Attachment of the dorsal margin of the outer lamella of the outer demibranchs to the inner surface of the mantle except at the posterior end of those demibranchs.

15. Attachment of the dorsal margin of the outer lamella of the outer demibranchs to the inner surface of the mantle for the entire length of those demibranchs.

16. Attachment of the dorsal margin of the inner lamella of the inner demibranchs to the visceral mass only at the anterior region of those demibranchs.

17. Attachment of the dorsal margin of the inner lamella of the inner demibranchs to the visceral mass for the entire length of the visceral mass.

18. Diaphragm (tissue separation of the suprabranchial and branchial components of the mantle cavity) complete, formed in part by the dorsal margin of the inner lamella of the inner demibranchs and in part by the siphonal musculature.

19. Diaphragm (tissue separation of the suprabranchial and branchial components of the mantle cavity) incomplete, with a single perforation in the siphonal musculature.

20. Diaphragm (tissue separation of the suprabranchial and branchial components of the mantle cavity) incomplete, formed only by the inner demibranchs.

21. Endobranchous (only the inner two gills) brooding.

22. Tetragenous (all four gills) brooding.

23. Ectobranchous (only the outer two gills) brooding.

24. Marsupial region of marsupial demibranchs seasonally extends below the ventral margin of the filaments.

25. Interlamellar space of all marsupial and nonmarsupial demibranchs divided into vertical watertubes by vertical, transverse interlamellar septa.

26. Relative number and spacing of transverse (primary) marsupial septa greater in marsupial than in nonmarsupial regions of marsupial and in nonmarsupial demibranchs.

27. Relative number and spacing of transverse (primary) marsupial septa similar in marsupial and nonmarsupial regions of marsupial and in nonmarsupial demibranchs.

28. Transverse (primary) septa are vertically perforated in marsupial but imperforate in nonmarsupial regions of marsupial demibranchs and throughout nonmarsupial demibranchs.

29. Transverse (primary) septa are vertically imperforate throughout marsupial and nonmarsupial demibranchs.

30. Tripartite water tubes.

31. Lateral ridges on primary septa approximating secondary vertical septa.

32. Glochidium larva.

33. Lasidium larva.

34. Haustorium larva.

35. Medioventral glochidial hooks.

36. Subtriangular glochidial shape (lateral view).

37. Subcircular glochidial shape (lateral view).

38. Subovate glochidial shape (lateral view).

39. Radial disk sculpture.

40. Reduced lateral teeth dentition.

41. Full lateral teeth dentition.

42. Pseudotaxodont lateral teeth dentition.

43. Reduced pseudocardinal teeth dentition.

44. Full pseudocardinal teeth dentition.

45. V-shaped lamellar-ligament fossette.

46. Double-looped beak sculpture.

47. Radial beak sculpture.

48. Concentric barred beak sculpture.

49. Zigzag beak sculpture.

50. Lateral muscle scars.

51. Triangular shaped labial palp.

52. Semicircular to kidney shaped labial palp.

53. Falciform shaped labial palp.

54. Elongate anterior adductor muscle shape.

55. Round anterior adductor muscle shape.

56. Marsupium restricted to middle of gill.

57. Marsupium restricted to posterior of gill.

Table 4.2

Bivalve species examined in this study

Taxonomic Position

Outgroup
Order Trigonioida
    Superfamily Trigonioidea
        Family Trigoniidae
            Neotrigonia margaritacea (Lamarck 1804)
Ingroup
Order Unionoida
    Superfamily Etherioidea
        Family Iridinidae
            Mutela dubia (Gmelin 1791)
            Mutela rostrata (Rang 1835)
        Family Mycetopodidae
            Anodontites guanarensis Marshall 1927
            Anodontites trigonus (Spix 1827)
            Monocondylaea minuana d'Orbigny 1835
    Superfamily Unionoidea
        Family Hyriidae
            Castalia stevensi (H. B. Baker 1930)
            Diplodon deceptus Simpson 1914
            Hyridella menziesi (Gray 1843)
            Lortiella rugata (Sowerby 1868)
            Velesunio angasi (Sowerby 1867)
        Family Margaritiferidae
            Cumberlandia monodonta (Say 1829)
            Margaritifera margaritifera (Linnaeus 1758)
        Family Unionidae
            Actinonaias ligamentina (Lamarck 1819)
            Amblema plicata (Say 1817)
            Anodonta cygnea (Linneaus 1758)
            Coelatura aegyptiaca (Cailliaud 1827)
            Cyrtonaias tampicoensis (Lea 1838)
            Elliptio dilatata (Rafinesque 1820)
            Fusconaia flava (Rafinesque 1920)
            Glebula rotundata (Lamarck 1819)
            Gonidea angulata (Lea 1838)
            Ligumia recta (Lamarck 1819)
            Pleurobema clava (Lamarck 1819)
            Potamilus alatus (Say 1817)
            Pyganodon grandis (Say 1829)
            Quadrula quadrula (Rafinesque 1820)
            Strophitus undulatus (Say 1817)
            Toxolasma lividus (Rafinesque 1831)
            Unio pictorum Linnaeus 1758
            Unio tumidus Retzius 1788

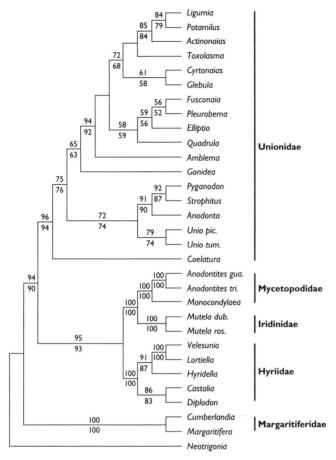

Figure 4.3. Single most parsimonious tree from a simultaneous analysis of 687 (243 parsimony-informative) unionoid morphological and molecular traits using successive weighting. Numbers represent bootstrap (above branch) and jackknife (below branch) percentages from 1,000 replicates using heuristic searches. Only percentages ≥50% are shown.

niques, could be used to evaluate the stability of the non-molecule–based topology (Figure 4.2). To that end, the 57-character presence/absence-coded morphology data matrix was combined with the 630-character transformed (only transversions coded at third positions) COI matrix, 186 parsimony-informative characters from Hoeh et al. (2001) and analyzed with successive weighting techniques using the parsimony algorithm in PAUP* v4.0b8. The single shortest tree obtained from this analysis is presented in Figure 4.3. Despite the detection of significant incongruence between the morphology and COI datasets ($p = 0.003$), using the character congruence (ILD) test (Farris et al. 1994; Chapter 3, this volume), with or without the deletion of invariant characters (Cunningham 1997a), the bootstrap support levels determined in the total evidence analysis for unionoid families and higher taxa are surprisingly high. The major unionoid family-level relationships indicated by the non-molecule–based analyses of Graf (2000) and Hoeh et al. (2001) are supported by the total evidence topology, except that here the Unionidae is monophyletic. This well-supported analysis represents the best-resolved, best-supported hypothesis of unionoid bivalve higher-order relationships produced to date. It is interesting to emphasize that despite the significant level of incongruence detected between the 57-character morphology dataset and the 630-character COI dataset, the total evidence analysis produced higher levels of bootstrap support for some of the same unionoid higher taxa clades supported in the morphology analysis. At least three of these clades were not evident in the strict consensus parsimony trees from both unweighted and successively weighted analyses of the transformed COI matrix alone (trees not shown). It seems reasonable to infer that congruence among the more consistent characters in the two datasets produced the high degree of support for the family and higher-level relationships obtained in the total evidence analysis.

If presence/absence character coding and successive weighting techniques are ultimately judged appropriate for these phylogenetic analyses, then the generally higher resolution and nodal support provided by them yield an increased level of confidence in the findings of the morphology and total evidence-based estimates of unionoid phylogeny presented here. Important evolutionary inferences regarding the Unionoida obtained from our total evidence analysis include the following: (1) tetragenous brooding is the ancestral condition, (2) the glochidium is the ancestral larval type, (3) endobranchous brooding evolved a single time from a tetragenous ancestor, and (4) ectobranchous brooding evolved at least twice from a tetragenous ancestor. Most of these hypotheses (except the fourth) are consistent with the classification of Ortmann (1912) and the unionoid tree of Heard and Guckert (1970: Fig. 1), but some run counter to findings from recent analyses based largely on mtDNA sequences as discussed above (Bogan and Hoeh 2000: Fig. 1; Graf and Ó Foighil 2000b: Fig. 3; Hoeh et al. 2001: Fig. 14.9).

Although a total evidence approach somewhat obviates their necessity, discussions regarding the existence and detection of incongruence among different data partitions are ongoing in systematics, as are discussions regarding the relative utility of morphological and molecular characters (e.g., Baker et al. 1998; Hillis and Wiens 2000; Yoder et al. 2001). The apparent incongruence between morphology and DNA-based estimates of unionoid phylogeny (e.g., Hoeh et al. 2001: Fig. 14.3) may be real in that single gene-based phylogenies need not accurately trace organismic phylogeny (e.g., Pamilo and Nei 1988;

Doyle 1992; Bull et al. 1993; Hoeh et al. 1997). Because animal mtDNA typically lacks recombination (but see Awadalla et al. 1999; Hagelberg et al. 1999; Ladoukakis and Zouros 2001), which causes the genes to be inherited as a single unit, simply adding sequences from other mitochondrial genes to our database will not counter this effect if it exists. To date, DNA-based analyses of unionoid phylogeny have largely been implemented using mtDNA sequences (e.g., Lydeard et al. 1996; Hoeh et al. 1998b, 2001; Roe and Lydeard 1998b; Bogan and Hoeh 2000; Graf and Ó Foighil 2000b; Roe et al. 2001; but see King et al. 1999; Graf and Ó Foighil 2000a; Graf 2002). Alternatively, the apparent incongruence between non-molecule–based and DNA-based estimates of phylogeny could be illusory: it may simply represent inadequate sampling of taxa and the relatively small morphological datasets used to date. Further sampling of additional morphological and molecular characteristics is necessary to address these alternative hypotheses. Phylogenetic analyses of the two independently inherited mtDNA genomes contained within the Unionoida (e.g., Hoeh et al. 1996; Liu et al. 1996) may offer a unique opportunity to further our understanding of the group's phylogeny. The increased resolution and strongly supported nodes generated using a total evidence approach combined with successive weighting suggest that significant advances in our understanding of unionoid systematics may be possible with the data currently available.

ACKNOWLEDGMENTS

We would like to thank two anonymous reviewers for their comments on an earlier version of this chapter. Arthur E. Bogan and Jeanne M. Serb provided valuable assistance and additional editorial comments on several aspects of the project. We also would like to thank the editors for the opportunity to contribute to this publication. K. Roe was a postdoctoral researcher at Saint Louis University and was supported by National Science Foundation grant DEB-0296162 during the preparation of this chapter.

REFERENCES

Avise, J. C. 1994. Molecular markers, natural history and evolution. Chapman and Hall, New York.

Awadalla, P., A. Eyre-Walker, and J. Maynard Smith. 1999. Linkage disequilibrium and recombination in hominid mitochondrial DNA. Science 286:2524–2525.

Baker, R. H., X. Yu, and R. DeSalle. 1998. Assessing the relative contribution of molecular and morphological characters in simultaneous analysis. Molecular Phylogenetics and Evolution 9:427–436.

Berg, D. J., E. G. Cantonwine, W. R. Hoeh, and S. I. Guttman. 1998. Genetic structure of *Quadrula quadrula* (Bivalvia: Unionidae): Little variation across large distances. Journal of Shellfish Research 17:1365–1373.

Bogan, A. E. 1993. Freshwater bivalve extinctions (Mollusca: Unionoida): A search for causes. American Zoologist 33:599–609.

Bogan, A. E. 1998. Freshwater molluscan conservation in North America: Problems and practices. Journal of Conchology Special Publication 2:223–230.

Bogan, A. E., and W. R. Hoeh. 2000. On becoming cemented: Evolutionary relationships among the genera in the freshwater bivalve family Etheriidae (Bivalvia: Unionoida). Pp. 159–168 in The Evolutionary Biology of the Bivalvia (E. M. Harper, J. D. Taylor and J. A. Crame, eds.). The Geological Society of London, London.

Bogan, A. E., and F. R. Woodward. 1992. A review of the higher classification of Unionoida (Mollusca: Bivalvia). Pp. 17–18 in 11th International Malacological Congress, Abstracts (F. Giusti and G. Manganelli, eds.). Vienna, Austria.

Bonetto, A. A. 1951. Acerca de las formas larvales de Mutelidae Ortmann. Jornadas Icticas (Sante Fé) 1:1–8.

Bonetto, A. A. 1997. The freshwater oysters (Muteloidea: Mutelidae): Their taxonomy and geographic distribution in the whole of the naiads of the world. Biociencias 5:113–142.

Boss, K. J. 1982. Mollusca. Pp. 945–1166 in Synopsis and Classification of Living Organisms (S. P. Parker, ed.). McGraw Hill, New York.

Bowden, J., and D. Heppell. 1968. Revised list of British Mollusca. 2. Unionacea–Cardiacea. Journal of Conchology 26:237–272.

Brooks, D. R., and D. A. McLennan. 1991. Phylogeny, Ecology, and Behavior: A Research Program in Comparative Biology. The University of Chicago Press, Chicago.

Buhay, J. E., J. M. Serb, C. R. Dean, Q. Parham, and C. Lydeard. 2002. Conservation genetics of two endangered unionid bivalve species: *Epioblasma florentina walkeri* and *Epioblasma capsaeformis* (Unionidae: Lampsilini). Journal of Molluscan Studies 68:385–391.

Burch, J. B. 1975. Freshwater Unionacean Clams (Mollusca: Pelecypoda) of North America (rev. ed.). Malacological Publications, Hamburg, MI.

Bull, J. J., J. P. Huelsenbeck, C. W. Cunningham, D. L. Swofford, and P. J. Waddell. 1993. Partitioning and combining data in phylogenetic analysis. Systematic Biology 42:384–397.

Buth, D. G. 1984. The application of electrophoretic data in systematic studies. Annual Review of Ecology and Systematics 15:501–522.

Butler, R. S. 1993. Results of a status survey for eight freshwater mussels (Bivalvia: Unionidae) endemic to eastern Gulf slope drainages of the Apalachicolan Region of southeast Alabama, southwest Georgia, and north Florida. Final Report. U.S. Fish and Wildlife Service, Jacksonville, FL.

Carpenter, J. M. 1988. Choosing among equally parsimonious cladograms. Cladistics 4:291–296.

Carpenter, J. M. 1994. Successive weighting, reliability and evidence. Cladistics 10:215–220.

Carpenter, J. M., J. E. Strassmann, S. Turillazzi, C. R. Hughes, C. R. Solis, and R. Cervo. 1993. Phylogenetic relationships among paper wasp social parasites and their hosts (Hymenoptera: Vespidae: Polistinae). Cladistics 9:129–146.

Cunningham, C. W. 1997a. Can three incongruence tests predict when data should be combined? Molecular Biology and Evolution 14:733–740.

Cunningham, C. W. 1997b. Is congruence between data partitions a reliable predictor of phylogenetic accuracy? Empirically testing an iterative procedure for choosing among phylogenetic methods. Systematic Biology 46:464–478.

Davis, G. M. 1983. Relative roles of molecular genetics, anatomy, morphometrics and ecology in assessing relationships among North American Unionidae (Bivalvia). Pp. 193–222 in Protein Polymorphism: Adaptive and Taxonomic Significance (G. S. Oxford and D. Rollinson, eds.). Systematics Association Special Volume 24, Academic Press, London.

Davis, G. M. 1984. Genetic relationships among some North American Unionidae (Bivalvia): Sibling species, convergence and cladistic relationships. Malacologia 25:629–648.

Davis, G. M., and S. L. H. Fuller. 1981. Genetic relationships among Recent Unionacea (Bivalvia) of North America. Malacologia 20:217–253.

Davis, G. M., W. H. Heard, S. L. H. Fuller, and C. Hesterman. 1981. Molecular genetics and speciation in *Elliptio* and its relationships to other taxa of North American Unionidae (Bivalvia). Biological Journal of the Linnean Society 15:131–150.

Dowling, T. E., C. Moritz, J. D. Palmer, and L. H. Rieseberg. 1996. Nucleic acids III: Analysis of fragments and restriction sites. Pp. 249–320 in Molecular Systematics, 2nd ed., (D. M. Hillis, C. Moritz, and B. K. Mable, eds.). Sinauer Associates, Sunderland, MA.

Doyle, J. J. 1992. Gene trees and species trees: Molecular systematics as one-character taxonomy. Systematic Botany 17:144–163.

Eldredge, N., and J. Cracraft. 1980. Phylogenetic Patterns and the Evolutionary Process. Columbia University Press, New York.

Farris, J. S. 1969. A successive approximations approach to character weighting. Systematic Zoology 18:374–385.

Farris, J. S., M. Kallersjo, A. G. Kluge, and C. Bult. 1994. Testing significance of incongruence. Cladistics 10:315–319.

Flemming, J. 1828. A History of British Animals. Bell and Bradfute, Edinburgh.

Folmer, O., M. Black, W. Hoeh, R. Lutz, and R. L. Vrijenhoek. 1994. DNA primers for amplification of mitochondrial cytochrome *c* oxidase subunit I from diverse metazoan invertebrates. Molecular Marine Biology and Biotechnology 3:294–299.

Graf, D. L. 2000. The Etherioidea revisited: A phylogenetic analysis of hyriid relationships (Mollusca: Bivalvia: Paleoheterodonta: Unionoida). Occasional Papers of the Museum of Zoology, Michigan, 1–21.

Graf, D. L. 2002. Molecular phylogenetic analysis of two problematic freshwater mussel genera (*Unio* and *Gonidea*) and a re-evaluation of the classification of Nearctic Unionidae (Bivalvia: Palaeoheterodonta: Unionoida. Journal of Molluscan Studies 68:65–71.

Graf, D. L., and D. Ó Foighil. 2000a. Molecular phylogenetic analysis of 28S rDNA supports a Gondwanan origin for Australasian Hyriidae (Mollusca: Bivalvia: Unionoida). Vie Et Milieu 50:245–254.

Graf, D. L., and D. Ó Foighil. 2000b. The evolution of brooding characters among the freshwater pearly mussels (Bivalvia: Unionoidea) of North America. Journal of Molluscan Studies 66:157–170.

Gray, J. E. 1840. Synopsis of the contents of the British Museum, London.

Gray, J. E. 1847. A list of the genera of Recent Mollusca, their synonyms and types. Proceedings of the Zoological Society of London 15:129–219.

Gutierrez, R. J., R. M. Zink, and S. Y. Yang. 1983. Genetic variation, systematic and biogeographic relationships of some Galliform birds. Auk 100:33–40.

Haas, F. 1940. A tentative classification of the Palearctic unionids. Field Museum of Natural History Zoology Series 24:115–141.

Haas, F. 1969a. Superfamilia Unionacea. Pp. x+663 in Das Tierreich, vol. 88, Walter de Gruyter and Co., Berlin.

Haas, F. 1969b. Superfamily Unionacea. Pp. 411–467 in Mollusca 6 (Bivalvia), vol. Part N. Treatise on Invertebrate Paleontology. The Geological Society of America and University of Kansas, Lawrence, KS.

Hafner, M. S., and S. A. Nadler. 1988. Phylogenetic trees support the coevolution of parasites and their hosts. Nature 332:258–259.

Hagelberg, E., N. Goldman, P. Lio, S. Whelan, W. Schiefenhovel, J. B. Clegg, and D. K. Bowden. 1999. Evidence for mitochondrial DNA recombination in a human population of island Melanesia. Proceedings of the Royal Society of London B 266:485–492.

Hannibal, H. 1912. A synopsis of the Recent and Tertiary freshwater Mollusca of the Californian Province, based upon an ontogenetic classification. Proceedings of the Malacological Society of London 10:112–211.

Heard, W. H., and R. H. Guckert. 1970. A re-evaluation of the Recent Unionacea (Pelecypoda) of North America. Malacologia 10:333–355.

Henderson, J. 1935. Fossil non-marine Mollusca of North America. Geological Society of America Special Papers, vol. 3, 313 pp.

Hillis, D. M. 1994. Homology in molecular biology. Pp. 339–367 in Homology: The Hierarchical Basis of Comparative Biology (B. K. Hall, ed.). Oxford University Press, New York.

Hillis, D. M., Mable, B. K., and Moritz, C. 1996b. Molecular Systematics, Sinauer Associates, Sunderland, MA.

Hillis, D. M. and J. J. Wiens. 2000. Molecules versus morphology in systematics: Conflicts, artifacts, and misconceptions. Pp. 1–19 in Phylogenetic Analysis of Morphological Data (J. J. Wiens, ed.). Smithsonian Institution Press, Washington, DC.

Hoeh, W. R. 1990. Phylogenetic relationships among Eastern North American *Anodonta* (Bivalvia: Unionidae). Malacological Review 23:63–82.

Hoeh, W. R., M. B. Black, R. G. Gustafson, A. E. Bogan, R. A. Lutz, and R. C. Vrijenhoek. 1998b. Testing alternative hypotheses of *Neotrigonia* (Bivalvia: Trigonioida) phylogenetic relationships using cytochrome *c* oxidase subunit I DNA sequences. Malacologia 40:267–278.

Hoeh, W. R., A. E. Bogan, and W. H. Heard. 2001. A phylogenetic perspective on the evolution of morphological and reproductive characteristics in the Unionoida. Pp. 257–280 in Evolution of the Freshwater Mussels (*Unionoida*), vol. 145. Ecological Studies (G. Bauer and K. Wachtler, eds.). Springer-Verlag, Berlin.

Hoeh, W. R., K. S. Frazer, E. Naranjo-Garcia, M. B. Black, D. J. Berg, and S. I. Guttman. 1998. Correlation between mating system and distribution of genetic variation in *Utterbackia* (Bivalvia: Unionidae). Journal of Shellfish Research 17:1383–1393.

Hoeh, W. R., K. S. Frazer, E. Naranjo-Garcia, and R. J. Trdan. 1995. A phylogenetic perspective on the evolution of simultaneous hermaphroditism in a freshwater mussel clade (Bivalvia: Unionidae: *Utterbackia*). Malacological Review 28:25–42.

Hoeh, W. R., D. T. Stewart, C. Saavedra, B. W. Sutherland, and E. Zouros. 1997. Phylogenetic evidence for role-reversals of gender-associated mitochondrial DNA genomes in *Mytilus* (Bivalvia: Mytilidae). Molecular Biology and Evolution 14:959–967.

Hoeh, W. R., D. T. Stewart, B. W. Sutherland, and E. Zouros. 1996. Multiple origins of gender-associated mitochondrial DNA lineages in bivalves. (Mollusca: Bivalvia). Evolution 50:2276–2286.

Hoggarth, M. A. 1988. The use of glochidia in the systematics of the Unionidae (Mollusca: Bivalvia). Ph.D. thesis, Ohio State University.

Hoggarth, M. A. 1999. Descriptions of some of the glochidia of the Unionidae (Mollusca: Bivalvia). Malacologia 41:1–118.

Johnson, R. I. 1970. The systematics and zoogeography of the Unionidae (Mollusca: Bivalvia) of the southern Atlantic Slope Region. Bulletin of the Museum of Comparative Zoology 140:263–449.

Johnson, R. I. 1978. Systematics and zoogeography of *Plagiola* (= *Dysnomia* = *Epioblasma*), an almost extinct genus of freshwater mussels (Bivalvia: Unionidae) from middle North America. Bulletin of the Museum of Comparative Zoology 148:239–320.

Kabat, A. R. 1997. Correct family names for the freshwater "muteloid" bivalves (Unionoida: Etherioidea). Occasional Papers on Mollusks, Harvard University 5:379–392.

Kandl, K. L., H.-P. Liu, R. Butler, W. R. Hoeh, and M. Mulvey. 2001. A genetic approach to resolving taxonomic ambiguity among *Pleurobema* (Bivalvia: Unionidae) of the eastern Gulf Coast. Malacologia 43:87–101.

Kat, P. W. 1983a. Patterns of electrophoretic and morphological variability in a widely distributed unionid: An initial survey. Netherlands Journal of Zoology 33:21–40.

Kat, P. W. 1983b. Conchiolin layers among the Unionidae and Margaritiferidae (Bivalvia): Microstructural characteristics and taxonomic implications. Malacologia 24:298–311.

Kat, P.W. 1983c. Genetic and morphological divergence among nominal species of North American *Anodonta* (Bivalvia: Unionidae). Malacologia, 23:361–374.

Kat, P.W. 1983d. Morphologic divergence, genetics, and speciation among *Lampsilis* (Bivalvia: Unionidae). Journal of Molluscan Studies, 49:133–145.

Kat, P. W. and G. M. Davis. 1984. Molecular genetics of peripheral populations of Nova Scotian Unionidae (Mollusca: Bivalvia). Biological Journal of the Linnean Society 22:157–185.

Kennard, A. S., A. E. Salisbury, and B. B. Woodward. 1925. Notes on the British post-Pliocene Unionidae with more special regard to the means of identification of fossil fragments. Proceedings of the Malacological Society of London 16:267–290.

King, T. L., M. S. Eackles, B. Gjetvaj, and W. R. Hoeh. 1999. Intraspecific phylogeography of *Lasmigona subviridis* (Bivalvia: Unionidae): Conservation implications of range discontinuity. Molecular Ecology 8 Supplement 1:S65–S78.

Kluge, A. G. 1989. A concern for evidence and a phylogenetic hypothesis of relationships among *Epicrates* (Boidae, Serpentes). Systematic Zoology 38:7–25.

Kluge, A. G. 1998. Total evidence or taxonomic congruence: Cladistics or consensus classification. Cladistics 14:151–158.

Knazek, E., S. Strahler, W. H. Heard, and W. R. Hoeh. 2001. The origin of simultaneous hermaphroditism in the freshwater mussel genus *Utterbackia* (Bivalvia: Anodontinae) as inferred from phylogenetic analysis of DNA sequences. Honors thesis, Department of Biological Sciences, Kent State University.

Ladoukakis, E. D., and E. Zouros. 2001. Direct evidence for homologous recombination in mussel (*Mytilus galloprovincialis*) mitochondrial DNA. Molecular Biology and Evolution 18:1168–1175.

Lamarck, J. B. P. A. 1805. Philosophie Zoologique, 328.

Lamarck, J. B. P. A. 1812. Extrait du Cours de Zoologique, 106.

Lamarck, J. B. P. A. 1830. Philosophie Zoologique, 318.

Lea, I. 1836. A synopsis of the family of naiades. Carey, Lea, and Blanchard, Philadelphia.

Lea, I. 1838. A synopsis of the family of naiades, enlarged and improved, 2nd ed. Carey, Lea, and Blanchard, Philadelphia.

Lea, I. 1852. A synopsis of the family of naiades, greatly enlarged and improved, 3rd ed. Blanchard and Lea, Philadelphia.

Lea, I. 1870. A synopsis of the family Unionidae, very greatly enlarged and improved, 4th ed. Henry C. Lea, Philadelphia.

Lee, D.-C., and H. N. Bryant. 1999. A reconsideration of the coding of inapplicable characters: Assumptions and problems. Cladistics 15:373–378.

Lieberman, B. S. 2000. Applying molecular phylogeography to test paleoecological hypotheses: A case study involving *Amblema plicata* (Mollusca: Unionidae). Pp. 83–103 in Evolutionary Paleoecology (W. D. Allmon and D. J Bottjer, eds.). Columbia University Press, New York.

Liu, H.-P., J. B. Mitton, and S.-K. Wu. 1996. Paternal mitochondrial DNA differentiation far exceeds maternal mitochondrial DNA and allozyme differentiation in the freshwater mussel, *Anodonta grandis grandis*. Evolution 50:952–957.

Lydeard, C., and R. L. Mayden. 1995. A diverse and endangered aquatic ecosystem of the southeast United States. Conservation Biology 9:800–805.

Lydeard, C., R. L. Minton, and J. D. Williams. 2000. Prodigious polyphyly in imperiled freshwater pearly-mussels (Bivalvia: Unionidae): A phylogenetic test of species and generic designations. Pp. 145–158 in The Evolutionary Biology of the Bivalvia (E. M. Harper, J. D. Taylor and J. A. Crame, eds.). Sp. pub. 177, The Geological Society of London.

Lydeard, C., M. Mulvey, and G. M. Davis. 1996. Molecular systematics and evolution of reproductive traits of North American freshwater unionacean mussels (Mollusca: Bivalvia) as inferred from 16S rRNA gene sequences. Philosophical Transactions of the Royal Society of London 351:1593–1603.

Lydeard, C., and K. J. Roe. 1998. Phylogenetic systematics: The missing ingredient in the conservation of freshwater unionid bivalves. Fisheries 23:16–17.

Master, L. L., S. R. Flack, and B. A. Stein. 1998. Rivers of Life: Critical Watersheds for Protecting Freshwater Biodiversity. The Nature Conservancy, Arlington, VA.

Mayden, R. L. 1988. Vicariance biogeography, parsimony, and evolution in North American freshwater fishes. Systematic Zoology 37:329–355.

McMichael, D. F., and I. D. Hiscock. 1958. A monograph of the freshwater mussels (Mollusca: Pelecypoda) of the Australian region. Australian Journal of Marine and Freshwater Research 9:372–503.

Meier, R. 1994. On the inappropriateness of presence/absence recoding, for non-additive multistate characters in computerized cladistic analyses. Zoological Anz 232:201–212.

Miyamoto, M. M. 1985. Consensus cladograms and general classification. Cladistics 1:186–189.

Modell, H. 1942. Das naturliche System der Najaden. Archiv für Molluskenkunde 74:161–191.

Modell, H. 1949. Das naturliche System der Najaden. 2. Archiv für Molluskenkunde 78:29–48.

Modell, H. 1964. Das naturliche System der Najaden. 3. Archiv für Molluskenkunde 93:71–126.

Moritz, C. 1996. Uses of molecular phylogenies for conservation. Pp. 203–214 in New Uses for New Phylogenies (P. H. Harvey, A. J. Leigh Brown, J. Maynard Smith, and S. Nee, eds.). Oxford University Press, Oxford.

Morrison, J. P. E. 1973. The families of the pearly freshwater mussels. Bulletin of the American Malacological Union, Inc. 39:45–46.

Mulvey, M., C. Lydeard, D. L. Pyer, K. M. Hicks, J. Brim-Box, J. D. Williams, and R. S. Butler. 1997. Conservation genetics of North American freshwater mussels *Amblema* and *Megalonaias*. Conservation Biology 11:868–878.

Nagel, K.-O., and G. Badino. 2001. Population genetics and systematics of European Unionoidea. Pp. 51–80 in Ecology and Evolution of the Freshwater Mussels Unionoida (G. Bauer and K. Wachtler, eds.). Springer-Verlag, Berlin.

Neves, R. J., A. E. Bogan, J. D. Williams, S. A. Ahlstedt, and P. W. Hartfield. 1997. Status of aquatic mollusks in the southeastern United States: A downward spiral of diversity. Pp. 43–85 in Aquatic Fauna in Peril: The Southeastern Perspective (G. W. Benz and D. E. Collins, eds.). Sp. pub. 1, Southeast Aquatic Research Institute, Lenz Design and Communications, Decatur, GA.

Ortmann, A. E. 1910. A new system of the Unionidae. The Nautilus 23:114–120.

Ortmann, A. E. 1911. The anatomical structure of certain exotic naiades compared with that of the North American forms. The Nautilus 24:103–108, 114–120, 127–131.

Ortmann, A. E. 1912. Notes upon the families and genera of the Najades. Annals of the Carnegie Museum 8:222–365.

Ortmann, A. E. 1916. The anatomical structure of *Gonidea angulata* (Lea). The Nautilus 30:50–53.

Ortmann, A. E. 1921. South American Naiades; A contribution to the knowledge of the freshwater mussels of South America. Memoirs of the Carnegie Museum 8:451–670.

Pamilo, P., and M. Nei. 1988. Relationships between gene trees and species trees. Molecular Biology and Evolution 5:568–583.

Parmalee, P. W., and A. E. Bogan. 1998. The freshwater mussels of Tennessee. The University of Tennessee Press, Knoxville.

Parodiz, J. J., and A. A. Bonetto. 1963. Taxonomy and zoogeographic relationships of the South American naiades (Pelecypoda: Unionacea and Mutelacea). Malacologia 1:179–213.

Pimentel, R. A., and R. Riggins. 1987. The nature of cladistic data. Cladistics 3:201–209.

Pleijel, F. 1995. On character coding for phylogeny reconstruction. Cladistics 11:309–315.

Quesada, H., D. A. G. Skibinski, and D. O. F. Skibinski. 1996. Sex-biased heteroplasmy and mitochondrial DNA inheritance in the mussel *Mytilus galloprovincialis* Lmk. Current Genetics 29:423–426.

Rafinesque, C. S. 1820. Mongraphie des coquilles bivalves fluviatiles de la Riviere Ohio, contenant douze genre et soixante-huit especes. Ann. Gen. Sci. Phys. Bruxelles 5:287–322.

Rafinesque, C. S. 1831. Continuation of a monograph of the bivalve shells of the river Ohio, and other rivers of the western states, Philadelphia.

Rafinesque, C. S. 1832. *Odatelia* N. G. of N. American bivalve fluviatile shell. Atlantic Journal and Friend of Knowledge 4:154.

Roe, K. J., P. D. Hartfield, and C. Lydeard. 2001. Phylogeographic analysis of the threatened and endangered superconglutinate-producing mussels of the genus *Lampsilis* (Bivalvia: Unionidae). Molecular Ecology 10:2225–2234.

Roe, K. J., and C. Lydeard. 1998a. Species delineation and the identification of evolutionary significant units: Lessons from the freshwater mussel genus *Potamilus* (Bivalvia: Unionidae). Journal of Shellfish Research 17:1359–1363.

Roe, K. J., and C. Lydeard. 1998b. Molecular systematics of the freshwater mussel genus *Potamilus* (Bivalvia: Unionidae). Malacologia 39:195–205.

Saiki, R. K., S. Scharf, F. Faloona, K. B. Mullis, G. T. Horu, H. A. Erlich, and N. Arnheim. 1985. Enzymatic amplification of β-globin genomic sequences and restriction site analysis for diagnosis of sickle cell anemia. Science 230:1350–1354.

Sanderson, M. J., B. G. Baldwin, G. Bharathan, C. S. Campbell, C. von Dohlen, D. Ferguson, J. M. Porter, M. F. Wojciechowski, and M. J. Donoghue. 1993. The growth of phylogenetic information and the need for a phylogenetic database. Systematic Biology 42:562–568.

Seitz, V., S. O. Garcia, and A. Liston. 2000. Alternative coding strategies and the inapplicable data coding problem. Taxon 49:47–54.

Simpson, C. T. 1896. The classification and geographical distribution of the pearly freshwater mussels. Proceedings of the U.S. National Museum 18:295–343.

Simpson, C. T. 1900. Synopsis of the naiades, or pearly fresh-water mussels. Proceedings of the U.S. National Museum 22:501–1044.

Simpson, C. T. 1914. A Descriptive Catalogue of the Naiades, or Pearly Freshwater Mussels. Bryant Walker, Detroit, MI.

Smith, D. G. 1980. Anatomical studies on *Margaritifera margaritifera* and *Cumberlandia monodonta* (Mollusca: Pelecypoda: Margaritiferidae). Zoological Journal of the Linnean Society 69:257–270.

Smith, D. G. 1986. The stomach anatomy of some eastern North American Margaritiferidae (Unionoida: Unionacea). American Malacological Bulletin 4:13–19.

Smith, D. G. 2000. On the taxonomic placement of *Unio ochraceus* Say, 1817 in the genus *Ligumia* (Bivalvia: Unionidae). The Nautilus 114:155–160.

Smith, D. G. 2001. Systematics and distribution of the Recent Margaritiferidae. Pp. 33–49 in Ecology and Evolution of the Freshwater Mussels Unionoida, vol. 145. Ecological Studies (G. Bauer and K. Wachtler, eds.). Springer-Verlag, Berlin.

Starobogatov, Y. I. 1970. Mollusk fauna and zoogeographical partitioning of continental water reservoirs of the world. Zoologischeskii Instituti Nauka, Akademiya Nauk SSSR, Leningrad.

Strong, E. E. and D. Lipscomb. 1999. Character coding and inapplicable data. Cladistics 15:363–371.

Swainson, W. 1840. A treatise on malacology; or the natural classification of shells and shellfish. Longman, Brown, Green and Longmans, London.

Swofford, D. L. 2002. PAUP*. Phylogenetic Analysis Using Parsimony (*and Other Methods). Version 4. Sinauer Associates, Sunderland, MA.

Swofford, D. L., and S. H. Berlocher. 1987. Inferring evolutionary trees from gene frequency data under the principle of maximum parsimony. Systematic Zoology 36:293–325.

Swofford, D. L., G. J. Olsen, P. J. Wadel, and D. M. Hillis. 1996. Phylogenetic inference. Pp. 407–514 in Molecular Systematics, 2nd ed. (D. M. Hillis, C. Moritz, and B. K. Mable, eds.). Sinauer Associates, Sunderland, MA.

Templeton, A., R. E. Routman, and C. A. Phillips. 1995. Separating population structure from population history: A cladistic analysis of the geographical distribution of mitochondrial DNA haplotypes in the tiger salamander, *Ambystoma tigrinum*. Genetics 140:767–782.

Thiele, J. 1935. Handbuch der Systematischen Weichtierkunde. 2:780–1022.

Troschel, F. H. 1847. Ueber die Brauchbarkeit der Mundlappen und Kiemen zur Familienunterscheidung und uber die Familie der Najaden. Archiv für Naturgeschichte 13:257–274.

Turgeon, D. D., J. F. Quinn, Jr., A. E. Bogan, E. V. Coan, F. G. Hochberg, W. G. Lyons, P. M. Mikkelsen, R. J. Neves, C. F. E. Roper, G. Rosenberg, B. Roth, A. Scheltema, F. G. Thompson, M. Vecchione, and J. D. Williams. 1998. Common and scientific names of aquatic invertebrates from the United States and Canada: Mollusks, 2nd ed. Sp. pub. 26, American Fisheries Society, Bethesda, MD.

Turner, T. F., J. C. Trexler, J. L. Harris, and J. L. Haynes. 2000. Nested cladistic analysis indicates population fragmentation shapes diversity in freshwater mussels. Genetics 154:777–785.

Vaught, K. C. 1989. A classification of the living Mollusca. American Malacologists, Melbourne, FL.

von Ihering, H. 1893. Najaden von S. Paulo und die geographische Verbreitung der Suss-wasser-Faunen von Sudamerika. Archiv für Naturgeschichte 1893:45–140.

Wachtler, K., M. C. Dreher-Mansur, and T. Richter. 2001. Larval types and early post-larval biology in naiads (Unionoida). Pp. 93–125 in Ecology and Evolution of the Freshwater Mussels Unionoida, vol. 145. Ecological Studies (G. Bauer and K. Wachtler, eds.). Springer-Verlag, Berlin.

Walker, K. F., M. Byrne, C. W. Hickey, and D. S. Roper. 2001. Freshwater mussels (Hyri-idae) of Australasia. Pp. 5–31 in Ecology and Evolution of the Freshwater Mussels Unionoida, vol. 145. Ecological Studies (G. Bauer and K. Wachtler, eds.). Springer-Verlag, Berlin.

Waller, T. R. 1990. The evolution of ligament systems in the Bivalvia. Pp. 49–71 in The Bivalvia: Proceedings of a Memorial Symposium in Honour of Sir Charles Maurice Yonge (B. Morton, ed.). Hong Kong University Press, Edinburgh.

Waller, T. R. 1998. Origin of the molluscan class Bivalvia and a phylogeny of major groups. Pp. 1–45 in Bivalves: An Eon of Evolution (P. A. Johnston and J. W. Haggart, eds.). University of Calgary Press, Calgary, Alberta.

Watters, G. T. 1994. Form and function of unionoidean shell sculpture and shape (Bivalvia). American Malacological Bulletin 11:1–20.

Welsh, J., and M. McClelland. 1990. Fingerprinting genomes with arbitrary primers. Nucleic Acids Research 18:7213–7218.

White, L. R., B. A. McPheron, and J. R. Stauffer. 1994. Identification of freshwater mussel glochidia on host fishes using restriction fragment length polymorphisms. Molecular Ecology 1994:183–185.

White, L. R., B. A. McPheron, and J. R. Stauffer. 1996. Molecular genetic identification tools for the unionids of French Creek, Pennsylvania. Malacologia 38:181–202.

Wiens, J. J. 2000. Phylogenetic analysis of morphological data. (J. J. Wiens, ed.). Smithsonian Institution Press, Washington, DC.

Wiens, J. J. 2001. Character analysis in morphological phylogenetics: Problems and solutions. Systematic Biology 50:689–699.

Wilkinson, M. 1995. A comparison of two methods of character construction. Cladistics 11:297–308.

Williams, J. D., and M. Mulvey. 1994. Recognition of freshwater mussel taxa: A conservation challenge. Pp. 57–58 in Principles of Conservation Biology (G. K. Meffe and C. R. Carroll, eds.). Sinauer Associates, Sunderland, MA.

Williams, J. D., M. L. Warren, Jr., K. S. Cummings, J. L. Harris, and R. J. Neves. 1993. Conservation status of the freshwater mussels of the United States and Canada. Fisheries 18:6–22.

Yoder, A. D., J. A. Irwin, and B. A. Payseur. 2001. Failure of the ILD to determine data combinability for slow loris phylogeny. Systematic Biology 50:408–424.

Zelditch, M. L., D. L. Swiderski, and W. L. Fink. 2000. Discovery of phylogenetic characters in morphometric data. Pp. 37–83 Phylogenetic Analysis of Morphological Data (J. J. Wiens, ed.). Smithsonian Institution Press, Washington, DC.

GERHARD STEINER AND PATRICK D. REYNOLDS

5

MOLECULAR SYSTEMATICS OF THE SCAPHOPODA

Scaphopods constitute one of the smaller molluscan classes, with approximately 520 validly described extant species (Steiner and Kabat 2001, and in press), which is composed exclusively of marine benthic infaunal "tusk shells." Although relatively minor contributors to benthic community species diversity, the group has a world-wide geographic distribution and is found from the intertidal zone (Lamprell and Healy 1998) to depths of 7,000 meters (Knudsen 1964). Current classification identifies two orders, Dentaliida (Palmer 1974) and Gadilida (Starobogatov 1974), 14 families, and 60 genera (46 extant) (Steiner and Kabat 2001) (Table 5.1).

The earliest scaphopod identified with certainty in the fossil record is a member of the Dentaliida from the Mississippian Carboniferous (362.5 million years; Yochelson 1999). However, several disputed records would place the first appearance considerably earlier; these include the Ordovician *Plagioglypta iowensis* and *Rhytiodentalium kentuckyensis* (Bretsky and Bermingham 1970; Pojeta and Runnegar 1979; Engeser and Riedel 1996; Lamprell and Healy 1998; Yochelson 1999) and several Devonian species that are also questionable (Ludbrook 1960; Emerson 1962; Yochelson 1999). The Gadilida appear much later, in the Tertiary (65 million years; Emerson 1962), but there are also earlier disputed reports, from the Permian (Yancey 1973; Yochelson 1999).

SCAPHOPOD MOLECULAR DATA

Scaphopod molecular data, as published in GenBank, has been scant compared with that available for larger molluscan classes. Scaphopoda have received only

123

Table 5.1

Classification of Scaphopoda (Steiner and Kabat 2001: Table 1 and text)

Taxa

Class SCAPHOPODA Bronn 1862
Order DENTALIIDA da Costa 1776
    Family ANULIDENTALIIDAE Chistikov 1975
        Genus *Anulidentalium* Chistikov 1975
        Genus *Epirhabdoides* Steiner 1999
    Family BALTODENTALIIDAE Engeser & Riedel 1992
        Genus: *Baltodentalium* Engeser & Riedel 1992 [F][a]
    Family CALLIODENTALIIDAE Chistikov 1975
        Genus *Calliodentalium* Habe 1964
    Family DENTALIIDAE [Children 1834]
        Genus *Antalis* H. & A. Adams 1854
        Genus *Coccodentalium* Sacco 1896
        Genus *Compressidentalium* Habe 1963
        Genus *Dentalium* Linnaeus 1758
        Genus *Eodentalium* Medina & del Valle 1985 [F]
        Genus *Eudentalium* Cotton & Godfrey 1933
        Genus *Fissidentalium* Fischer 1885
        Genus *Graptacme* Pilsbry & Sharp 1897
        Genus *Paleodentalium* Gentile 1974 [F]
        Genus *Paradentalium* Cotton & Godfrey 1933
        Genus *Pictodentalium* Habe 1963
        Genus *Plagioglypta* Pilsbry in Pilsbry & Sharp 1898
        Genus *Schizodentalium* Sowerby 1894
        Genus *Striodentalium* Habe 1964
        Genus *Tesseracme* Pilsbry & Sharp 1898
    Family FUSTIARIIDAE Steiner 1991
        Genus *Fustiaria* Stoliczka 1868
    Family GADILINIDAE Chistikov 1975
      Subfamily EPISIPHONINAE Chistikov 1975
        Genus *Episiphon* Pilsbry & Sharp 1897
      Subfamily GADILININAE Chistikov 1975
        Genus *Gadilina* Foresti 1895
      Subfamily LOBANTALINAE Chistikov 1975
        Genus *Lobantale* Cossmann 1888 [F]
    Family LAEVIDENTALIIDAE Palmer 1974
        Genus *Laevidentalium* Cossmann 1888
        Genus *Pipadentalium* Yoo 1988 [F]
        Genus *Rhytiodentalium* Pojeta & Runnegar 1979 [F]
        Genus *Scissuradentalium* Yoo 1988 [F]
    Family OMNIGLYPTIDAE Chistikov 1975
        Genus *Omniglypta* Kuroda & Habe in Habe 1953
    Family PRODENTALIIDAE Starobogatov 1974
        Genus: *Prodentalium* Young 1942 [F]
    Family RHABDIDAE Chistikov 1975
        Genus *Rhabdus* Pilbry & Sharp 1897

Table 5.1 continued

Taxa

Dentaliida, *incertae sedis*
      Genus *Progadilina* Palmer 1974 [F].
      Genus *Suevidontus* Engeser, Riedel & Bandel 1993 [F]
Order GADILIDA Starobogatov 1974
  Suborder ENTALIMORPHA Steiner 1992
  Family ENTALINIDAE Chistikov 1979
    Subfamily BATHOXIPHINAE Chistikov 1983
      Genus *Bathoxiphus* Pilsbry & Sharp 1897
      Genus *Rhomboxiphus* Chistikov 1983
      Genus *Solenoxiphus* Chistikov 1983
    Subfamily ENTALININAE Chistikov 1979
      Genus *Entalina* Monterosato 1872
    Subfamily HETEROSCHISMOIDINAE Chistikov 1982
      Genus *Costentalina* Chistikov 1982
      Genus *Entalinopsis* Habe 1957
      Genus *Heteroschismoides* Ludbrook 1960
      Genus *Pertusiconcha* Chistikov 1982
      Genus *Spadentalina* Habe 1963
  Suborder GADILIMORPHA Steiner 1992
  Family GADILIDAE Stoliczka 1868
    Subfamily GADILINAE Stoliczka 1868
      Genus *Bathycadulus* Scarabino 1995
      Genus *Cadulus* Philippi 1844
      Genus *Gadila* Gray 1847
      Genus *Sulcogadila* Moroni & Ruggieri 1981 [F]
    Subfamily SIPHONODENTALIINAE Tryon 1884
      Genus *Dischides* Jeffreys 1867
      Genus *Polyschides* Pilsbry & Sharp 1898
      Genus *Sagamicadulus* Sakurai & Shimazu 1963
      Genus *Siphonodentalium* M. Sars 1859
      Genus *Striocadulus* Emerson 1962
    Gadilidae, *incertae sedis*
      Genus *Calstevenus* Yancey 1973 [F]
      Genus *Gadilopsis* Woodring 1925 [F]
  Family PULSELLIDAE Scarabino in Boss 1982
      Genus *Annulipulsellum* Scarabino 1986
      Genus *Pulsellum* Stoliczka 1868
      Genus *Striopulsellum* Scarabino 1995
  Family WEMERSONIELLIDAE Scarabino 1986
      Genus *Chistikovia* Scarabino 1995
      Genus *Wemersoniella* Scarabino 1986
  Gadilida, *incertae sedis*
      Genus *Compressidens* Pilsbry & Sharp 1897
      Genus *Megaentalina* Habe 1963
  Scaphopoda, *incertae sedis*
      Genus *Cyrtoconella* Patrulius 1996 [F]

[a] F = fossil taxa.

marginal attention in molecular phylogenetic studies until recently, serving either as outgroups in gastropod or bivalve phylogenies or among representative molluscan taxa in phylum-level analyses; none of these studies shed light on the origin or the relationships among scaphopods. More recently, focus has been brought to bear on relationships among scaphopods and has therefore increased the representation of scaphopod taxa in the published database, at least for a limited number of genes.

Of protein-coding genes, only sequences for *engrailed* and mitochondrial cytochrome *c* oxidase subunit I (COI) have been published for scaphopods. In their comparison of molluscan partial *engrailed* sequences, Wray et al. (1995) were the first to deposit scaphopod (*Graptacme eborea* and *Cadulus fusiformis*) sequences of any description in GenBank, although they were not analyzed cladistically. In contrast, several scaphopod species have been used as outgroup taxa for bivalve COI analyses: *Dentalium* sp. (most probably *Rhabdus rectius*) (Hoeh et al. 1998), *Antalis pilsbryi* and *Rhabdus rectius* (Chapter 3, this volume), and *Episiphon subrectum* (Matsumoto 2002). All are partial sequences of about 690 bp near the 5′ end of the gene except that of *E. subrectum*, which is 903 bp long and starts farther downstream. Some of these sequences are also included in a phylogenetic analysis of an extended COI dataset, including 17 scaphopod species, presented below.

Scaphopod sequences of rDNA are limited to 28S and 18S genes. Two short partial 28S rDNA sequences of *Antalis vulgaris* are available: Chenuil et al. (1997) investigated the evolution of a protein-binding site in the D7 region, and Rosenberg et al. (1997) used the D6 region for molluscan phylogeny. Giribet and Distel (Chapter 3, this volume) added the D3 region of *Antalis pilsbryi* and *Rhabdus rectius* to the database.

The first 18S rDNA sequence was that of *A. vulgaris*, produced by Winnepenninckx et al. (1996) in their analysis of molluscan phylogeny. Subsequently, Steiner and Hammer (2000) added near-complete 18S sequences of *A. inaequicostata* and *A. perinvoluta* (as *Dentalium bisexangulatum*) as outgroup taxa in a bivalve phylogeny; the sequence of *Dentalium laqueatum* in Steiner and Hammer (2000) was subsequently found to be from a sipunculid inhabiting the scaphopod shell. In the same year, Giribet et al. (2000) used the 18S sequences of *A. pilsbryi* and *Rhabdus rectius* among molluscan representatives in the assessment of lower metazoan relationships. The first published molecular phylogeny of scaphopods, by Steiner and Dreyer (in press), presents an 18S dataset containing 17 scaphopod sequences of sufficient systematic range for an initial assessment of ingroup relationships, and 4 to 17 representatives of the other higher conchiferan classes (Bivalvia, Gastropoda, Cephalopoda) to examine interclass relationships.

RELATIONSHIPS AMONG SCAPHOPODS AND OTHER MOLLUSCAN CLASSES

Scaphopoda is the most recent molluscan class to appear in the fossil record. Although relationships among the earliest lineages of the Mollusca, established by the late Cambrian (e.g., Neomeniomorpha–Solenogastres, Caudofoveata, Polyplacophora, Conchifera), have been difficult to resolve, there has been greater divergence of opinion on the placement of the scaphopods among conchiferan classes. These hypotheses can be broadly categorized as either supporting the Rostroconchia–Diasoma concept (†Rostroconchia + Bivalvia + Scaphopoda), or the Helcionelloida–Cyrtosoma concept (†Helcionelloida + Cephalopoda + Scaphopoda + Gastropoda).

Several factors contribute to the variety of hypotheses and lack of consensus of morphology-based analyses of scaphopod relationships within the Mollusca. First, the fossil record provides no conclusive connections to other classes. Despite much study on the early development and morphogenesis of scaphopod embryos, little information has been available until recently on their later larval development. Furthermore, the absence of several organ systems found in other conchiferan classes, such as osphradia, ctenidia, and auricles, and conflicting interpretation of the orientation of the body between the elongated dorsoventral axis and functional the posterior–anterior axis (Edlinger 1991; Steiner 1992; Shimek and Steiner 1997; Waller 1998) has helped obscure scaphopod affinities.

ROSTROCONCHIA–DIASOMA HYPOTHESIS

The Diasoma–Cyrtosoma concept of conchiferan relationships, proposed by Runnegar and Pojeta (1974), argued for the derivation of Bivalvia and Scaphopoda from the Rostroconchia, a Paleozoic group of pseudo-bivalved mollusks, the three classes placed within the subphylum Diasoma, and the remaining conchiferan classes placed in the Cyrtosoma. The sister relationship of scaphopods and rostroconchs is argued by Pojeta and Runnegar (1985) with reference to the trend of anterior shell elongation found in all rostroconchs (both conocardioids and ribeirioids). In particular, the ribeirioid *Pinnocaris,* with its roughly tusk-shaped shell, served as a model for the derivation of the Scaphopoda, despite being posteriorly, rather than anteriorly, elongated. Furthermore, Pojeta and Runnegar (1979) and Runnegar (1996) present the hypothesis of a Scaphopoda–Ribeiroida sister group (i.e., scaphopods as a clade within Rostroconchia). In contrast, Engeser and Riedel (1996) argue that

Scaphopoda are most likely sister to the rostroconch order Conocardioida, based on similarities between the scaphopod larval shell and conocardioid adult shell and a putative heterochronic evolutionary event.

The Diasoma hypothesis (=Loboconcha; Salvini-Plawen 1985) was supported by a cladistic-based morphological analysis by Salvini-Plawen and Steiner (1996), who found support for a Scaphopoda–Bivalvia sister grouping among Recent taxa, noting Rostroconchia as intermediates. Synapomorphies supporting this sister grouping were the common formation of the laterally elongated mantle and shell, the similar, anteriorly extended foot, and the epiathroid nervous system with identical innervation areas. The absence of a ventral sole gland was presented as another synapomorphy for the grouping.

No molecular data to date support the Diasoma hypothesis. Steiner and Dreyer (in press) show that with 18S rDNA data, the best parsimony and maximum likelihood trees with a constrained Diasoma topology are significantly longer or less likely than the optimal trees based on the nonconstrained analyses.

HELCIONELLOIDA–CYRTOSOMA HYPOTHESIS

The hypothesis that Scaphopoda is most closely related to the Cephalopoda and Gastropoda, rather than Bivalvia, was extensively discussed by Waller (1998), who proposed including Scaphopoda and Helcionelloida in the Cyrtosoma as defined by Wingstrand (1985) (=Visceroconcha; Salvini-Plawen 1985). Morphological features supporting this grouping include a dorsoventral growth axis, with U-shaped gut. Support for a Scaphopoda + Cephalopoda + Helcionelloida clade includes endogastric coiling; Waller (1998) argues that the similarities in mantle and pedal morphology and in the nervous system cited in the Diasoma hypothesis may result from convergent adaptation to the infaunal lifestyle of Scaphopoda and Bivalvia. Haszprunar (2000), in an analysis of recent Mollusca using a significantly different set of character coding from Waller (1998), nevertheless also found support for a sister relationship with a Gastropoda + Cephalopoda clade. The arguments on potential synapomorphies and convergences are summarized in Steiner and Dreyer (in press).

Recent developmental evidence supports the cephalopod + gastropod + scaphopod clade. Wanninger and Haszprunar (2002), studying the development of musculature in the larva of *Antalis entalis,* have found cephalic retractor muscles as found in gastropods and cephalopods but not in bivalves, and argue the similarities between the musculo-hydrostat system of the scaphopod foot and captacula and that of cephalopod arms and prosobranch cephalic tentacles. They further argue that secondary losses in scaphopods may account for differences in prototroch or velum retractor muscles among these groups.

Further evidence in support of this hypothesis is provided by the single shell field of the scaphopod larva (Lacaze-Duthiers 1856–57; Kowalevsky 1883), recently confirmed by the expression of *engrailed* in marginal cells of the embryonic shell field (Wanninger and Haszprunar 2002). The scaphopod shell is, therefore, univalved throughout morphogenesis. The expression pattern is similar to that in the gastropod *Ilyanassa* (Moshel et al. 1998), whereas two centers of calcification are indicated by *engrailed* expression in bivalves (Jacobs et al. 2000).

The 18S rDNA data of Steiner and Dreyer (in press) robustly support the monophyly of the Scaphopoda and Cephalopoda and their sister-group relationship (Figure 5.1A), and Gastropoda closer to the scaphopod + cephalopod clade than Bivalvia. With a bootstrap value of 86, the branch connecting Scaphopoda and Cephalopoda is the only interclass branch with a bootstrap value above 50, the only significant signal in the lower part of the tree. This result is also confirmed by a likelihood puzzling test (Figure 5.2) where 51.6% of the quartets support the Scaphopoda + Cephalopoda topology, but only 19.5% and 2.3% support the alternative topologies. Spectral analysis further corroborates the (Gastropoda (Scaphopoda, Cephalopoda)) topology. The six top-ranked splits are those of the four cephalopod species, the scaphopod orders Dentaliida and Gadilida, and the Polyplacophora. Split 7 unites Cephalopoda and Scaphopoda, whereas the scaphopod split ranks tenth.

RELATIONSHIPS WITHIN SCAPHOPODA

Morphology-Based Analyses

A series of morphology-based cladistic analyses of scaphopod phylogeny (Steiner 1992, 1998, 1999; Reynolds 1997; Reynolds and Okusu 1999), using overlapping but not identical data, produced largely congruent trees and classifications (e.g., Figure 5.3, A and B). There is support for the monophyly of the two scaphopod orders, Dentaliida and Gadilia, although that of the Dentaliida is not as robust.

All analyses agree that the basal branch within the Gadilida is the suborder Entalimorpha, containing the single family Entalinidae. Steiner (1998) pointed out that Entalimorpha/Entalinidae may be paraphyletic because they, like the Dentaliida, share plesiomorphic traits of the gadilid clade. Because of its radula characters, the entalinid *Heteroschismoides subterfissum* appears more closely related to the Gadilimorpha in the species-based analysis of Steiner (1998) (Figure 5.3B). In this analysis, the Gadilimorpha have a basal polytomy of the families Pulsellidae, Wemersoniellidae, and Gadilidae, but in the family-level

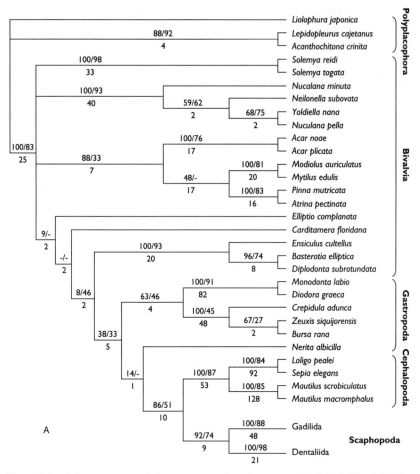

Figure 5.1. Strict consensus of three most parsimonious trees (L = 3206, CI = 0.5221, RC = 0.3935) with support indices, 18S rDNA. Bootstrap and puzzling values are above, and decay index below branches. (A) Subtree showing relationships of Scaphopoda with other molluscan taxa. (B) Subtree of Scaphopoda (from Steiner and Dreyer, in press) (next page).

analysis of Reynolds and Okusu (1999), Pulsellidae are sister to a Wemersoni-ellidae + Gadilidae clade. The Gadilidae subfamilies, Gadilinae and Siphono-dentaliinae, have been treated by Scarabino (1995) as families, but are supported in the above analyses as sister groups.

The relationships of the Dentaliida families are less clear and more contro-versial. The investigation of numerous organ systems (e.g., dorsoventral and pedal musculature, mantle margins, radula, and shell microstructure) has re-

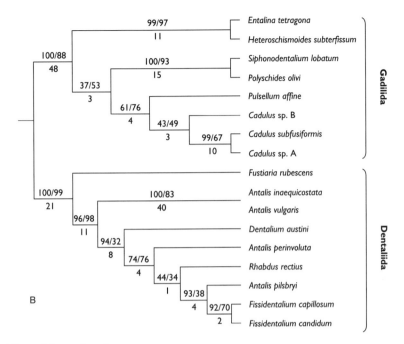

Figure 5.1 continued

vealed character incongruency and (potential) homoplasy (Reynolds and Okusu 1999; Steiner 1999) and, as a result of different taxon samples and character representations, the phylogenetic hypotheses in Reynolds and Okusu (1999) and Steiner (1999) differ. The main discrepancy concerns the position of the Fustiariidae, being the sister taxon to Dentaliidae in the former but sister to the Dentaliidae + Laevidentaliidae + Rhabdidae + Calliodentaliidae in the latter. Both studies agree, however, in placing the Gadilinidae, as a potentially paraphyletic taxon, at the base of the dentaliid tree, and a clade Laevidentaliidae + Rhabdidae + Calliodentaliidae in a relatively derived position.

Molecular-Based Analyses

Molecular-based analyses have to date been limited to the genes 18S rDNA and COI mtDNA. The 18S rDNA data of Steiner and Dreyer (in press) fully support the monophyly of the major scaphopod orders Dentaliida and Gadilida (Figure 5.1B). The Gadilida are characterized by an increased sequence length of 100 to 150 bp caused by inserts in the helices E23_1 and E23_2 to E23_5 of the V4 region, referring to the secondary structure model in Wuyts et al. (2002). The sequences of these inserts are surprisingly similar, which suggests their homology.

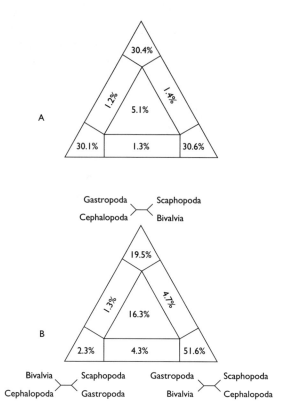

Figure 5.2. Maximum likelihood mapping using Tree-Puzzle 5.0. Areas at the corners of the triangle represent fully resolving one of the three possible four-taxon (quartet) topologies, those along the edges represent partly resolved quartets for which it is not possible to decide between two possible topologies. The central area represents unresolved quartets. Figures give percentages of 10,000 randomly chosen quartets in each area. (A) General likelihood mapping showing 91.1% of all quartets being fully resolved and only 5.1% unresolved. This indicates high phylogenetic information contents of the 18S rDNA data. (B) Four-cluster likelihood mapping of the conchiferan classes testing their phylogenetic relationships. The corners of the triangle are labeled with the corresponding unrooted tree topology. The Scaphopoda–Cephalopoda topology receives greatest support with 51.6% of all quartets, compared to 19.5% and 2.3% for the competing topologies. Note that the topology at the top corner does not necessarily represent the Diasoma–Cyrtosoma concept because the root can also be placed at the Bivalvia branch resulting in the modified Visceroconcha concept of Haszprunar (2000).

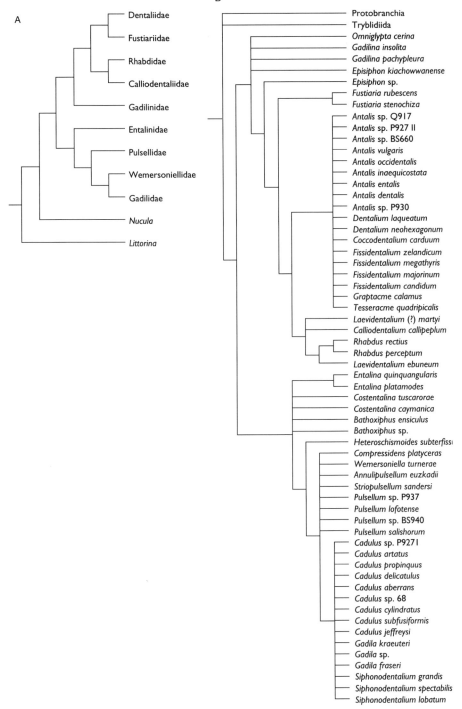

Figure 5.3. Morphology-based analyses of scaphopod phylogeny. (A) Single most parsimony tree for family representatives (L = 61, CI = 0.83, RC = 0.72) from Reynolds and Okusu (1999). (B) Strict consensus tree of species-level analysis (length = 91, CI = 0.527, RC = 0.444) from Steiner (1998).

The restricted taxon sample of this study allows only certain tests of relationships among Scaphopoda. Contrasting with morphological analyses, monophyly of the suborder Entalimorpha is well supported (*Heteroschismoides subterfissum* + *Entalina tetragona*). Branch support for the Gadilimorpha is weak. There is considerable signal, however, that Siphonodentaliidae and Gadilidae are not sister taxa but separated by the Pulsellidae. This result indicates that the anterior constriction of the shell arose independently in Siphonodentaliidae and Gadilidae. Considering that some species of *Siphonodentalium* (e.g., *S. lobatum*) have no such constriction, this possibility appears likely, but increased taxon sampling is required to confirm this. The two siphonodentaliid species form a robust clade, whereas the monophyly of the three *Cadulus* species representing the Gadilidae is not well supported.

Only three of the seven dentaliid family taxa, Dentaliidae, Fustiariidae, and Rhabdidae, are represented in the 18S rDNA dataset. The basal position of *Fustiaria rubescens* is in accordance with the morphological analysis of Steiner (1999) (Figure 5.3B). The Dentaliidae do not appear monophyletic, because *Rhabdus rectius* is included. The species of *Fissidentalium* form a clade, but the four *Antalis* species are widely separated, although the effects of limited taxon sampling are unclear (e.g., whether the deep rooting species-pair of *Antalis inaequicostata* and *A. vulgaris* is artificial because of the relatively long branch). Nuclear 18S rDNA delineates Gadilida well from Dentaliida because of inserts, but yields low interfamily branch support within Gadilida. The high support of dentaliid branches may indicate suitability of this gene for family relationships in this order.

Some of the results from 18S analyses are also reflected in the COI dataset (Reynolds et al., Hamilton College, NY, unpublished data), but there are significant differences. For this analysis, 24 partial sequences (forward and reverse strands; 687 aligned bp) in 17 species were sequenced using the Folmer et al. (1994) primers, aligned using the Clustal W program with manual adjustment using MacClade 4.0. Maximum parsimony and maximum likelihood analyses were carried out with PAUP* 4.0b10 (Swofford 2002) using the GTR+G+I substitution model derived from Modeltest v3.06 (Posada and Crandall 1998). For maximum likelihood analysis, 1,000 bootstrap replicates were run to assess node support. New sequences are deposited in GenBank (accession numbers AY260813 to AY260833). The analysis presented below also includes GenBank sequences U56843 (*Dentalium* sp.) and AB084110 (*Episiphon subrectum*); the latter is an incomplete sequence starting at position 214 bp. Outgroups used were the GenBank sequences for the cephalopods *Loligo pealei* (AF120629), *Sepia officinalis* (AF120630), and *Octopus bimaculoides* (AF377967).

COI nucleotides, like morphological and 18S data, strongly support the

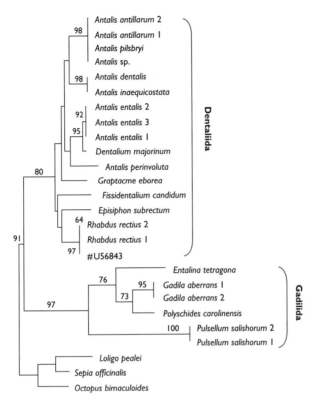

Figure 5.4. Maximum likelihood tree of scaphopod relationships based on COI mtDNA nucleotide sequences (–ln L = 6153.28516); node values: bootstrap values >50%.

monophyly of the orders Dentaliida and Gadilida. Within the Dentaliida, relationships among families and most genera are not well resolved; a clade of *Rhabdus rectius* (Rhabdidae), *Episiphon subrectum* (Gadilinidae), and *Fissidentalium candidum* (Dentaliidae), rendering the Dentaliidae paraphyletic, has poor bootstrap support (the GenBank sequence U56834, *Dentalium* sp., is likely *Rhabdus rectius;* Figure 5.4). Relationships among *Antalis* and *Dentalium* representatives have stronger support; the paraphyly of *Antalis* confirms the result from 18S data and mirrors the morphological uncertainty regarding the validity of the genus *Antalis* (compare Scarabino 1995, Lamprell and Healy 1998). Within Gadilida, relationships among genera and among families are well supported; Gadilidae monophyly is supported, but the suborder Gadilimorpha is rendered paraphyletic in this analysis, as the Entalinidae/ Entalimorpha representative *Entalina tetragona* is sister to Gadilidae (Figure 5.4).

CONCLUSIONS

Molecular phylogenetics of the Scaphopoda is in its earliest stages. Nevertheless, scaphopod molecular data hold significant potential for addressing important questions in conchiferan evolution that morphological data has been unable to resolve to date. On the other hand, both morphological and molecular-based analyses have demonstrated that relationships within the Dentaliidae, and particularly the genus *Antalis,* are far from understood, and these taxa are likely paraphyletic. Further taxonomic sampling is required to examine the extent of paraphyly and to understand relationships in these groups. Further sampling across other genera and families may well reveal other shortcomings in morphology-based classification within the class. Although genes such as 18S and COI may address questions at disparate levels, such as the placement of Scaphopoda within the Conchifera and the validity of *Antalis,* other genes appropriate to examining family-level relationships have yet to be developed in the scaphopods.

ACKNOWLEDGMENTS

We would like to thank Charles Lydeard for the invitation to contribute to this treatise. We would particularly like to thank several students who have worked on molecular phylogeny in scaphopods in our labs, including Hermann Dreyer, Sabine Hammer, Joanna Peters, Melissa Vetter, Anne Stires, Rebecca Hamm, and Akiko Okusu. We greatly appreciate the efforts of those who have provided scaphopod specimens to us or helped in their collection: James Woods, James Nybakken, Craig Staude, Woody Lee, and Hugh Reichardt. This study was funded in part by the Austrian Science Fund (FWF project no. P11846-GEN) and the National Science Foundation (DEB–9707400).

REFERENCES

Bretsky, P.W., and J. J. Bermingham. 1970. Ecology of the Paleozoic scaphopod genus *Plagioglypta* with special reference to the Ordovician of eastern Iowa. Journal of Paleontology 44:908–924.

Chenuil, A., M. Solignac, and M. Bernard. 1997. Evolution of the large-subunit ribosomal RNA binding site for protein L23/25. Molecular Biology and Evolution 14:578–588.

Edlinger, K. 1991. Zur Evolution der Scaphopoden-Konstruktion. Natur und Museum 121:116–122.

Emerson, W. K. 1962. A classification of the scaphopod molluscs. Journal of Paleontology 36:461–482, pls. 76–80.

Engeser, T. S., and F. Riedel. 1996. The evolution of the Scaphopoda and its implications for the systematics of the Rostroconchia (Mollusca). Mitteilungen aus dem Geologisch-Paläontologischen Institut der Universität Hamburg 79:117–138.

Folmer, O., M. B. Black, W. R. Hoeh, R. A. Lutz, and R. C. Vrijenhoek. 1994. DNA primers for the amplification of mitochondrial cytochrome *c* oxidase subunit I from diverse metazoan invertebrates. Molecular Marine Biology and Biotechnology 3:294–299.

Giribet, G., D. L. Distel, M. Polz, W. Sterrer, and W. Wheeler. 2000. Triploblastic relationships with emphasis on the acoelomates and the position of Gnathostomulida, Cycliophora, Plathelminthes, and Chaetognatha: A combined approach of 18S sequences and morphology. Systematic Biology 49:539–562.

Haszprunar, G. 2000. Is the Aplacophora monophyletic? A cladistic point of view. American Malacological Bulletin 15:115–130.

Hoeh, W. R., M. B. Black, R. Gustafson, A. E. Bogan, R. A. Lutz, and R. C. Vrijenhoek. 1998. Testing alternative hypotheses of *Neotrigonia* (Bivalvia: Trigonioida) phylogenetic relationships using cytochrome *c* oxidase subunit I DNA sequences. Malacologia 40:267–278.

Jacobs, D. K., C. G. Wray, C. J. Wedeen, R. Kostriken, R. DeSalle, J. L. Staton, R. D. Gates, and D. R. Lindberg. 2000. Molluscan *engrailed* expression, serial organization, and shell evolution. Evolution and Development 2:340–347.

Kowalevsky, A. 1883. Étude sur l'embryogénie du Dentale. Annales de la Museum d'Histoire Naturelle de Marseille 1:1–54.

Knudsen, J. 1964. Scaphopoda and Gastropoda from depths exceeding 6000 meters. Galathea Report 7:125–136.

Lamprell, K. L. and J. M. Healy. 1998. A revision of the Scaphopoda from Australian waters (Mollusca). Records of the Australian Museum, Supplement 24:1–189.

Lacaze-Duthiers, H. 1856–57. Histoire de l'organisation et du développement du Dentale. Annales des Sciences Naturelles, Quatrième Serie, *Paris* Tome 6, 225–281, pls. 8–10; 319–385, pls. 11–13; Tome 7, 5–51, pls. 2–4; 171–255, pls. 5–9.; Tome 8, 18–44.

Ludbrook, N. H. 1960. Scaphopoda. Pp. I37–I41 in Treatise on Invertebrate Paleontology, Part I, Mollusca 1 (R. C. Moore, ed.). Geological Society of America and University of Kansas Press, New York and Lawrence, KS.

Matsumoto, M. 2002. Direct submission to GenBank, accession number AB084110.

Moshel, S. M., M. Levine, and J. R. Collier. 1998. Shell differentiation and *engrailed* expression in the *Ilyanassa* embryo. Development, Genes and Evolution 208:135–141.

Palmer, C. P. 1974. A supraspecific classification of the scaphopod Mollusca. The Veliger 17:115–123.

Pojeta, J. Jr., and B. Runnegar. 1979. *Rhytiodentalium kentuckyensis,* a new genus and new species of Ordovician scaphopod, and the early history of scaphopod mollusks. Journal of Paleontology 53:530–541, pls. 1–3.

Pojeta, J. Jr., and B. Runnegar. 1985. The early evolution of diasome molluscs. Pp. 295–336 in The Mollusca, Vol. 10, Evolution (E. R. Trueman and M. R. Clark, eds.). Academic Press, New York.

Posada, D., and K. A. Crandall. 1998. MODELTEST: Testing the model of DNA substitution. Bioinformatics 14 (9):817–818.

Reynolds, P. D. 1997. The phylogeny and classification of Scaphopoda (Mollusca): An assessment of current resolution and cladistic reanalysis. Zoologica Scripta 26:13–26.

Reynolds, P. D., and A. Okusu. 1999. Phylogenetic relationships among families in the Class Scaphopoda (Phylum Mollusca). Zoological Journal of the Linnean Society 126:131–154.

Rosenberg, G., S. Tillier, A. Tillier, G. S. Kuncio, R. T. Hanlon, M. Masselot, and C. J. Williams. 1997. Ribosomal RNA phylogeny of selected major clades in the Mollusca. Journal of Molluscan Studies 63:301–309.

Runnegar, B. N. 1996. Early evolution of the Mollusca: The fossil record. Pp. 77–87 in Origin and Evolutionary Radiation of the Mollusca (J. D. Taylor, ed.), Oxford University Press, Oxford.

Runnegar, B. N., and J. Pojeta. 1974. Molluscan phylogeny: The paleontological viewpoint. Science 186:311–317.

Salvini-Plawen, L. von. 1985. Early evolution and the primitive groups. Pp. 59–150 in The Mollusca vol. 10, Evolution (E. R. Trueman and M. R. Clarke, eds.). Academic Press, London.

Salvini-Plawen, L. von, and G. Steiner. 1996. Synapomorphies and plesiomorphies in higher classification of Mollusca. Pp. 29–51 in Origin and Evolutionary Radiation of the Mollusca (J. D. Taylor, ed.). Oxford University Press, Oxford.

Scarabino, V. 1995. Scaphopoda of the tropical Pacific and Indian Oceans, with descriptions of 3 new genera and 42 new species. Pp. 189–379 in Résultats des Campagnes Musorstom, vol.14 (P. Bouchet, ed.), *Mémoires* du Muséum national d'Histoire naturelle 167:189–379.

Shimek, R. L., and G. Steiner. 1997. Scaphopoda. Pp. 719–781 in Microscopic Anatomy of Invertebrates, Vol. 6B, Mollusca II (F.W. Harrison and A. J. Kohn, eds.). Wiley-Liss, New York.

Starobogatov, Y. I. 1974. Xenoconchias and their bearing on the phylogeny and systematics of some molluscan classes. Paleontological Journal of the American Geological Institute 8:1–13.

Steiner, G. 1992. Phylogeny and classification of Scaphopoda. Journal of Molluscan Studies 58:385–400.

Steiner, G. 1998. Phylogeny of Scaphopoda (Mollusca) in light of new anatomical data on the Gadilinidae and some problematica, and a reply to Reynolds. Zoologica Scripta 29:73–82.

Steiner, G. 1999. A new genus and species of the family Annulidentaliidae (Scaphopoda: Dentaliida) and its systematic implications. Journal of Molluscan Studies 65:151–161.

Steiner, G., and H. Dreyer. (in press). Molecular phylogeny of Scaphopoda (Mollusca) inferred from 18S rDNA sequences: Support for a Scaphopoda–Cephalopoda clade. Zoologica Scripta.

Steiner, G., and S. Hammer. 2000. Molecular phylogeny of Bivalvia (Mollusca) inferred from 18S rDNA sequences with particular reference to the Pteriomorphia. Pp. 11–29 in The Evolutionary Biology of the Bivalvia (E. M. Harper, J. D. Taylor, and J. A. Crame, eds.). Sp. pub. 77, Geological Society of London.

Steiner, G., and A. R. Kabat. 2001. Catalogue of supraspecific taxa of Scaphopoda (Mollusca). Zoosystema 23:433–460.

Steiner, G., and A. R. Kabat. (in press). Catalogue of species-group taxa of Scaphopoda (Mollusca). Zoosystema.

Swofford, D. L. 2002. PAUP*. Phylogenetic Analysis Using Parsimony (*and other methods). Version 4. Sinauer Associates, Sunderland, MA.

Thompson, J. D., D. G. Higgins, and T. J. Gibson. 1994. CLUSTAL W: Improving the sensitivity of progressive multiple sequence alignment through sequence weighting, positions-specific gap penalties and weight matrix choice. Nucleic Acids Research 22:4673–4680.

Waller, T. R. 1998. Origin of the Molluscan class Bivalvia and a phylogeny of major groups. Pp. 1–45 in Bivalves: An Eon of Evolution (P. A Johnston and J. W. Haggart, eds.). University of Calgary Press, Calgary, Alberta.

Wanninger, A., and H. Haszprunar. 2002. The expression of an engrailed protein during embryonic shell formation of the tusk-shell, *Antalis entalis* (Mollusca, Scaphopoda). Evolution and Development 3:312–321.

Wingstrand, K. G. 1985. On the anatomy and relationships of recent Monoplacophora. Galathea Report 16:1–94.

Winnepenninckx, B., T. Backeljau, and R. De Wachter. 1996. Investigation of molluscan phylogeny on the basis of 18S rRNA sequences. Molecular Biology and Evolution 13:1306–1317.

Wray, C. G., D. K. Jacobs, R. Kostriken, A. P. Vogler, R. Baker, and R. DeSalle. 1995. Homologues of the engrailed gene from five molluscan classes. FEBS Letters 365:71–74.

Wuyts, J., Y. Van de Peer, T. Winkelmans, and R. De Wachter. 2002. The European database on small subunit ribosomal RNA. Nucleic Acid Research 30:183–185.

Yancey, T. E. 1973. A new genus of Permian siphonodentalid scaphopods, and its bearing on the origin of Siphonodentaliidae. Journal of Paleontology 47:1062–1064.

Yochelson, E. L. 1999. Rejection of Carboniferous *Quasidentalium* Shimansky, 1974, from the phylum Mollusca. Journal of Paleontology 73:63–65.

ANDREW G. McARTHUR AND M. G. HARASEWYCH

6

MOLECULAR SYSTEMATICS OF THE MAJOR LINEAGES OF THE GASTROPODA

Gastropod mollusks are among the oldest and most evolutionarily successful groups of animals to inhabit the earth. With origins in the Cambrian (Erwin and Signor 1991; Tracey et al. 1993), gastropods have come to inhabit nearly all marine, freshwater, and terrestrial habitats and have achieved a diversity at all taxonomic levels that is exceeded only by the Insecta. Because of their long and extensive fossil record, Gastropoda have figured prominently in paleontological studies spanning the entire Phanerozoic. By virtue of their abundance and diversity in the Recent fauna, gastropods have been used in studies spanning numerous biological disciplines, including ecology, physiology, developmental biology, population genetics, biodiversity, biomechanics, biogeography, and molecular evolution. Gastropods are also of considerable economic importance as sources of food, ornament, and pharmacological compounds, and as agricultural and mariculture pests and vectors of disease (see Cheng 1967; Faust et al. 1968; Abbott 1972; Olivera 1997, and references therein).

The higher classification of gastropods had become entrenched for much of the twentieth century, being based on an arrangement advanced by Thiele (1929–31) and only slightly modified by subsequent authors (Wenz 1938–44; Moore 1960; Boss 1982; Brusca and Brusca 1990). Thiele (1929–31) adopted Milne-Edwards's (1848) division of Gastropoda into Prosobranchia, Opisthobranchia, and Pulmonata, but further subdivided the Prosobranchia into the anagenic series Archaeogastropoda, Mesogastropoda, and Stenoglossa (renamed Neogastropoda by Wenz 1938–1944). Thorough, historical reviews of gastropod classification are provided in Bieler (1992) and Ponder and Lindberg (1997).

The dramatically divergent classification of prosobranch gastropods proposed by Golikov and Starobogatov (1975) catalyzed a renewed interest in gastropod phylogeny. This interest was fueled by the discoveries of numerous new, higher taxa, especially at hydrothermal vents and sulphide seeps; the rapid increase in the amount of new data provided by new morphological and molecular techniques; and, perhaps most importantly, by the application of increasingly rigorous methodologies for analyzing the data and generating phylogenetic hypotheses. Although discrepancies remain among phylogenies based on morphological (e.g., Salvini-Plawen 1980; Haszprunar 1988; Ponder and Lindberg 1996, 1997), DNA–RNA sequence (e.g., Tillier et al. 1992, 1994; Rosenberg et al. 1994, 1997; Winnepenninckx et al. 1994, 1998; Harasewych et al. 1997a, b, 1998; McArthur and Koop 1999; Colgan et al. 2000; Harasewych and McArthur 2000), and paleontological (e.g., Wagner 1995; Bandel 1997) datasets, these recent studies converge on a broad outline of gastropod evolution that differs significantly from those advocated by Thiele and Wenz. Virtually all modern classifications divide the Gastropoda into the following monophyletic groups: Patellogastropoda, Cocculinoidea, Lepetelloidea, Neritopsina, Neomphalina, Vetigastropoda (generally, but not always including the Pleurotomarioidea), Caenogastropoda, and Heterobranchia. However, differences arise regarding the rank and relationships among these clades (see Figure 6.1).

Some features, such as the basal position of the patellogastropod limpets within Gastropoda, which had been recognized in earlier classifications (e.g., Lister 1678; Troschel 1861; Pelseneer 1906), have been reaffirmed by morphological (e.g., Golikov and Starobogatov 1975; Haszprunar 1988; Ponder and Lindberg 1997; Sasaki 1998) and molecular (e.g., Harasewych et al. 1997a; Harasewych and McArthur 2000) studies. Similarly, the monophyly of the Apogastropoda and the sister-group relationship of Caenogastropoda and Heterobranchia are strongly confirmed by multiple datasets. However, the relationships of a number of basal taxa, among them the cocculiniform superfamilies Cocculinoidea and Lepetelloidea, the vetigastropod superfamily Pleurotomarioidea, and the order Neritopsina, are poorly resolved. Their positions have been strongly influenced by such factors as data type (morphological versus molecular), taxon sampling, and outgroup selection.

Over the last decade, many studies have been published in which sequence data have been used to investigate phylogenetic relationships within Gastropoda at varying hierarchical levels (e.g., Tillier et al. 1992, 1994; Rosenberg et al. 1994, 1997; Winnepenninckx et al. 1994, 1996, 1998; Harasewych et al. 1997a, 1997b, 1998; Colgan et al. 2000; Harasewych and McArthur 2000). In general, these studies have readily identified the major gastropod lineages (outlined

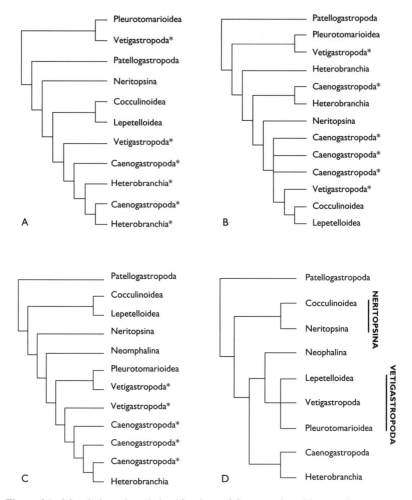

Figure 6.1. Morphology-based classifications of Gastropoda, with trees drawn to represent relationships among the nine clades (see text). Taxon names and ranks have been modified for the sake of uniformity and clarity. Hypothesis of nonmonophyly for the modern major groupings are marked with an asterisk. (A) Thiele (1929–31), represented as a phylogenetic tree (modified from Ponder and Lindberg 1996). (B) Golikov and Starobogatov (1975). (C). Haszprunar (1988). (D). Ponder and Lindberg (1997).

above), but have had much difficulty in resolving basal relationships. In this chapter, we examine the growing consensus concerning the major lineages of the Gastropoda and use the largest alignment of 18S (small subunit) ribosomal RNA sequences yet assembled as an independent assessment of confidence.

EXAMINING CONSENSUS USING SUPERTREE ANALYSIS

With so many independent studies, each using a different dataset, it is difficult to obtain an overall view of our current understanding of gastropod phylogeny. In particular, it is often hard to compare the results of molecular phylogenetic investigations (based on sequences from individual specimens) and parsimony analysis of anatomical characters (often based on summaries of character states for higher-order taxonomic groups). Even among studies using the same kind of data, differences in taxonomic sampling, the exact characters considered, and how the analysis was performed can make comparison difficult. Table 6.1 shows the data used by a metaanalysis of 11 studies of gastropod phylogeny, ranging from the precladistic hypotheses of Thiele (1929–31) and Golikov and Starobogatov (1975) to the recent molecular phylogenetic investigation of Colgan et al. (2000). We used the supertree method of Baum (1992) and Ragan (1992) (for a review, see Sanderson et al. 1998), in which the phylogenetic trees of each study are recoded as new metadata by the matrix representation using parsimony (MRP) method. In essence, the new metadata matrix reflects the presence or absence (or lack of data) of all of the possible clades of gastropod groups found in these studies. The resulting matrix was subjected to a cladistic analysis under parsimony to provide a consensus supertree in which shared results among original studies are recovered and differences in branching order among the original studies are resolved using the most parsimonious explanation. The advantage of this method is that the studies do not have to be compatible—each can use a different selection of taxa. For example, the detailed neontological study of Ponder and Lindberg (1997) includes the architaenioglossan families Cyclophoridae and Ampullariidae, whereas the molecular study of McArthur and Koop (1999) includes representatives of Ampullariidae and Viviparidae.

The metadata were compiled by recoding each of the original phylogenetic hypotheses to the taxonomic names used by Ponder and Lindberg (1997), with the addition of a few taxa present in one or more of the other studies and some changes to more inclusive higher-order taxa. Because some molecular studies had multiple representatives of a single taxonomic group (e.g., multiple representatives of the Trochidae), we trimmed these phylogenetic trees to single representatives, albeit with some difficult choices, when molecular trees did not support monophyly. Some taxonomic groups were not included because they were not informative about overall gastropod phylogeny. The resulting dataset included 33 taxa and 151 parsimony-informative characters. Because not all of the original studies included bootstrapping or other measures of internal con-

Table 6.1

Sources of data used for the supertree metaanalyses

Source	Type of Data	Type of Analysis	Number of MRP Characters
Thiele (1929–31)	Morphology	Nonphylogenetic	16
Golikov and Starobogatov (1975)	Morphology	Nonphylogenetic	19
Haszprunar (1988)	Anatomy	Cladistic	17
Ponder and Lindberg (1997)	Anatomy and Morphology	Phylogenetic	30
Tillier et al. (1994)	28S rDNA	Phylogenetic	9
Harasewych et al. (1997a)	18S rDNA	Phylogenetic	12
Harasewych et al. (1998)	18S rDNA	Phylogenetic	11
Winnepenninckx et al. (1998)	18S rDNA	Phylogenetic	6
McArthur and Koop (1999)	28S rDNA	Phylogenetic	14
Colgan et al. (2000)	28S rDNA and Histone H3	Phylogenetic	16
Harasewych and McArthur (2000)	18S rDNA	Phylogenetic	11

Note: Trees were recoded as binary characters by the matrix representation using parsimony (MRP) method (Baum 1992; Ragan 1992).

fidence, each of the new metadata characters was given equal weight in the final analysis.

We subjected the metadata to analysis under the parsimony optimality criteria using the computer program PAUP* (Swofford 2001). We used 100 random taxa addition replicates with tree bisection and reconnection (TBR) branch swapping to find the set of most parsimonious trees. We assessed confidence of the results by performing bootstrap analysis on 100 random bootstrap replicate datasets, with 10 random taxa addition replicates per bootstrap dataset and a limit of 100 trees in memory for TBR swapping. Eighty equally parsimonious reconstructions of the metadata were found; the majority-rule consensus tree is presented in Figure 6.2. The resulting phylogenetic hypothesis was quite conventional, with Ponder and Lindberg's (1996) hypothesized basal position of the Patellogastropoda and monophyly of all of the major groups, although this was assumed for the Patellogastropoda (based on Harasewych and McArthur 2000), Neritopsina (consistent with all of the studies), Neomphalina (based on McArthur and Koop 1999), and Pulmonata. As with recent studies, bootstrapping revealed negligible support for the basal branching order of the Gastropoda. Monophyly of groups such as the Cocculiniformia (Cocculinoidea + Lepetelloidea), Caenogastropoda, Vetigastropoda, and Heterobranchia was not statistically supported. However, there was strong support for monophyly of the Apogastropoda (Caenogastropoda + Heterobranchia) and a clade containing the Architectonicoidea, Pulmonata, and a monophyletic Opisthobranchia. Resolu-

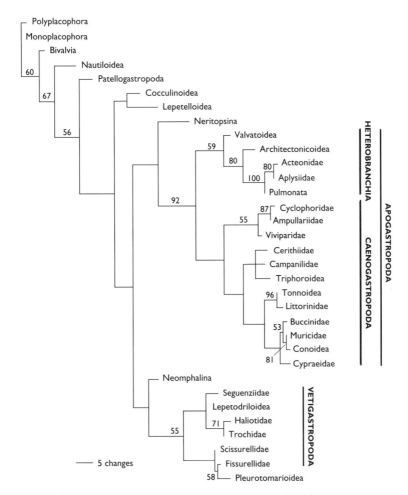

Figure 6.2. Majority-rule consensus tree of the 80 most parsimonious supertrees found using all of the studies listed in Table 6.1. Bootstrap proportions where higher than 50% are shown.

tion of evolutionary relationships within the Prosobranchia did not improve when the studies of Thiele (1929–31), Golikov and Starobogatov (1975), and Haszprunar (1988) were excluded and the data reanalyzed (Figure 6.3). This second result reflected common themes found in molecular studies: unexpected placement of the root of the Gastropoda (Neritopsina as the basal clade?), lack of resolution of the prosobranch grade, yet good support for monophyly of the Apogastropoda, Caenogastropoda, and Heterobranchia. As did the molecular studies, this method suggested shared ancestry of the Cocculinoidea, Lepetelloidea, and Patellogastropoda.

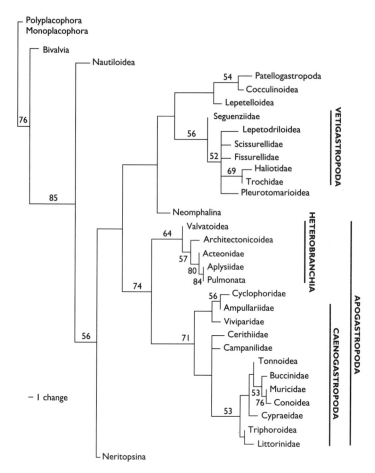

Figure 6.3. Majority-rule consensus tree of the 20 most parsimonious supertrees found using all the studies listed in Table 6.1, except those of Thiele (1929–31), Golikov and Starobogatov (1975), and Haszprunar (1988). Bootstrap proportions where higher than 50% are shown.

A PRELIMINARY BAYESIAN PERSPECTIVE

Molecular investigations of gastropod phylogeny have mainly used the DNA–RNA sequences of the small (18S) and large (28S) subunits of ribosomal RNA genes. Every gene examined in molecular systematics has its limits. The consistently poor resolution of early gastropod phylogeny suggests that ribosomal sequences have a difficult time "reaching back" to the early Paleozoic and Cambrian. This is contrary to expectation, because these gene sequences have been used successfully to examine the Tree of Life and have helped elucidate the existence of the three domains of life—the Bacteria, Archaea, and Eukaryota (Woese et al. 1990; Embley et al. 1994). Our lack of significant progress in the Gastropoda is more likely a reflection of sampling effort. Outside of the Euthyneura, most gastropod rRNA sequences are partial. Statistical resolution improves with the number of characters sequenced. Similarly, representative taxon sampling can be very important in resolving major phylogenetic patterns because crucial taxa can split long internal branches, reducing artifacts and providing more phylogenetic signal about early evolutionary events. We have attempted to address these concerns by building as large an 18S rDNA dataset as possible. To that end, an 18S rDNA alignment of 163 gastropod taxa has been constructed (available on request, including source information). This alignment includes all partial and full-length 18S rRNA sequences available in GenBank, aligned according to the secondary structure model of the Ribosomal Database Project (Olsen et al. 1992) using an iterative application of ClustalW (Thompson et al. 1994) and subsequent manual editing using MacClade (Maddison and Maddison 2000). The alignment also includes 21 new sequences (Table 6.2) determined using the methods outlined in Harasewych and McArthur (2000). These new sequences attempt to improve the taxon sampling throughout the Prosobranchia, although most were partial sequences because preserved specimens were used.

To incorporate maximum likelihood, phylogenetic uncertainty, and easy examination of competing hypotheses, we analyzed this large alignment using a Bayesian statistical procedure, as implemented by the computer program MrBayes (Huelsenbeck and Ronquist 2001). MrBayes performs a Metropolis-coupled Markov chain Monte Carlo (MCMCMC) estimation of posterior probabilities (see Shoemaker et al. 1999; Huelsenbeck et al. 2001; Lewis 2001b). Bayesian methods have the advantage of incorporating maximum likelihood, complex substitution models, and fast analyses. Their goal is not to find the best tree, but to instead sample the "cloud of best trees." Bayesian methods recognize that there is uncertainty in phylogenetic analyses owing to the finite amount of data sampled. Instead of finding the single best tree (a very difficult procedure), Bayesian methods instead use their sampling of "best trees" to estimate posterior probabilities of relationships between taxa. Any possible clade can be assigned an easy-to-interpret posterior probability. For example, Bayesian methods make it easy to ask, "What is the probability that the sister taxon to the Caenogastropoda is the Heterobranchia versus the probability that it is the Neritopsina?" The consensus of all trees sampled from the "cloud of best trees" provides an overview of the most probable phylogeny. In practice, this is often the same tree found using maximum likelihood heuristic searches. As our dataset was too large for maximum likelihood heuristic searching, we performed MCMCMC estimation of posterior probabilities using noninformative prior

Table 6.2

New 18S rDNA sequences included in this study, localities, preservation means, vouchers, sequence lengths, and GenBank accession numbers

Taxon	Specimen Details	GenBank Accession No.
Caenogastropoda		
Viviparus georgianus	Lake Talquin, Tallahassee, Florida (Frozen, USNM 1003901)	AY090794 (1797 bp)
Cocculinoidea		
Coccopigya hispida	On wood, off Cape Palliser, 41°45.2'S 175°26.8'E, 1,039–1,077 m (EtOH, Marshall 87048)	AY090795 (537 bp)
Cocculina messingi	On deployed wood, Bahamas, 26°37.30'N 78°58.55'W, 1,372 ft. (Frozen, USNM 888655)	AY090796 (1752 bp)
Heterobranchia		
Rissoella caribaea	Fiesta Key, Florida (Frozen, USNM 881221)	AY090797 (2123 bp)
Lepetelloidea		
Caymanabyssia fosteri	On deployed wood, 11°51.00'N 103°50.00'W, East Pacific Rise, 2,700 m (EtOH, USNM 784765)	AY090798 (145 bp)
Copulabyssia gradata	On wood, off Cape Egmont 38°58.5'S 172°10.2'E, 1,045–1,055 m (EtOH, USNM 888730)	AY090799 (530 bp)
Mesopelex zelandica	On kelp holdfast, Chatham Rise, 42°53'S 176°04'E, 370–420 m (EtOH, Marshall 118916)	AY090800 (547 bp)
Pyropelta musaica	Hydrothermal vents, 45°57.00'N 130°01.00'W, Juan de Fuca Ridge, 1,546 m (EtOH, USNM 858229)	AY090801 (149 bp)
Tentaoculus haptricola	On algal holdfast, off Chatham Island, 42°50'S 176°30'W, 945 m (EtOH, USNM 888731)	AY090802 (141 bp)
Neomphalina		
Cyathermia naticoides	Hydrothermal vents, 20°49.9'N 109°06.0'W, East Pacific Rise, 2,615 m (Frozen, Lutz A2232)	AY090803 (526 bp)
Depressigyra globulus	Hydrothermal vents, 44°59.43'N, 130°12.08'W, Juan de Fuca Ridge, 2,249 m (Frozen, Tunn. HYS202)	AY090804 (561 bp)
Melanodrymia aurantiaca	Hydrothermal vents, 20°47.0'N 109°08.9'W, East Pacific Rise, 2,577 m (Frozen, Lutz A2233)	AY090805 (490 bp)
Neomphalus fretterae	Hydrothermal vents, Oyster Bed and Garden of Eden, Galapagos Rift (EtOH, USNM 784638)	AY090806 (576 bp)
Peltospira operculata	Hydrothermal vents, 00°48'N 86°13'W, Galapagos Rift, 2,462 m (Frozen, Lutz A2010)	AY090807 (538 bp)
Symmetromphalus regularis	Hydrothermal vents, 18°12.36'N 144°42.24'E, Mariana back-arc basin, 3,640 m (EtOH, USNM 784763)	AY090808 (538 bp)
Pleurotomarioidea		
Entemnotrochus adansonianus	Guadeloupe (Frozen, USNM 888647)	AY090809 (1993 bp)

Table 6.2 continued

Taxon	Specimen Details	GenBank Accession No.
Vetigastropoda		
Bathymargarites symplector	Hydrothermal vents, 20°49.9'N 109°06.0'W, East Pacific Rise, 2,615 m (Frozen, Lutz A2232)	AY090810 (534 bp)
Cittarium pica	Jamaica (Frozen, USNM 888661)	AY090811 (403 bp)
Lepetodrilus fucensis	Hydrothermal vents, 40°58'N 129°05.5'W, Juan de Fuca Ridge, 2,125 m (Frozen, Tunn. F20-A2413)	AY090812 (529 bp)
Sinezona confusa	Long Key, Florida (Frozen, USNM 888716)	AY090813 (530 bp)
Temnocinclis euripes	Hydrothermal vents, 44°56'N 130°15'W, Juan de Fuca Ridge, 2282 m. (EtOH, Tunn. A2078-1452)	AY090814 (541 bp)

Note: Sequences for *Cocculina messingi*, *Rissoella caribaea*, and *Entemnotrochus adansonianus* are extensions of previously published sequences.

Preservation method: EtOH = fixation in formalin, followed by storage in ethanol.

Source codes: Lutz = collection of R. Lutz, Rutgers University, U.S.A.; Marshall = collection of B. Marshall, Museum of New Zealand, New Zealand; Tunn. = collection of V. Tunnicliffe, University of Victoria, Canada; USNM = Mollusk collection, National Museum of Natural History, Smithsonian Institution.

probabilities, the TrN+I+Γ substitution model with inclusion of unequal nucleotide frequencies (see Swofford et al. 1996), and four incrementally heated Markov chains. The TrN+I+Γ substitution model was selected using the Likelihood Ratio Test as implemented in the computer program ModelTest (Posada and Crandall 1998). The Markov chains were run for 150,000 generations, with sampling of topologies every 100 generations. Posterior probabilities of topologies, clades, and parameters were estimated from the sampled topologies after removal of MCMCMC burn-in. We included all taxa in the alignment, except that we used representative taxon sampling for the Caenogastropoda and Heterobranchia, because both of these groups included a heavy sampling of closely related genera. Regions of poor or uncertain alignment were excluded from analyses and the final dataset included 81 taxa and 1,431 characters (899 of which were constant). Based on our experience with these data, we did not include a nongastropod outgroup because the preponderance of partial sequences at the base of the Gastropoda make rooting extremely unreliable. Instead, we used the Patellogastropoda as a visual root when presenting trees. As such, all discussion of monophyletic and sister relationships from our results should be treated as conditional on the final placement of the root of the Gastropoda.

A consensus of posterior probabilities as determined by Bayesian analysis

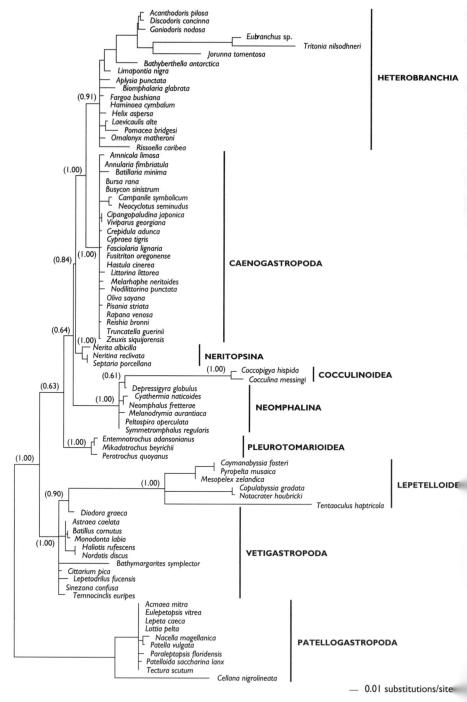

Figure 6.4. Majority-rule consensus tree of trees sampled from the "cloud of best trees" by MCMCMC (after removal of burn-in). Such a sampling is representative of posterior probabilities, which are given in parentheses for the major clades and a few internal nodes related to the placement of the Cocculinoidea and Lepetelloidea.

of the 81 taxa alignment is shown in Figure 6.4. Monophyly of the Hetero-branchia, Caenogastropoda, Apogastropoda, Neritopsina, and Patellogastropoda was supported. In contrast, the Cocculiniformia was divided into two groups— a monophyletic Cocculinoidea appeared as a member of the hydrothermal vent Neomphalina and a monophyletic Lepetelloidea appeared as a member of the Vetigastropoda. Both of these results had strong support. As found previously, the Pleurotomarioidea branched separately from the Vetigastropoda. Unfortu-nately, the sampling used to determine these posterior probabilities was very small. The MCMCMC was only able to run for 150,000 generations before MrBayes ran out of memory, and the first 100,000 generations had to be dis-carded as burn-in. Only 500 trees were sampled. As in heuristic searching, in which the number of random addition replicates needed to find the best tree is unknown, the number of generations needed by MCMCMC to obtain an unbi-ased estimate of posterior probabilities is also unknown. However, because our sampling ran only half as long as the burn-in, the posterior probabilities pre-sented in Figure 6.4 are almost certainly inaccurate estimates. As such, we must view Figure 6.4 as a preliminary Bayesian estimate of gastropod phylogeny.

REPRESENTATIVE TAXON SAMPLING

Although our goal in using Bayesian methods was to obtain an estimate of gas-tropod phylogeny using the largest possible sampling of taxa, we ran aground of computational limits. As such, we decided to take a smaller, representative sampling of gastropod phylogeny from our 163 taxon alignment and examine gastropod phylogeny under maximum parsimony, minimum evolution, maxi-mum likelihood, and Bayesian perspectives. Based on our broader Bayesian analysis (Figure 6.4), we sampled representative sequences from the Patel-logastropoda, Vetigastropoda, and Apogastropoda and included all sequences available for the Lepetelloidea, Pleurotomarioidea, Neomphalina, Cocculi-noidea, and Neritopsina. In all, 33 taxa and 1,431 characters (1,070 constant) were included. We searched for the best tree under the maximum likelihood op-timality criterion using 10 random taxon addition replicates, TBR branch swap-ping, and the TrN+I+Γ substitution model with inclusion of unequal nucleotide frequencies (determined by ModelTest). Bootstrap measures of internal confi-dence were determined using 100 bootstrap replicates with 10 (maximum par-simony / minimum evolution) or 2 (maximum likelihood) random addition

Figure 6.4 continued
Branch lengths represent the amount of evolutionary change. The posterior probabili-ties shown are probably inaccurate estimates caused by the overly short Markov chain lengths (see text).

replicates each. All searches and bootstrapping were restricted to 20 trees in memory during branch swapping. In addition, we performed Bayesian estimation of posterior probabilities using noninformative prior probabilities, the TrN+I+Γ substitution model with inclusion of unequal nucleotide frequencies, and four incrementally heated Markov chains. The Markov chains were run for a million generations, with sampling of topologies every 100 generations. Posterior probabilities of topologies, clades, and parameters were estimated from the sampled topologies after removal of MCMCMC burn-in.

The best tree found under maximum likelihood (Figure 6.5) agreed in general with the tree found with the broader Bayesian analysis. There was strong bootstrap and posterior probability support for the major clades, with the exception of the Vetigastropoda/Lepetelloidea grouping. All analyses found strong support for the Apogastropoda, but disagreed on the Neritopsina being its sister taxa (the maximum likelihood bootstrap value was low). There was strong support under maximum likelihood and by posterior probabilities that the Cocculinoidea shared common ancestry with the hydrothermal vent Neomphalina. The placement of the Lepetelloidea within the Vetigastropoda is merely an unsupported suggestion because bootstrap values and posterior probabilities were negligible for the placement of vetigastropod taxa, although support was strong for independent origins of the Pleurotomarioidea.

The combination of new discoveries such as the novel Neomphalina from hydrothermal vents, new tools such as electron microscopy and DNA sequencing, and new perspectives such as parsimony and maximum likelihood have all combined to revolutionize our understanding of gastropod phylogeny. As shown in our metaanalyses, neontological and molecular data do not conflict for much of gastropod phylogeny. We have clear evidence that traditional groupings such as the Prosobranchia, Archaeogastropoda, Mesogastropoda, and Streptoneura are phylogenetically meaningless or are grades of organization. We now understand that a great deal of gastropod diversity, including marine, terrestrial, and aquatic forms with a variety of lifestyles is the product of a single phylogenetic lineage—the Apogastropoda. This clade, with its internal sister relationship between the Heterobranchia and Caenogastropoda, finds strong support in all neontological and molecular studies. However, neontological and molecular studies are not in agreement on the remainder of gastropod phylogeny. The most notable difference is that neontological studies find resolved (i.e., bootstrap support) branching patterns for basal gastropod phylogeny (e.g., Ponder and Lindberg 1997), whereas molecular studies often resolve the major clades, but cannot resolve the relationships among them. The differences in basal branching pattern and consistent poor basal bootstrap support in our two supertrees (Figures 6.2 and 6.3) are due in part to introduction of basal phylogenetic noise by molecular studies. This noise, within the molecular studies

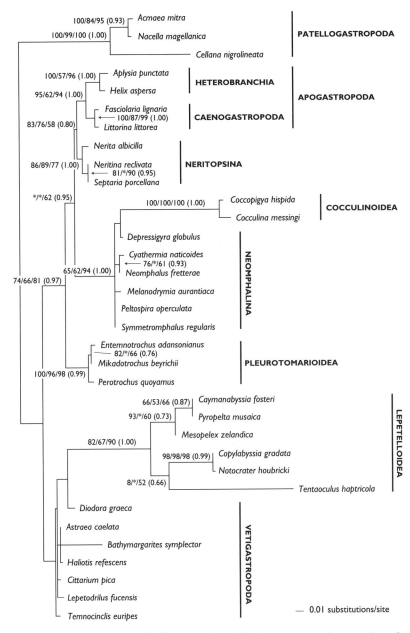

Figure 6.5. Best maximum likelihood tree found for the representative sampling of the Gastropoda. Bootstrap values are shown where higher than 50% (maximum parsimony / minimum evolution / maximum likelihood). Bayesian posterior probabilities are given in parentheses where higher than 50%. * represent bootstrap values lower than 50%. Branch lengths represent amount of evolutionary change.

themselves or in our supertree analyses, produces phylogenetic artifacts such as the basal placement of the Neritopsina (Figure 6.3). Most notably, basal position of the Patellogastropoda finds considerable, very believable support in studies of gastropod anatomy and ultrastructure (e.g., Ponder and Lindberg 1996, 1997), but long branch problems essentially make the Patellogastropoda a rogue taxon in ribosomal sequence investigations (McArthur and Koop 1999; Harasewych and McArthur 2000). Thus, molecular investigations of gastropod phylogeny need to pay careful attention to sources of variation, noise, and bias.

Molecular investigations of gastropod phylogeny provide a very powerful tool for independent assessment of neontological hypotheses. Our preliminary Bayesian investigation and thorough maximum likelihood analysis confirm the monophyly of the Patellogastropoda, Neritopsina, Cocculinoidea, Lepetelloidea, Pleurotomarioidea, Caenogastropoda, Heterobranchia, and Apogastropoda. Use of 18S rDNA sequences confirms the earlier finding of McArthur and Koop (1999) that the hydrothermal vent endemic Neomphalina represents a major gastropod lineage, but additionally supports common ancestry with the Cocculinoidea, another group of deep-sea sulphophiles known from hydrothermal vents, hydrocarbon seeps, whale falls, and sunken wood. We also find considerable evidence for origins of the Lepetelloidea independent of the Cocculinoidea, with a statistically unsupported association of the Lepetelloidea with the Vetigastropoda. The lack of support for monophyly of the Vetigastropoda (less the Pleurotomarioidea) contrasts with previous molecular investigations and illustrates some concerns with these data. Long internal branches separate the Patellogastropoda, Cocculinoidea, and Lepetelloidea from their closest relatives. Long branches can obscure phylogenetic history through introduction of localized noise, although maximum likelihood handles this artifact quite well (Swofford et al. 1996). Although the Cocculinoidea does not appear to be acting as a rogue taxon (i.e., high support for monophyly of Neomphalina + Cocculinoidea), the combination of very long branches in the Lepetelloidea and drastic loss of support for the Vetigastropoda makes the association of these two taxa suspect. The separation of the Pleurotomarioidea from the Vetigastropoda has been found in previous molecular investigations (e.g., Harasewych and McArthur 2000). The Pleurotomarioidea has novel 18S rDNA inserts of considerable size associated with elevated sequence variation in conservative regions of the alignment (Harasewych et al. 1997a; Harasewych and McArthur 2000). On a smaller scale, the same is true for the Patellogastropoda and Lepetelloidea (Harasewych and McArthur 2002; alignment used in this study). Models of nucleotide substitution used by maximum likelihood analyses assume that the rules of evolution are homogeneous for all lineages within the sampled phylogenetic tree (Swofford et al. 1996). The opposite-elevated or differing patterns of variation in some of the sampled lineages (covarion pattern of variation) can confuse tree reconstruction considerably (Lockhart et al. 1998). Thus, although the Pleurotomarioidea does not exhibit long branches, examinations of the rDNA alignments suggest it may be contributing a covarion structure to the data (as may be the Patellogastropoda and Lepetelloidea in association with long branch problems). Although Ponder and Lindberg's (1997) detailed investigation of gastropod anatomy supported a monophyletic Vetigastropoda that included the Lepetelloidea and Pleurotomarioidea, a covarion structure in the rDNA data may be degrading our ability to accurately examine the Vetigastropoda and its relatives.

Consistent with other molecular investigations, we do not find a clear resolution of basal gastropod phylogeny. Although Bayesian methods hold promise for future resolution of basal gastropod phylogeny, we were unable to run sufficiently long Markov chains in our dataset with a large taxon sampling (Figure 6.4). We suspect that the lack of support for the Vetigastropoda + Lepetelloidea in the maximum likelihood analysis (Figure 6.5) could also in part be the product of restricted taxon sampling. Overall, there was a general agreement between maximum likelihood bootstrap values and Bayesian posterior probabilities in the smaller taxon sampling (Figure 6.5), with the exception of two of the deeper nodes in the tree (position of the Neritopsina and placement of the Neomphalina + Cocculinoidea). Posterior probabilities support a sister relationship between the Neritopsina and the Apogastropoda, with this grouping in a sister relationship with the Neomphalina + Cocculinoidea clade. There is no current understanding of the relationship between bootstrap values and posterior probabilities. That the two are in conflict for these two hypothetical sister relationships means these relationships should be taken as preliminary hypotheses testable by additional data and future advances in our understanding of the bootstrap–posterior probability relationship.

Instead of focusing on the search for the best tree, molecular systematic investigations can often be more fruitful if they are considered opportunities to examine important hypotheses. One of the advantages of Bayesian methods is that posterior probabilities can be determined for any number of phylogenetic hypotheses. A series of important hypotheses and their associated posterior probabilities are presented in Table 6.3. These data do not support monophyly of the Cocculiniformia (Cocculinoidea + Lepetelloidea; $p = 0.00$), nor their joint association with another major lineage, such as the Patellogastropoda ($p = 0.00$) or Neomphalina ($p = 0.00$). Although the Cocculinoidea appear to share common ancestry with the Neomphalina ($p = 1.00$), the data are equivocal on whether this could be a sister relationship ($p = 0.38$). These data also reject common ancestry of the Neomphalina and Vetigastropoda ($p = 0.00$). In fact, any

Table 6.3

A Bayesian look at several alternative phylogenetic hypotheses

Hypothesis	Posterior Probability [a,b]
Monophyly of Cocculinoidea + Lepetelloidea	0.00
Monophyly of Cocculinoidea + Neomphalina (exists in Figure 6.5)	1.00
Cocculinoidea as sister taxon to the Neomphalina only	0.38
Monophyly of Lepetelloidea + Vetigastropoda (exists in Figure 6.5)	0.30
Lepetelloidea as sister taxon to the Vetigastropoda only	0.17
Monophyly of the Vetigastropoda (regardless of Lepetelloidea)	0.17
Monophyly of Vetigastropoda + Pleurotomarioidea	0.00
Monophyly of Patellogastropoda + Cocculinoidea + Lepetelloidea	0.00
Monophyly of Neomphalina + Vetigastropoda	0.00
Monophyly of Cocculinoidea + Neritopsina	0.00
Monophyly of Neomphalina + Cocculinoidea + Lepetelloidea	0.00

[a]Posterior probabilities are from the Bayesian analysis shown in Figure 6.5.
[b]Rounded to two decimal places.

hypothesis that does not place the Lepetelloidea, Vetigastropoda (not including Pleurotomarioidea), and Patellogastropoda basal to all other gastropod lineages is strongly rejected ($p = 0.97$, maximum likelihood bootstrap = 81). However, because Bayesian approaches are a recent introduction to molecular systematics and many aspects of their strengths, weaknesses, and biases have yet to be investigated, we should exercise some caution in interpreting these results.

It should be noted that the long branches associated with the Lepetelloidea in our analyses are in part due to use of some very short sequences (Table 6.2). Short sequences increase error in branch length estimates. Resolution of the position of the Lepetelloidea, plus improved resolution of basal relationships, could improve with use of longer sequences. Overall, our large 18S rDNA alignment contained little variation for the taxa studied (532 variable sites for Figure 6.4, 361 variable sites for Figure 6.5). Given the dramatically decreased cost of DNA sequencing and development of long PCR techniques, future studies should attempt to amplify and sequence the entire ribosomal operon when examining overall gastropod phylogeny. Combined analysis of both large and small subunit ribosomal sequences is proving quite powerful in resolving deep animal relationships (Medina et al. 2001; Mallatt and Winchell 2002), so amplifying the entire operon would be the most efficient and cost effective way to obtain these data. Statistical and probabilistic approaches to gastropod phylogeny are proving to be powerful and, as our emphasis shifts from finding the very best tree to recognizing phylogenetic uncertainty, we are certain to expand our use of morphological and molecular data for phylogenetic hypothesis testing. For example, Lewis (2001a) recently described a maximum likelihood method for discrete morphological characters that could be applied to our rich knowledge of gastropod morphology, anatomy, and ultrastructure. Because statistical and model-based methods have associated error, it will be important to minimize random error by using the longest sequences and largest datasets possible. Given the expanding use of diverse approaches such as electron microscopy, examination of development, DNA sequencing, comparative genomics, and broad taxon sampling, the vigorous malacological research community is sure to rise to the challenge.

ACKNOWLEDGMENTS

We are grateful to Richard Lutz (Rutgers University, NJ), Bruce Marshall (Museum of New Zealand), and Verena Tunnicliffe (University of Victoria, Canada) for providing living or preserved material for this study. We thank Sarah Pacocha (Marine Biological Laboratory) for considerable assistance with construction of the alignments. A. McArthur was supported by a Smithsonian Institution Postdoctoral Fellowship (Office of Fellowships and Grants) and the National Museum of Natural History's Laboratory of Molecular Systematics. Computational resources were provided by the Josephine Bay Paul Center for Comparative Molecular Biology and Evolution (Marine Biological Laboratory) through funds provided by The W. M. Keck Foundation and The G. Unger Vetlesen Foundation.

REFERENCES

Abbott, R. T. 1972. Kingdom of the seashell. Crown Publishers, New York.

Bandel, K. 1997. Higher classification and pattern of evolution of the Gastropoda. Courier Forschungsinstitut Senckenberg 201:57–81.

Baum, B. R. 1992. Combining trees as a way of combining data sets for phylogenetic inference, and the desirability of combining gene trees. Taxon 41:3–10.

Bieler, R. 1992. Gastropod phylogeny and systematics. Annual Review of Ecology and Systematics 23:311–338.

Boss, K. 1982. Mollusca. Pp. 945–1161 in Synopsis and Classification of Living Organisms (S. P. Parker, ed.). McGraw Hill, New York.

Brusca, R. C., and G. J. Brusca. 1990. Invertebrates, Sinauer Associates, Sunderland, MA.

Cheng, T. C. 1967. Marine Molluscs as Hosts for Symbiosis: With a Review of Known Parasites of Commercially Important Species. Academic Press, New York.

Colgan, D. J., W. F. Ponder, and P. E. Eggler. 2000. Gastropod evolutionary rates and phylogenetic relationships assessed using partial 28S rDNA and histone H3 sequences. Zoologica Scripta 29:29–63.

Embley, T. M., R. P. Hirt, and D. M. Williams. 1994. Biodiversity at the molecular level: The domains, kingdoms and phyla of life. Philosophical Transactions of the Royal Society of London, B 345:21–33.

Erwin, D. H., and P. W. Signor. 1991. Extinction in an extinction–resistant clade: The evolutionary history of the Gastropoda. Pp. 152–160 in The Unity of Evolutionary Biology. Proceedings of the 4th International Congress of Systematic and Evolutionary Biology, 1 (E. C. Dudley, ed.). Dioscorides Press, Portland, OR.

Faust, E. C., P. C. Beaver, and R. C. Jung. 1968. Animal Agents and Vectors of Human Disease, 3rd ed. Lea and Febiger, Philadelphia.

Golikov, A. N., and Y. I. Starobogatov. 1975. Systematics of prosobranch gastropods. Malacologia 15:185–232.

Harasewych, M. G., S. L. Adamkewicz, J. A. Blake, D. Saudek, T. Spriggs, and C. J. Bult. 1997a. Phylogeny and relationships of pleurotomarid gastropods (Mollusca: Gastropoda): an assessment based on partial 18S rDNA and cytochrome oxidase I sequences. Molecular Marine Biology and Biotechnology 6:1–20.

Harasewych, M. G., S. L. Adamkewicz, J. A. Blake, D. Saudek, T. Spriggs, and C. J. Bult. 1997b. Neogastropod phylogeny: A molecular perspective. Journal of Molluscan Studies 63:327–351.

Harasewych, M. G., S. L. Adamkewicz, M. Plassmeyer, and P. M. Gillevet. 1998. Phylogenetic relationships of the lower Caenogastropoda (Mollusca, Gastropoda, Architaenioglossa, Campaniloidea, Cerithioidea) as determined by partial 18S rDNA sequences. Zoologica Scripta 27:361–372.

Harasewych, M. G., and A. G. McArthur. 2000. A molecular phylogeny of the Patellogastropoda (Mollusca: Gastropoda). Marine Biology 137:183–194.

Haszprunar, G. 1988. On the origin and evolution of major gastropod groups, with special reference to the Streptoneura. Journal of Molluscan Studies 54:367–441.

Huelsenbeck, J. P., and F. Ronquist. 2001. MrBayes: Bayesian inference of phylogenetic trees. Bioinformatics 17:754–755.

Huelsenbeck, J. P., F. Ronquist, R. Nielsen, and J. P. Bollback. 2001. Bayesian inference of phylogeny and its impact on evolutionary biology. Science 294:2310–2314.

Lewis, P. O. 2001a. A likelihood approach to estimating phylogeny from discrete morphological character data. Systematic Biology 50:913–925.

Lewis, P. O. 2001b. Phylogenetic systematics turns over a new leaf. Trends in Ecology and Evolution 16:30–37.

Lister, M. 1678. Historiae Animalium Angliae Trest Tractatus. III. Cochleis Marinis, Royal Society, London.

Lockhart, P. J., M. A. Steel, A. C. Barbrook, D. H. Huson, M. A. Charleston, and C. J. Howe. 1998. A covariotide model explains apparent phylogenetic structure of oxygenic photosynthetic lineages. Molecular Biology and Evolution 15:1183–1188.

Maddison, W. P., and D. R. Maddison. 2000. MacClade, version 4.0b10. Sinauer Associates, Sunderland, MA.

Mallatt, J., and C. J. Winchell. 2002. Testing the new animal phylogeny: First use of combined large–subunit and small–subunit rRNA gene sequences to classify the protostomes. Molecular Biology and Evolution 19:289–301.

McArthur, A. G., and B. F. Koop. 1999. Partial 28S rDNA sequences and the antiquity of hydrothermal vent endemic gastropods. Molecular Phylogenetics and Evolution 13:255–274.

Medina, M., A. G. Collins, J. D. Silberman, and M. L. Sogin. 2001. Evaluating hypotheses of basal animal phylogeny using complete sequences of large and mall subunit rRNA. Proceedings of the National Academy of Sciences U.S.A. 98:9707–9712.

Milne-Edwards, H. 1848. Note sur la classification naturelle chez Mollusques Gasteropodes. Annales des Sciences Naturalles 9:102–112.

Moore, R. C. 1960. Summary of classification. Pp. I157–I169 in Treatise on Invertebrate Paleontology, Part I (R. C. Moore, ed.). Geological Society of America and University of Kansas Press, Lawrence, KS.

Olsen, G. J., R. Overbeek, N. Larsen, T. L. Marsh, M. J. McCaughey, M. A. Maciukenas, W.-M. Kuan, T. J. Macke, Y. Xing, and C. R. Woese. 1992. The ribosomal database project. Nucleic Acids Research Supplement 2199–2200.

Olivera, B. 1997. *Conus* venom peptides, receptor and ion channel targets, and drug design: 50 million years of neuropharmacology. Molecular Biology and Cell 8:2101–2109.

Pelseneer, P. 1906. A Treatise on Zoology. V. Mollusca, A and C Black, London.

Ponder, W. F., and D. R. Lindberg. 1996. Gastropod phylogeny. Challenges for the 90s. Pp. 135–154 in Origin and Evolutionary Radiation of the Mollusca (J. D. Taylor, ed.). Oxford University Press, Oxford.

Ponder, W. F., and D. R. Lindberg. 1997. Towards a phylogeny of gastropod molluscs: An analysis using morphological characters. Zoological Journal of the Linnean Society 119:83–265.

Posada, D., and K. Crandall. 1998. Modeltest: Testing the model of DNA substitution. Bioinformatics 14:817–818.

Ragan, M. A. 1992. Phylogenetic inference based on matrix representation of trees. Molecular Phylogenetics and Evolution 1:53–58.

Rosenberg, G., G. S. Kuncio, G. M. Davis, and M. G. Harasewych. 1994. Preliminary ribosomal RNA phylogeny of gastropod and unioidean bivalve mollusks. The Nautilus, Supplement 2:111–121.

Rosenberg, G., S. Tillier, A. Tillier, G. Kuncio, R. T. Hanlon, M. Masselot, and C. J. Williams. 1997. Ribosomal RNA phylogeny of selected major clades in the Mollusca. Journal Molluscan Studies 63:301–309.

Salvini-Plawen, L. von. 1980. A reconsideration of systematics in the Mollusca (phylogeny and higher classification). Malacologia 19:249–278.

Sanderson, M. J., A. Purvis, and C. Henze. 1998. Phylogenetic supertrees: Assembling the trees of life. Trends in Ecology and Evolution 13:105–109.

Sasaki, T. 1998. Comparative anatomy and phylogeny of the Recent Archaeogastropoda (Mollusca: Gastropoda). Bulletin, The University Museum, The University of Tokyo 38:1–224.

Shoemaker, J. S., I. S. Painter, and B. S. Weir. 1999. Bayesian statistics in genetics. Trends in Genetics 15:354–357.

Swofford, D. L. 2001. PAUP*. Phylogenetic Analysis Using Parsimony (*and other methods), v. 4.0b8. Sinauer Associates, Sunderland, MA.

Swofford, D. L., G. J. Olsen, P. J. Waddell, and D. M. Hillis. 1996. Phylogenetic inference. Pp. 407–514 in Molecular Systematics (D. M. Hillis, C. Moritz, and B. K. Mable, eds.). Sinauer Associates, Sunderland, MA.

Thiele, J. 1929–31. Handbuch de systematischen weichtierkunde, Gustav Fischer Verlag, Jena.

Thompson, J. D., D. G. Higgins, and T. J. Gibson. 1994. CLUSTAL W: Improving the sensitivity of progressive multiple sequence alignment through sequence weighting, position-specific gap penalties and weight matrix choice. Nucleic Acids Research 22:4673–4680.

Tillier, S., M. Maselot, J. Guerdox, and A. Tillier. 1994. Monophyly of major gastropod taxa tested from partial 28S rRNA sequences, with emphasis on Euthyneura and hot–vent limpets Peltospiroidea. The Nautilus, Supplement 2:122–140.

Tillier, S., M. Masselot, P. Herve, and A. Tillier. 1992. Phylogénie moléculaire des Gastropoda (Mollusca) fondée sur le séquençage partiel de l'ARN ribosomique 28S. Comptes Rendus Academie de Sciences (Paris) 134:79–85.

Tracey, S., J. A. Todd, and D. H. Erwin. 1993. Mollusca: Gastropoda. Pp. 137–167 in The Fossil Record 2 (M. J. Benton, ed.). Chapman and Hall, London.

Troschel, F. H. 1861. Das Gebiss der Schnecken, zur Begründung einer natürlichen Classification, Nicolaische Verlagsbuchhandlung, Berlin.

Wagner, P. J. 1995. Diversity patterns among early gastropods: Contrasting taxonomic and phylogenetic descriptions. Paleobiology 21:410–439.

Wenz, W. 1938–44. Gastropoda, Teil 1: Allgemeiner Teil und Prosobranchia. in Handbuch der Paläozoologie (O. H. Schindewolf, ed.). Gebrüder Borntraeger, Berlin.

Winnepenninckx, B., T. Backeljau, and R. De Wachter. 1994. Small ribosomal subunit RNA and the phylogeny of Mollusca. The Nautilus, Supplement 2:98–110.

Winnepenninckx, B., T. Backeljau, and R. De Wachter. 1996. Investigation of molluscan phylogeny on the basis of 18S rRNA sequences. Molecular Biology and Evolution 13:1306–1317.

Winnepenninckx, B., G. Steiner, T. Backeljau, and R. De Wachter. 1998. Details of gastropod phylogeny inferred from 18S rRNA sequences. Molecular Phylogenetics and Evolution 9:55–63.

Woese, C. R., O. Kandler, and M. L. Wheelis. 1990. Towards a natural system of organisms: Proposal for the domains Archaea, Bacteria and Eucarya. Proceedings of the National Academy of Sciences U.S.A. 87:4576–4579.

7

GOALS AND LIMITS OF PHYLOGENETICS

The Euthyneuran Gastropods

In addition to providing classifications, the main goals of phylogenetics are to give insight into the evolutionary history of characters and the evolution of taxa to assess their biogeography and their first occurrence in the fossil record, and to provide data for conservation programs. In this context, obtaining phylogenetic trees is more a preliminary step than a final goal. Phylogenetic relationships shown on a tree (i.e., status of taxa in terms of monophyly, paraphyly, and polyphyly) are most interesting for their evolutionary implications. Because conclusions based on poorly resolved relationships may be totally erroneous, it is crucial to have a clear idea of which relationships are strongly supported in a tree and which are not. More specifically, we should not be afraid of accepting absence of resolution as a result. Absence of resolution is not necessarily the result of a failure in the method of tree construction, but may accurately represent patterns of evolution. In the case of poorly supported nodes or absence of resolution, what can we say about evolution of characters, according to what we know and what we do not know?

Euthyneura, the crown taxon of Gastropoda, is also among the most diverse in ecological terms. For example, terrestrial and freshwater adaptations that are uncommon within the noneuthyneuran gastropods (i.e., the prosobranch grade) are widely successful within the euthyneurans. Euthyneura is also diverse in morphological terms. For example, complete loss of the shell is only known in the genus *Titiscania* and some internal parasitic eulimids outside of Euthyneura, but has occurred several times within Euthyneura. Monophyly of Euthyneura was first suggested by Mörch (1865) by the name "Androgyna" and has always been supported by anatomy (Spengel 1881; Pelseneer 1894; Plate 1895;

161

Boettger 1954; Johansson 1954; Robertson 1985; Salvini-Plawen and Steiner 1996; Dayrat and Tillier 2002) and recent molecular studies (e.g., Wade and Mordan 2000; Dayrat et al. 2001).

Euthyneuran gastropods classically comprise two groups, Opisthobranchia and Pulmonata, which naturalists can easily distinguish in most cases based on their respective habitats. All of the opisthobranchs live in marine waters, except for a few acochlidiacean species that live in tropical freshwaters. Pulmonates are principally terrestrial and freshwater. Even marine pulmonates, except some species of the genus *Williamia* that live permanently submerged, live in the upper littoral zone or part time out of water. Morphology is also helpful: all pulmonates have a pallial cavity that is closed by a contractile pneumostome and with a suprapallium that is hypervascularized by pulmonary vessels (except for species of *Siphonaria* that have a pallial cavity without pulmonary vessels and closed by a noncontractile pneumostome).

Monophyly of Pulmonata is well supported by anatomy (e.g., Salvini-Plawen and Steiner 1996; Dayrat and Tillier 2002), and molecular studies (Tillier et al. 1996; Wade and Mordan 2000). The situation differs somewhat for the opisthobranchs. As a result of their important morphological diversity (e.g., Gosliner and Ghiselin 1984; Gosliner 1991), opisthobranchs do not share any diagnostic character, unless the absence of some organs via secondary loss is considered. This approach has already been criticized by Dayrat and Tillier (2000) regarding pentaganglioneury. When characters are coded as they are observed, without any preconceived notions, no synapomorphy can be found for Opisthobranchia (Dayrat and Tillier 2002). As a result, molecular studies are crucial for understanding the evolutionary relationships of opisthobranchs, but, unfortunately and as discussed further, they have been of only limited use.

The uncertain monophyletic status of Opisthobranchia requires the examination of characters in a global dataset that includes all relevant euthyneurans. Several authors have discussed affinities between opisthobranchs and pulmonates (Johansson 1954; Morton 1955; Duncan 1960; Salvini-Plawen and Steiner 1996). Robertson (1985, 1993) studied the variation of four important characters within Euthyneura and some of their closely related outgroups (heterostrophy, pallial raphes, chalaze, and pigmented mantle organ). To discuss their evolutionary history, Dayrat and Tillier (2002) reevaluated all of the euthyneuran morphological characters available in a global cladistic analysis that included opisthobranch and pulmonate taxa, and obtained globally unresolved relationships. Not surprisingly, they have been able to confirm monophyly of many opisthobranch subtaxa: Acteonoidea, Acochlidioidea, Sacoglossa, Umbraculoidea, Pleurobranchoidea, Nudibranchia, Thecosomata, and Gymnosomata. Four higher euthyneuran clades were also found to be well supported:

Pulmonata, Geophila (including Stylommatophora and Systellommatophora), "Cephalaspidea" (including Anaspidea, but excluding Runcinidae and Acteonoidea), and a clade that includes bulloids and Anaspidea.

Quite a few molecular studies that involved euthyneuran species have been published in the last 10 years. Those that take an insufficient number of nucleotides and/or sampled euthyneuran taxa into account are not discussed here: Emberton et al. 1990; Tillier et al. 1992, 1994; Harasewych et al. 1997; McArthur and Koop 1997; Colgan et al. 2000; Wollscheid-Lengeling et al. 2001. Six other studies included more characters and taxa (Table 7.1): Thollesson (1999); Wollscheid and Wägele (1999); Wade and Mordan (2000); Dayrat et al. (2001); Dutra-Clarke et al. (2001); and Wade et al. (2001). Only ribosomal genes have been used for euthyneuran relationships (see Table 7.1): partial mitochondrial 16S rDNA (Thollesson 1999), complete nuclear 18S rDNA (Wollscheid and Wägele 1999; Dutra-Clarke et al. 2001), partial nuclear 28S rDNA (Wade and Mordan 2000; Dayrat et al. 2001; Wade et al. 2001), and partial nuclear 5.8S rDNA (Wade and Mordan 2000; Wade et al. 2001). Other recent contributions using genomic data (mitochondrial gene order) will be discussed as potentially good markers for euthyneuran phylogenetics (Kurabayashi and Ueshima 2000; Grande et al. 2001; Medina et al. 2001; Ueshima 2001). However, because only a few complete mitochondrial genomes are available for Euthyneura, these results were not used for constructing a global euthyneuran tree.

Our aim in this chapter is to propose a consensus tree for euthyneuran relationships based on combining several molecular trees with the morphological results obtained from our former work (Dayrat and Tillier 2002). The fully resolved phylogeny of Euthyneura proposed by Salvini-Plawen and Steiner (1996) is not used for constructing the consensus tree because of methodological problems associated with character coding (see Dayrat and Tillier 2000, 2002). After constructing the consensus tree, we discuss euthyneuran classification and evolution of key characters. The fossil record has been discussed elsewhere (Dayrat et al. 2001), and unfortunately, discussion of biogeographic patterns would be meaningless at this level. Our main goal is to discuss overall euthyneuran relationships and evolution of characters. Chapter 8 focuses more specifically on select opisthobranch taxa.

HANDLING LOW RESOLUTION IN MULTIPLE DATASETS AND TREES

Distinct phylogenetic data can be combined at two different levels. The first approach entails combining the different datasets into a single data matrix and

Table 7.1

Studies used to construct the consensus tree for euthyneuran gastropods shown in Figure 7.2

Source	Dataset	Number of Characters	Opisthobranchia Species	Pulmonata Species	Phylogenetic Analysis Methods
(1) Thollesson 1999	Partial mitochondria 16S rRNA	427 positions (231 informative sites)	10 Nudibranchia; 2 shell-less Sacoglossa; 1 Pleurobranchoidea; 1 Anaspidea; 3 Cephalaspidea s. str.; 1 Acteonoidea; 1 Gymnosomata	2 Hygrophila; 2 Stylommatophora	MP; ME
(2) Wollsheid and Wägele 1999	Complete 18S rRNA	2,556 positions (631 informative sites)	19 Nudibranchia; 1 shell-less Sacoglossa; 3 *Aplysia*; (Anaspidea) 1 Cephalaspidea s. str.	10 Hygrophila; 1 Onchidioidea; 1 Soleolifera; 5 Stylommatophora	NJ; MP
(3) Wade and Mordan 2000	3' end of 5.8S rRNA; ITS-2 region; Complete 5' end of 28S rRNA	661 positions (184 variable sites)	2 Nudibranchia; 1 *Aplysia*	1 *Siphonaria*; 3 Ellobiidae; 2 Soleolifera; 8 Stylommatophora	NJ; FM; ML; M
(4) Wade et al. 2001	3' end of 5.8S rRNA; Complete ITS-2 region; 5' end of 28S rRNA	843 positions	None	*Siphonaria*; 3 Ellobiidae; 2 Soleolifera; 105 Stylommatophora	NJ; FM; ML; MP (only NJ published)
(5) Dayrat and Tillier 2001	5' end of 28S rRNA	1,100 positions (254 informative sites)	1 Nudibranchia; 2 Sacoglossa; 2 Anaspidea; 1 Pleurobranchoidea; 1 Umbraculoidea; 1 Runcinoidea; 5 Cephalaspidea s. str.; 2 Acteonoidea; 1 Gymnosomata; 1 Thecosomata	1 *Siphonaria*; 1 *Amphibola*; 5 Ellobiidae; 2 Hygrophila (including *Chilina*); 1 Onchidioidea;2 Soleolifera; 2 Stylommatophora	ML; MP; (only MP published)
(6) Dutra-Clarke et al. 2001	Complete 18S rRNA	1,926 positions (176 informative sites)	1 *Aplysia*	1 *Siphonaria*; 9 Hygrophila; 1 Onchidioidea;1 Soleolifera; 6 Stylommatophora	ML; MP
(7) Dayrat and Tillier 2002	Anatomy	75 characters	39 genera sampled from all high taxa	22 genera from all high taxa	MP

Note: Phylogenetic analysis methods: FM = Fitch-Margoliash; ME = maximum evolution; ML = maximum likelihood; MP = maximum parsimony; NJ = neighbor-joining.

searching for the most parsimonious tree or trees (so-called character congruence or total evidence; Kluge 1998). The second approach treats each dataset independently and the resultant trees are compared and combined on the basis of taxonomic congruence (Mickevich 1978). The first approach is considered constrained because identical taxa must be sampled in the different datasets, datasets should be congruent as measured by a test such as the character congruence (ILD) test (Farris et al. 1995), and it is often argued, although we do not agree, that it is presumptuous to mix together such characters as anatomical features, nucleotides, and gene order. Taxonomic congruence, however, is convenient for combining several trees based on different samples of taxa, such as the phylogenetic studies that include euthyneuran taxa mentioned above. Here, we use taxonomic congruence in an approach referred to as *combinable component consensus* (Bremer 1990).

All types of consensus tree (Adams 1972; Sokal and Rohlf 1981; Bremer 1990; Swofford 1991) are based on the study of contradiction and congruence. When the trees combined are fully resolved, contradiction and congruence are clear cut. In the absence of resolution, there is neither positive congruence nor positive contradiction. The consensus method used here is a combinable component consensus in which unresolved nodes are considered to be combinable with resolved nodes. Only positively incompatible nodes are not combined in the final consensus tree. This method avoids the loss of positive phylogenetic resolution without passing over the limits of phylogenetics. For example, if we take two trees (Figure 7.1, a and b), a strict consensus tree would be totally unresolved (Figure 7.1c). However, two combinable component consensus trees can be obtained: one is fully unresolved (Figure 7.1c) if absence of resolution is considered as a contradiction (i.e., an unresolved position is not considered to be combinable with a resolved position). However, a partly resolved consensus tree is obtained (Figure 7.1d), if absence of resolution is taken into account by considering that an unresolved position may be combined with a resolved position. In other words, we consider the two clades (A, B, C) and (D, E, F, G) as "combinable" even if they are found in only one tree (Figure 7.1b), because absence of resolution for the taxa A to E in the other tree (Figure 7.1a) does not contradict these two clades. Nevertheless, the topology remains unresolved within the clade (D, E, F, G) because (F, G) is either monophyletic (Figure 7.1a) or polyphyletic (Figure 7.1b). Let us take a new tree (Figure 7.1e) that consists of two clades (A, B, C) and (D, E, F) and in which the position of G is not resolved. Its combination with the first tree (Figure 7.1a) may provide a resolved tree (Figure 7.1f) because only the resolved position of G is taken into account (Figure 7.1a) and is not contradicted by the unresolved position of G (Figure 7.1e).

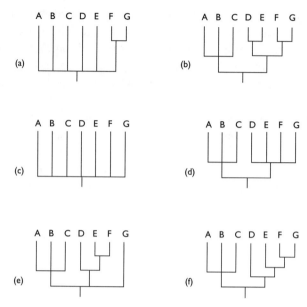

Figure 7.1. Combinable component consensus trees constructed by either taking into account the absence of resolution or not. (a) Unresolved tree, except for the monophyly of the taxon that includes F and G. (b) Resolved tree, except for the relationships within the clade that includes the taxa A, B, and C. (c) Consensus for trees (a) and (b): fully unresolved strict consensus or combinable component consensus that does not take into account the absence of resolution. (d) Consensus for trees (a) and (b): partly resolved combinable component consensus tree that takes into account the absence of resolution. (e) Partly unresolved tree. (f) Consensus for trees (a) and (e): partly resolved combinable component consensus tree that takes into account the absence of resolution.

Lack of resolution should also influence the vocabulary that we use to describe phylogenetic results. Phylogenetic relationships of taxa that are part of an unresolved clade should not be referred to as monophyletic, paraphyletic, or polyphyletic. The taxon (D, E, F) can be monophyletic (Figure 7.1e), paraphyletic (Figure 7.1f), or polyphyletic (Figure 7.1b) only if the relationships are resolved, but not if they are unresolved (Figure 7.1, a, c, and d). In the latter case, the status of (D, E, F) should be said to be partly unresolved, and not further characterized.

Before constructing a consensus tree for euthyneuran gastropods, it is necessary to determine which nodes are supported enough to be retained in the consensus tree. The robustness of nodes is usually evaluated by the Bremer support index (Bremer 1988) or by bootstrap proportions (Felsenstein 1985), both of which have a statistical basis. Dayrat and Tillier (2002) used an empirical method that tests many possible character matrices consisting of alternative coding, or interpretation, of ambiguous characters. Dayrat and Tillier retained a strict consensus tree from all of the trees obtained from the repeated phylogenetic analyses. This results in a consensus tree that includes only the nodes that are strongly supported. A node is most reliable when obtained from independent datasets, whether morphological, molecular, or different gene trees. In our subjective classification, bootstrap values higher than 90% means nodes are strongly supported; 80% to 90%, nodes are well supported; 70% to 80%, nodes are interesting but hypothetical; and below 70%, nodes are not resolved.

A CONSENSUS TREE FOR EUTHYNEURAN GASTROPODS

The sister group of Euthyneura should be found among allogastropods (Valvatidae, Glacidorbidae, Rissoellidae, Omalogyridae, Pyramidelloidea, Architectonicoidea, etc.), a paraphyletic grade that constitutes, together with Euthyneura, the clade Heterobranchia. However, the particular lineage within the allogastropods is not yet precisely identified (Roberston 1985; Haszprunar 1988; Ponder and Lindberg 1997; Dayrat and Tillier 2002). Pyramidelloids could be either sister group to Euthyneura (Salvini-Plawen and Steiner 1996) or included within Euthyneura (e.g., Fretter and Graham 1949; Thompson 1976; Dayrat and Tillier 2002). The architectonicoids may be the closest sister group to Euthyneura, but could also be rooted more deeply within Heterobranchia, depending on the relative importance given to sperm characters and heterostrophy (Roberston 1985; Haszprunar 1988; Healy 1993; Salvini-Plawen and Steiner 1996; Ponder and Lindberg 1997; Dayrat and Tillier 2002). Mitochondrial gene order may also be phylogenetically informative in resolving the sister-group status of Euthyneura as more complete mitochondrial genomes are sequenced from allogastropods (e.g., Kurabayashi and Ueshima 2000).

Monophyly of Euthyneura is well supported, especially by molecular studies—the strongest support has been obtained by Wade and Mordan (2000) based on nuclear ribosomal gene data. However, no strong morphological synapomorphy has been found even after reevaluation of characters. One possible synapomorphy already proposed by Dayrat and Tillier (2002) is the loss of the operculum, but this character needs to be discussed more fully, which we have done under "Evolutionary History of Characters."

Monophyly of Pulmonata (Figure 7.2) is well supported by at least three anatomical characters (Dayrat and Tillier 2002): pallial cavity opening by a

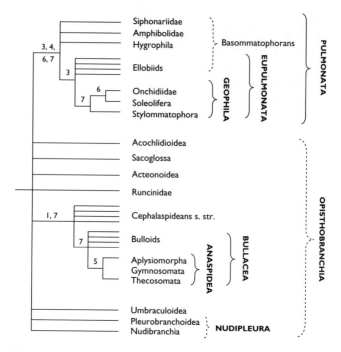

Figure 7.2. Consensus tree for euthyneuran gastropods, based on seven different molecular and morphological analyses. The numbers refer to the analyses in Table 7.1, in which the nodes are well supported.

pneumostome, a procerebrum, and cerebral bodies in the central nervous system. The best molecular support has been obtained with partial 28S rDNA sequences by Wade and Mordan (2000), who have also shown the basal position of *Siphonaria* toward Eupulmonata. Positions of other marine taxa such as Amphibolidae, Trimusculidae, Otinidae, or Smeagolidae remain unresolved. Monophyly of Hygrophila (including Chilinidae) has recently been shown by Dayrat et al. (2001), but its position remains uncertain.

Monophyly of Eupulmonata (*sensu* Morton 1955 and Tillier et al. 1996, not *sensu* Wade and Mordan 2000), which includes ellobiids, Systellommatophora, and Stylommatophora, has been hypothesized with unconvincing results by Tillier et al. (1996), but has been strongly supported by more recent studies (Wade and Mordan 2000). No morphological synapomorphy has been discovered (Dayrat and Tillier 2002). Relationships of ellobiids (excluding Otinidae and Smeagolidae) are also unresolved, but they most probably constitute a paraphyletic group at the base of Eupulmonata.

Monophyly of Geophila is supported by morphological synapomorphies only (Dayrat and Tillier 2002): anterior gland lying free on the floor of the visceral cavity, eyes at the tip of posterior tentacles, loss of paired lateral jaws (unpaired dorsal jaw ulteriorly lost in Onchidiidae), and loss of gizzard pouch. Monophyly of Systellommatophora, including Onchidiidae and Soleolifera, is well supported by recent molecular studies (Dutra-Clarke et al. 2001).

Monophyly of Stylommatophora, which includes the typical land snails and slugs, is strongly supported by several molecular (Wade and Mordan 2000; Dayrat et al. 2001; Dutra-Clarke et al. 2001; Wade et al. 2001) and morphological studies (Tillier 1989; Nordsieck 1990; Dayrat and Tillier 2002). The two strong synapomorphies of Stylommatophora are a long pedal gland placed beneath a membrane and retractile tentacles (not contractile as in Soleolifera). Occurrence of a secondary ureter could also be considered as a synapomorphic character (Dayrat and Tillier 2002), but embryology of the ureters of the systellommatophorans needs further observation (Fretter 1943; Delhaye and Bouillon 1972). Several anatomical classifications have been proposed within Stylommatophora (e.g., Tillier 1989; Emberton et al. 1990; Nordsieck 1990), but a general cladistic review of all characters is still lacking. According to molecular studies (Tillier et al. 1994; Wade et al. 2001; Christopher Wade, Natural History Museum, London, personal communication), the relationships within Stylommatophora remain mostly unresolved except for a few groups of which the monophyly has been substantiated or discovered by Wade et al. (2001): Elasmognatha (including Heterurethra and Athoracophoridae), endodontoids (Punctidae, Charopidae, Otoconchidae, but not Discidae), Helicoidea (excluding Camaenidae), Limacoidea, Clausiliidae, and Orthurethra (excluding Chondrinidae).

Within Stylommatophora, Wade et al. (2001) also proposed a basal divergence between an "achatinoid clade" (strongly supported) and a "non-achatinoid clade" (bootstrap value, 65%). Previously, Tillier et al. (1996) had also found an indication of early divergence of a clade that included Achatinidae (*Achatina*), Streptaxidae (*Gonaxis*), and Limacidae (*Limax*). The early divergence of Limacidae is not confirmed by Wade et al. (2001), whose sample, however, does not include representatives of the genus *Limax* itself. Although it is poorly supported, this clade should retain our attention because the "non-achatinoid clade," including some Limacidae, has been found repetitively in independent analyses performed with new taxonomic samples by Wade (personal communication); for instance, an analysis including 30 taxa provided bootstrap proportions of 69% in maximum parsimony, 78% in neighbor-joining, and 90% in Fitch-Margoliash. This basic distinction between "achatinoid" and "non-achatinoid" stylommatophorans would mean that the sigmurethran kidney constitutes the plesiomorphic condition, or that the character is much more labile than previously expected. As shown by Tillier et al. (1996) and by Wade et al.

(2001), the higher classical classification of Stylommatophora proposed by Pilsbry (1900) and Baker (1955) and comprising Orthurethra, Sigmurethra, Mesurethra, and Heterurethra, is not supported by molecular studies, except for Heterurethra, which only consists of succineids.

Monophyly of Opisthobranchia (Figure 7.2) was not confirmed by a global cladistic analysis performed previously (Dayrat and Tillier 2002). Paraphyly of Opisthobranchia is supported by two molecular studies (Thollesson 1999; Wollscheid and Wägele 1999); however, the latter focused primarily on a single opisthobranch group—the Nudibranchia. Recently, more extensive molecular datasets support the monophyly of the Opisthobranchia (Medina et al. 2001; Ueshima 2001). Mitochondrial gene order varies within mollusks (Boore and Brown 1994) and gastropods (Kurabayashi and Ueshima 2000), and recent studies indicate synapomorphic rearrangements supporting the monophyly of opisthobranchs (Grande et al. 2001, 2002; Medina et al. 2001; Ueshima 2001). Because three of four of these studies are published abstracts or have limited sampling, we will consider the monophyly of Opisthobranchia as unsubstantiated.

Except for a few nodes, relationships among higher opisthobranch taxa are unresolved. Because phylogenetic relationships among opisthobranch taxa are discussed in more depth in the Chapter 8, here we provide only a summary of some main results obtained from our global euthyneuran approach (see also Dayrat et al. 2001).

For many years (e.g., Pelseneer 1894; Plate 1895), because they seem to be the "most primitive" opisthobranchs, the cephalaspideans and especially Acteonoidea have been considered as an ideal "stem group." Dayrat and Tillier (2000) have shown that this view largely influenced previous phylogenetic works. Actually, the relationships of Opisthobranchia are so poorly resolved that it is still not possible to delineate the higher group "Cephalaspidea" (Figure 7.2). The phylogenetic study of Cephalaspidea *sensu lato* (in the former sense) by Mikkelsen (1996), despite providing interesting data on character coding, used an arbitrary taxon sampling of some shell-bearing opisthobranchs that prevented objective conclusions concerning the status of "Cephalaspidea." A phylogenetic definition of the name "Cephalaspidea" in accordance with the principles of phylogenetic taxonomy (de Queiroz and Gauthier 1992, 1994) would be illusory at the moment. It appears that a group overlapping with the fuzzy concept of Cephalaspidea includes at least some taxa like *Ringicula, Cylichna, Retusa, Scaphander,* or *Philine* that can be called "cephalaspideans *sensu stricto*" and of which the phylogenetic status is still unresolved. These cephalaspideans *sensu stricto* constitute the sister group to a clade we call "Bullacea," which includes a few taxa like *Haminoea, Bulla,* or *Smaragdinella,* of which the relationships are also still unresolved, and the clade "Anaspidea." The latter clade includes the Akeridae and the Aplysioidea (sea-hares), two taxa that we call "Aplysiomorpha," of which the monophyly is well supported by molecular data (Medina and Walsh 2000). The status and the relationships of Aplysiomorpha, Gymnosomata, and Thecosomata are still unknown. No exclusive synapomorphy can be proposed for all of the taxa previously included under the name "Cephalaspidea" because the corresponding characters (cephalic shield, Hancock's organs, pallial caecum, pallial raphes, plicatidium, esophageal gizzard; see below for discussion) are found in several other taxa in which the phylogenetic position remains unresolved (like Acteonoidea, Runcinidae, Sacoglossa, *Siphonaria,* or *Chilina*).

The name "Nudipleura" was introduced by Wägele and Willan (2000) for a clade that included Nudibranchia and Pleurobranchoidea, which were formerly proposed as monophyletic (e.g., Schmekel 1985; Salvini-Plawen 1990). The Pleurobranchoidea were traditionally included in Notaspidea with Umbraculoidea. In total, six synapomorphies (androdiaulic system, loss of osphradium, possession of a blood gland, lack of albumen gland, loss of the gizzard, haploid chromosome number = 12) have been proposed by Schmekel (1985) or Salvini-Plawen (1990) and partly discussed by Wägele and Willan (2000), who accepted three of them (androdiaulic system, loss of osphradium, possession of a blood gland). Reevaluation of variation of these characters (Dayrat and Tillier 2002) has shown that only one character (posterior migration of the male genital aperture close to the female aperture) could be considered as a possible synapomorphy for this taxon—variation of other characters is much higher than expected (see below). The migration of the male genital aperture is therefore in contradiction with another important character, presence of a lateral pinnate gill, that was classically supposed to characterize all of the notaspideans (Willan 1987). Monophyly of Nudipleura or Notaspidea, therefore, depends mainly on the relative importance of these two characters.

Monophyly of Sacoglossa, a clade that consists of shell-bearing and shell-less species, is strongly supported by the presence of an "ascus" in the radular apparatus, a character that they share exclusively (Jensen 1996; Dayrat and Tillier 2002). Monophyly is also well supported by molecular data (Dayrat et al. 2001). Monophyly of Acteonoidea (including *Acteon, Pupa, Hydatina, Micromelo,* etc.) has been confirmed by recent morphological results (Dayrat and Tillier 2002). Only monophyly of the Acteonidae (*Acteon* and *Pupa*) has been confirmed by molecular data (Dayrat et al. 2001). Phylogenetic position of Runcinidae, most often considered as derived cephalaspideans, is still unknown. The position of the fascinating Acochlidioidea, the only group of opisthobranchs that includes freshwater species, also remains a mystery.

From these data, a general classification can be proposed for the Euthyneura

(Table 7.2) that takes into account all of the resolved and unresolved relation-ships. Even if a few clades are well supported, like Bullacea, Anaspidea, Pulmonata, Eupulmonata, Geophila, or Stylommatophora, the principles of phylogenetic taxonomy (de Queiroz and Gauthier 1992, 1994) may hardly be applied as a result of the generally poor resolution of the tree. For instance, the position of the marine pulmonates has to be resolved to know which group con-stitutes the most basal branch before giving a definition of "Pulmonata."

EVOLUTIONARY HISTORY OF CHARACTERS

An important goal of phylogenetics is to examine the evolution of characters. Because the evolutionary history of characters is discussed elsewhere (Dayrat et al. 2001; Dayrat and Tillier 2002), we provide comments on some characters for which the evolutionary history is important for euthyneuran evolution or which may have altered because of later phylogenetic studies.

Heterostrophy

Heterostrophy, in which the larval shell ("protoconch") and the adult shell ("teleoconch") coil around two different axes (left and right), was first described for pyramidellids (Plate 1895). This character is also shared by Architectoni-coidea and Euthyneura except Geophila and Hygrophila. Larval coiling is also equivocal for several euthyneuran genera (Robertson 1985, 1993; Dayrat and Tillier 2002). Evolutionary history of heterostrophy depends directly on the po-sition of Architectonicoidea relative to the other heterobranch gastropods. One or two acquisitions and one or two losses (because Hygrophila could be sister group to Geophila) have occurred.

Operculum

In adult euthyneurans, an operculum is present in Amphibolidae, Acteonidae (part of Acteonoidea, the rest lack an operculum), and two genera of coiled the-cosomatous pteropods (*Peraclis* and *Limacina*). Because the operculum appears in the early developmental stages of nearly all euthyneurans, we can conclude that the operculum has been lost secondarily through regulatory gene action. Moreover, opercula of Acteonidae and thecosomatous pteropods seem to be identical (Lalli and Gilmer 1989) and might be homologous. Because the po-sition of Acteonidae and Amphibolidae is still unknown, no precise evolution-ary history can be given concerning this character. The adult operculum could have been lost once (as proposed by Dayrat and Tillier 2002) and secondarily

Table 7.2

Temporary classification of Euthyneura, taking into account the unresolved positions of several taxa *(incertae sedis)*[a][b]

EUTHYNEURA
Pulmonata
    Siphonariidae *(inc. sed.)*
    Trimusculidae *(inc. sed.)*
    Amphibolidae *(inc. sed.)*
    Otinidae *(inc. sed.)*
    Smeagolidae *(inc. sed.)*
    Hygrophila *(inc. sed.)*
        Chilinidae
        Lymnaeidae
        Planorbidae
        Physidae
        Ancylidae
    Eupulmonata
        Ellobiidae
        Geophila
        Systellommatophora
                Onchidiidae
                Soleolifera
          Stylommatophora
Opisthobranchia
    Acteonoidea *(inc. sed.)*
    Acochlidioidea *(inc. sed.)*
    Runcinidae *(inc. sed.)*
    Sacoglossa *(inc. sed.)*
    Umbraculoidea *(inc. sed.)*
    Pleurobranchoidea *(inc. sed.)*
    Nudibranchia *(inc. sed.)*
    "Cephalaspidea" *(inc. sed.)*
        Ringiculidae *(inc. sed.)*
        Scaphandridae *(inc. sed.)*
        Philinidae *(inc. sed.)*
        Diaphanidae *(inc. sed.)*
        Retusidae *(inc. sed.)*
        Gastropteridae *(inc. sed.)*
        Aglajidae *(inc. sed.)*
        Bullacea
           Bullidae *(inc. sed.)*
           Hamineidae *(inc. sed.)*
           Smaragdinellidae *(inc. sed.)*
           Anaspidea
               Aplysiomorpha (inc. sed.)
               Gymnosomata (inc. sed.)
               Thecosomata (inc. sed.)

---

[a]The taxa of which the status is unknown are italicized.

[b]The well-supported clades are indicated in bold font.

regained three times. This would mean that the genetic tools directing the development of the operculum could have been "turned on" again three times. Alternatively, and if Acteonidae and Amphibolidae are the most basal branch(es) of Euthyneura (or Opisthobranchia and Pulmonata), their operculum could be primary and could never have been "turned off" and lost in adults; however, this hypothesis is complicated by the absence of an operculum in other adult acteonoids (like *Micromelo* or *Hydatina*). Every combination of the two previous hypotheses is also acceptable. The operculum possessed by *Peraclis* and *Limacina* cannot be primary because of the phylogenetic position of these genera among other operculum-less Anaspidea. This means that at least one (perhaps two) developmental change(s) happened in the common ancestors of these genera for maintaining the development of the operculum. For this reason, and because the shell of *Limacina* and *Peraclis* is sinistrally coiled, some authors proposed that the cosomes could be neotenic opisthobranchs (e.g., Lemche 1948), but this view has recently been criticized (Lalli and Gilmer 1989). Unfortunately, the reasons for such plasticity in the development of the adult operculum during evolution, as well as its corresponding genetic mechanisms, are unknown.

Cephalic Shield and Hancock's Organs

A cephalic shield, which is an organ related to burrowing, is present in Acteonoidea, cephalaspideans *sensu stricto,* Bullacea (except for sea-hares and pteropods), and some shelled sacoglossans (*Cylindrobulla, Ascobulla,* etc.). If Sacoglossa and Acteonoidea are included in Cephalaspidea, as supposed by Mikkelsen (1996), one common acquisition for all of the cephalaspideans *sensu lato* (i.e., including cephalaspideans *sensu stricto,* Acteonoidea, Sacoglossa, bulloids, sea hares, and pteropods), and at least two independent losses (at least one for the pteropods and one for the sacoglossans) can be inferred. Hancock's organs, a pair of sensory rows of leaves located under each side of the cephalic shield, co-occur with the cephalic shield, except in the shell-bearing sacoglossans that do not possess these organs. The evolutionary history of Hancock's organs is then similar to that of the cephalic shield. They have been lost at least once by Anaspidea (sea-hares and pteropods).

Osphradium

The osphradium, a sensory organ at the entrance of the pallial cavity involved in perception of water quality, shows a complex pattern of variation in Heterobranchia (Dayrat and Tillier 2002). This variation is clearly related to the presence/absence of a pallial cavity for opisthobranchs (a pallial cavity is absent in notaspideans, shell-less sacoglossans, Acochlidioidea, Nudibranchia, and Gymnosomata), and most probably with the presence/absence of an active water current for pulmonates (an osphradium is present in *Chilina, Amphibola,* and *Siphonaria* only). Because its occurrence is plesiomorphic for Heterobranchia, it seems to have been lost more than once. Wägele and Willan (2000) proposed the loss of the osphradium as a synapomorphy for Nudipleura (i.e., Nudibranchia and Pleurobranchoidea). However, this synapomorphy could not be confirmed from a data matrix that included all of the euthyneurans, as must be done because it varies in all of them. This does not imply that the loss of the osphradium could not be a synapomorphy of Nudipleura (i.e., an historical event that is unique for this taxon, even if it also happened in other taxa), but it means only that an analysis that includes only nudipleuran taxa does not allow this conclusion.

Pallial Cavity and Raphes

The pallial cavity of several opisthobranchs, such as cephalaspideans, acteonoids, or sacoglossans, shares affinities with that of pulmonates (*Chilina* and *Siphonaria*). For this reason, it has long been proposed that *Chilina* and *Siphonaria* were the "most primitive" pulmonates and were closely related to cephalaspideans (e.g., Pelseneer 1894; Plate 1895). Because the phylogenetic position of all of these taxa is not well known (especially for the opisthobranchs), it is difficult to give a precise evolutionary history for the characters related to the pallial cavity (Dayrat and Tillier 2002). This difficulty is complicated by the fact that the pallial cavity has been lost in various groups of opisthobranchs (notaspideans, shell-less sacoglossans, Nudibranchia, Acochlidioidea). The main concern is the homology or analogy of pallial organs present in opisthobranchs and pulmonates. The pallial caecum is surely analogous in *Chilina* and the opisthobranchs in which it is present (i.e., Acteonoidea, cephalaspideans [not all of them], bulloids, and *Akera*), because *Chilina* is included in the clade Hygrophila. The caecum has been lost at least once in sea-hares and pteropods.

Evolutionary history of pallial raphes, organs involved in water current inside the pallial cavity, is more complicated because they are present in Acteonoidea, cephalaspideans, bulloids, shell-bearing sacoglossans, *Siphonaria,* and *Chilina*. In pallial raphes, the uncertainty is due to the relationships of Acteonoidea and Sacoglossa with cephalaspideans, and the position of *Siphonaria* within Pulmonata. Because *Siphonaria* could be the first emerging lineage of Pulmonata, we cannot exclude homology of pallial raphes among Euthyneura. Pallial raphes have been lost at least once in sea-hares and pteropods.

The plicatidium shows the same variation as the pallial raphes, except that it is absent in both shelled sacoglossans (only in *Ascobulla, Cylindrobulla,* and *Volvatella*) and *Chilina.* Moreover, in sacoglossans and *Siphonaria,* it is only one-sided and not two-sided. Again, a better phylogenetic resolution is needed to draw conclusions about its evolutionary history and homology among Euthyneura.

Jaws

The occurrence of a pair of lateral jaws is plesiomorphic for Heterobranchia (Ponder and Lindberg 1997; Dayrat and Tillier 2002), but is not observed in many opisthobranchs (Sacoglossa, Acochlidioidea, Gymnosomata, several cephalaspideans, etc.) and pulmonates (*Amphibola* and all Geophila). Thus, we can at least conclude that these two lateral jaws have been lost several times in Euthyneura and that their evolution is complex within opisthobranchs. Another jaw, dorsal and unpaired, is also shared by Pulmonates except Onchidiidae and *Amphibola,* which have no jaw at all. If Amphibolidae is not the most basal taxon of Pulmonata, an unpaired jaw could have been acquired once in the common ancestral branch of Pulmonates and secondarily lost in the common ancestral branch of Amphibolidae (acquisition of the dorsal jaw could be a synapomorphy for Pulmonata). If Amphibolidae is the most basal taxon of Pulmonata, acquisition of the unpaired dorsal jaw could have occurred in the ancestral branch of a clade that included all Pulmonata less Amphibolidae. Because Systellommatophora (Onchidiidae + Soleolifera) is sister group to Stylommatophora, it is likely that the two lateral jaws have been lost in the common ancestral branch of Geophila and the unpaired dorsal jaw has been lost in the common ancestral branch of Onchidiidae only.

Gizzard Pouch

A gizzard pouch is present in the stomach of all basommatophoran pulmonates (i.e., all pulmonates excluding Systellommatophora and Stylommatophora) except for few genera like *Siphonaria* or *Acroloxus.* Evolutionary history of this character depends directly on the position of *Siphonaria:* (1) If *Siphonaria* constitutes the most basal branch of pulmonates, the gizzard pouch could have been first acquired on the common ancestral branch of the clade that includes all of the pulmonates less *Siphonaria* and secondarily lost by Geophila. Of course, in this case, the acquisition of a gizzard pouch could be a synapomorphy for the clade that includes all of the pulmonates less *Siphonaria,* and its secondary loss could be a synapomorphy for Geophila. (2) If *Siphonaria* is not the most basal branch of pulmonates, the acquisition of the gizzard pouch could be a synapomorphy for Pulmonata.

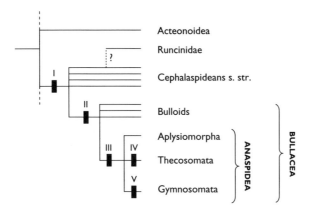

Figure 7.3. Evolutionary history of the esophageal gizzard plates and spines in Euthyneura. (I) Acquisition of three calcified plates, and subsequent losses for cephalaspideans that do not have them (e.g., *Aglaja* or *Gastropteron*). (II) Acquisition of chitinous spines. (III) Transformation into several calcified plates distinct from the three previous plates. (IV) Loss of the calcified plates and chitinous spines, acquisition of chitinous plates. (V) Loss of all the esophageal gizzard structures.

Esophageal Gizzard

Esophageal gizzard plates and spines, distinct from the stomach gizzard pouch, are present in several cephalaspideans, in the clade we refer to as Bullacea (except for Gymnosomata), and in Runcinidae. The phylogenetic importance of such characters has been suggested by Mikkelsen (1996). According to our consensus tree (Figure 7.2), these structures are present in one large clade and in Runcinidae only. The position of runcinids has not been resolved because of its simplified and yet derived morphology (Dayrat and Tillier 2002), nor by molecular data (Dayrat et al. 2001). The fact that runcinids share three calcified plates identical to those observed in cephalaspideans (like *Scaphander*, *Cylichna*, *Retusa* or *Philine*) is indeed a good reason to consider them as belonging to cephalaspideans. Unfortunately, this has not yet been demonstrated by any cladistic analysis. A hypothesis of evolutionary history for this character has five major steps, as illustrated in Figure 7.3.

Male Copulatory Organ

A male copulatory organ is present in all Euthyneura. It is free and external in some (Acteonoidea, *Ringicula*, Umbraculoidea, *Siphonaria*) or internal and embedded in a sheath in the others. Because a free male copulatory organ is also present in some allogastropods (e.g., *Valvata*, *Rissoella*, *Glacidorbis*), an internal organ should be apomorphic. Internalization then happened at least once; nevertheless, we still do not know if it happened once for all Euthyneura, or independently for pulmonates and opisthobranchs—nor do we know if free male copulatory organs observed in Euthyneura are really plesiomorphic or secondary. Posterior migration of the male genital aperture close to the female aperture is the best synapomorphy that we have found for "Nudipleura," the clade that includes Nudibranchia and Pleurobranchoidea (Dayrat and Tillier 2002). Such a migration is found in these two taxa, and also in the minute marine pulmonate *Otina*.

LAND AND FRESHWATER INVASION

Because of lack of resolution within Pulmonata, it is difficult to give a precise history of adaptation to freshwater and terrestrial habitats in this taxon (Figure 7.4). Because most of the outgroups of euthyneurans are marine organisms, and because of the necessarily basal position (even if it is presently unresolved) of one of the marine pulmonate taxa, we can reasonably suppose that the first pulmonates were marine. Most of the so-called marine pulmonates live in supralittoral and upper littoral zones or mangroves, and rarely live fully submerged (some species of *Williamia* only). At the moment, we infer that colonization of freshwater habitat happened once because of the monophyly of Hygrophila (including Chilinidae). Because the phylogenetic position of Hygrophila is still unknown, we cannot determine whether this clade colonized freshwater from land or from the seashore. History of adaptation to the land cannot be precisely established because the relationships of ellobiids, especially the terrestrial ones (like *Carychium*), remain unresolved. Most of the onchidiids live on the seashore or in mangroves, but some are known to be truly terrestrial (Tillier 1984). Adaptation to living on land could have occurred once on the common ancestral branch of Geophila (in this case, another similar adaptation occurred for ellobiids) and the marine habitat for Onchidiidae could be secondary. Alternatively, land invasion could have occurred independently for some onchidiids, ellobiids, Soleolifera, and Stylommatophora, and then the marine habitat would be primary.

FINAL REMARKS

At present, molecular data seem far from reaching their phylogenetic potential in resolving opisthobranch evolutionary relationships and monophyly. The

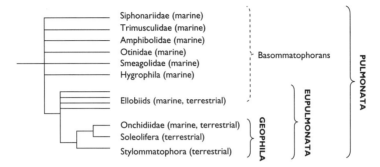

Figure 7.4. Proposed hypothesis of evolutionary history of the adaptation of Pulmonata to freshwater and terrestrial habitats.

problem is particularly serious for the deepest nodes (i.e., between the higher taxa of opisthobranchs and for all of the basal pulmonates). One solution may be the use of longer sequences, either by a complete sequencing of one gene (e.g., 28S rDNA) or by simultaneous addition of several genes. Even then, it is not certain that traditional markers like ribosomal genes will provide well-supported resolution. New genes—not just longer sequences—are obviously needed for Euthyneura. Taxon sampling is also an issue, not only because homoplasy may increase with the number of taxa, but also because our available molecular dataset does not provide good representation of the diversity of Euthyneura. Despite this shortcoming, molecular data have already provided nearly as many resolved nodes as the much richer and older morphological studies. In addition, we suspect, comparative genomics (we think particularly here of mitochondrial gene order comparison) will give us some hope for resolving euthyneuran relationships (Kurabayashi and Ueshima 2000; Grande et al. 2001; Medina et al. 2001; Ueshima 2001).

Resorting to combinable component consensus provides a relatively large number of unresolved nodes. Nevertheless, it is a crucial step for determining rigorously what we know and what we do not know about phylogenetic relationships. Sorting out the nodes that are strongly supported from those that are unresolved is necessary because discussing evolutionary history of characters or historical biogeography from a tree without considering the absence of resolution would imply positively erroneous conclusions. Even if it is an important limit for phylogenetics, and whether considered either as an artifact of data or as a result implying an evolutionary radiation, the absence of resolution must be taken into account. Any absence of resolution reflects our current knowledge. This is especially relevant for constructing consensus trees.

Our combinable component consensus tree for euthyneurans is based on the principle that resolved topologies are not contradicted by absence of resolution for other topologies derived from other datasets. This allows for the construction of a phylogenetic hypothesis with the most explanatory power, derived from several independent analyses with distinct taxon samples. Taxonomic congruence is particularly relevant when each independent analysis provides only one or two well-supported nodes, as is the case for the euthyneurans. We recommend a global phylogenetic analysis of opisthobranchs + pulmonates to fully ascertain the phylogenetic relationships of Euthyneura and to determine the monophyletic status of the Opisthobranchia.

ACKNOWLEDGMENTS

We are grateful to Marilyn Eversole, Rebecca Johnson, and Mónica Medina who reviewed the manuscript, Terry Gosliner, Robert Robertson, Christopher Wade, Serge Gofas, and an anonymous reviewer for their help and valuable comments. This contribution was supported by the National Science Foundation (PEET DEB-9978155), the California Academy of Sciences, the Centre National de la Récherche Scientifique (FR 1541), and the Muséum National d'Histoire Naturelle (Paris).

REFERENCES

Adams, E. N. I. 1972. Consensus techniques and the comparison of taxonomic trees. Systematic Zoology 21:390–397.

Baker, H. B. 1955. Heterurethrous and Aulacopod. Nautilus 58:109–112.

Boettger, C. 1954. Die systematik der euthyneuren Schnecken. Verhandlungen der Deutschen Zoologischen Geselschaft in Tübingen 18:253–280.

Boore, J. L., and W. M. Brown. 1994. Mitochondrial genomes and the phylogeny of Mollusks. The Nautilus, Supplement 108:61–78.

Bremer, K. 1988. The limits of amino–acid sequence data in angiosperm phylogenetic reconstruction. Evolution 42:795–803.

Bremer, K. 1990. Combinable component consensus. Cladistics 6:369–372.

Colgan, D. J., W. F. Ponder, and P. E. Eggler. 2000. Gastropod evolutionary rates and phylogenetic relationships assessed using partial 28S rDNA and histone H3 sequences. Zoologica Scripta 29:29–63.

Dayrat, B., and S. Tillier. 2000. Taxon sampling, character sampling and systematics: How gradist presuppositions created additional ganglia in euthyneuran taxa. Zoological Journal of the Linnean Society 129:403–418.

Dayrat, B., and S. Tillier. 2002. Evolutionary relationships of euthyneuran gastropods (Mollusca): A re-evaluation of morphological characters. Zoological Journal of the Linnean Society 135:403–470.

Dayrat, B., A. Tillier, G. Lecointre, and S. Tillier. 2001. New clades of euthyneuran gastropods (Mollusca) from 28S rRNA sequences. Molecular Phylogenetics and Evolution 19:225–235.

Delhaye, W., and J. Bouillon. 1972. L'évolution et l'adaptation de l'organe excréteur chez les Mollusques gastéropodes pulmonés. Biological Bulletin 106:45–77.

de Queiroz, K., and J. Gauthier. 1992. Phylogenetic taxonomy. Annual Review of Ecology and Systematics 23:449–480.

de Queiroz, K., and J. Gauthier. 1994. Toward a phylogenetic system of biological nomenclature. Trends in Ecology and Evolution 9:27–31.

Duncan, C. J. 1960. The evolution of the pulmonate genital system. Proceedings of the Zoological Society of London 134:601–609.

Dutra-Clarke, A. V. C., C. Williams, R. Dickstein, N. Kaufer, and J. R. Spotila. 2001. Inferences on the phylogenetic relationships of Succineidae (Mollusca, Pulmonata) based on 18S rRNA gene. Malacologia 43:223–236.

Emberton, K. C., G. S. Kuncio, G. M. Davis, S. M. Phillips, K. M. Monderewicz, Y. H. Guo. 1990. Comparison of recent classifications of Stylommatophoran land-snail families, and evaluation of large ribosomal-RNA sequencing for their phylogenetics. Malacologia 31:327–352.

Farris, J. S., M. Kallersjö, A. G. Kluge, and C. Bult. 1995. Testing significance of incongruence. Cladistics 10:313–319.

Felsenstein, J. 1985. Confidence limits on phylogenies: An approach using bootstrap. Evolution 39:783–791.

Fretter, V. 1943. Studies in the functional morphology and embryology of *Onchidella celtica* (Forbes and Hanley) and their bearing on its relationships. Journal of Marine Biology Association, U.K. 25:685–720.

Fretter, V., and A. Graham. 1949. The structure and mode of life of the Pyramidellidae, parasitic opisthobranchs. Journal of Marine Biology Association, U.K. 28:493–532.

Gosliner, T. M. 1991. Morphological parallelism in opisthobranch gastropods. Malacologia 32:313–327.

Gosliner, T. M., and M. T. Ghiselin. 1984. Parallel evolution in opisthobranch gastropods and its implications for phylogenetic methodology. Systematic Zoology 33:255–274.

Grande, C., J. Templado, J. L. Cervera, and R. Zardoya. 2001. Complete sequence of mitochondrial genome of a nudibranch, *Roboastra europaea* (Mollusca, Opisthobranchia). P. 130 in World Congress of Malacology (L. Salvini-Plawen, J. Woltzow, H. Sattman, and G. Steiner, eds.). Vienna, Austria, Unitas Malacologica.

Grande, C. J., J. Templado, J. L. Cervera, and R. Zardoya. 2002. The complete mitochondrial genome of the nudibranch *Roboastra europaea* (Mullusca, Gastropoda) supports the monophyly of opisthobranchs. Molecular Biology and Evolution 19:1672–1685.

Harasewych, M. G., S. L. Adamkewicz, J. A. Blake, S. Saudek, T. Spriggs, and C. J. Bult. 1997. Phylogeny and relationships of pleurotomariid gastropods (Mollusca: Gastropoda): An assessment based on partial 18S rDNA and cytochrome *c* oxydase I sequences. Molecular Marine Biology and Biotechnology 6:1–20.

Haszprunar, G. 1988. On the origin and evolution of major gastropod groups, with special reference to the Streptoneura (Mollusca). Journal of Molluscan Studies 54:367–441.

Healy, J. M. 1993. Comparative sperm ultrastructure and spermiogenesis in basal heterobranch gastropods (Valvatoidea, Architectonicidea, Rissoellidea, Omalogyridea, Pyramidelloidea) (Mollusca). Zoologica Scripta 22:263–276.

Jensen, K. R. 1996. Phylogenetic systematics and classification of the Sacoglossa (Mollusca, Gastropoda, Opisthobranchia). Philosophical Transactions of the Royal Society of London B 351:91–122.

Johansson, J. 1954. On the pallial gonoduct of *Actaeon tornatilis* (L.) and its significance for the phylogeny of the Euthyneura. Zoologiska Bidrag fran Uppsala 30:223–232.

Kluge, A. G. 1998. Total evidence or taxonomic congruence: Cladistics or consensus classification. Cladistics 14:151–158.

Kurabayashi, A., and Ueshima, R. 2000. Partial mitochondrial genome organization of the heterostrophan gastropod *Omalogyra atomus* and its systematic significance. Venus 59:7–18.

Lalli, C. M., and R. W. Gilmer. 1989. Pelagic snails. The biology of Holoplanktonic Gastropod Mollusks. Stanford University Press, CA.

Lemche, H. 1948. Northern and Arctic tectibranch gastropods. Det Kongelige Danske Videnskabernes Selskab, Biologiske Skifter 5:1–136.

McArthur, A. G., and B. F. Koop. 1997. Partial 28S rDNA sequences and the antiquity of hydrothermal vent endemic gastropods. Molecular Phylogenetics and Evolution 13:255–274.

Medina, M., Y. Vallès, T. Gosliner, and J. Boore. 2001. Mitochondrial evolution of crown gastropods: Insight from large subunit sequences and gene order data. P. 218 in World Congress of Malacology (L. Salvini-Plawen, J. Woltzow, H. Sattman, and G. Steiner, eds.). Vienna, Austria, Unitas Malacologica.

Medina, M., and P. J. Walsh. 2000. Molecular systematics of the Order Anaspidea based on mitochondrial DNA sequence (12S, 16S, and COI). Molecular Phylogenetics and Evolution 15:41–58.

Mickevich, M. F. 1978. Taxonomic congruence. Systematic Zoology 27:143–158.

Mikkelsen, P. M. 1996. The evolutionary relationships of Cephalaspidea s.l. (Gastropoda: Opisthobranchia): A phylogenetic analysis. Malacologia 37:375–442.

Mörch, O. A. L. 1865. On the systematic value of the organs which have been employed as fundamental characters in the classification of Mollusca. Annual Magazine of Natural History 25:1–16.

Morton, J. E. 1955. The evolution of the Ellobiidae with a discussion of the origin of the Pulmonata. Proceedings of the Zoological Society of London 125:127–168.

Nordsieck, H. 1990. Phylogeny and system of the Pulmonata (Gastropoda). Arch. Moll. 121:31–52.

Pelseneer, P. 1894. Récherche sur divers opisthobranches. Mémoirs of the Royal Academy of Sciences, Belgium 53:1–157.

Pilsbry, H. A. 1900. On the zoological position of *Partula* and *Achatinella*. Proceedings of the Philadelphia Academy of Sciences 52:561–567.

Plate, L. 1895. Bemerkungen Über die Phylogenie und die Enstehung der Asymmetrie der Mollusken. Zoologische Jahrbucher 9:162–206.

Ponder, W. F., and D. R. Lindberg. 1997. Towards a phylogeny of gastropod molluscs: An analysis using morphological characters. Zoological Journal of the Linnean Society 119:83–265.

Robertson, R. 1985. Four characters and the higher category systematics of gastropods. American Malacological Bulletin, sp. ed. 1:1–22.

Robertson, R. 1993. Snail handedness. National Geographic Research Exploration 9:104–119.

Roth, V. L. (1988). The biological basis of homology. Pp. 1–26 in Ontogeny and Systematics (C. J. Humphries, ed.). Columbia University Press, New York.

Salvini-Plawen, L. von. 1990. Origin, phylogeny and classification of the phylum Mollusca. Iberus 9:1–33.

Salvini-Plawen, L. von, and G. Steiner. 1996. Synapomorphies and plesiomorphies in higher classification of Mollusca. Pp. 29–51 in Origin and Evolutionary Radiation of the Mollusca (J. D. Taylor, ed.). Oxford University Press, London.

Schmekel, L. 1985. Aspects of evolution within the opisthobranchs. Pp. 221–267 in The Mollusca (K. M. Wilbur, ed.). Academic Press, London.

Sokal, R. R., and F. J. Rohlf. 1981. Taxonomic congruence in the Leptopodomorpha reexamined. Systematic Zoology 30:309–325.

Spengel, J. W. 1881. Die Geruchsorgane und das Nervensystem der Mollusken. Zeitschrift für Wissenschaftliche 35:333–383.

Swofford, D. L. 1991. When are phylogeny estimates from molecular and morphological data incongruent? Pp. 295–333 in Phylogenetic Analysis of DNA Sequences (M. M. Miyamoto and J. Cracraft, eds.). Oxford University Press, New York.

Thollesson, M. 1999. Phylogenetic analysis of Euthyneura (Gastropoda) by means of the 16S rRNA gene: Use of a "fast" gene for "higher–level" phylogenies. Proceedings of the Royal Society of London 266:75–83.

Thompson, T. E. 1976. Biology of Opisthobranch Molluscs, I. Ray Society, London.

Tillier, S. 1984. A new mountain *Platevindex* from Philippine Islands. Journal of Molluscan Studies, supplement 12A:198–202.

Tillier, S. 1989. Comparative morphology, phylogeny and classification of land snails and slugs (Gastropoda: Pulmonata: Stylommatophora). Malacologia 30:1–303.

Tillier, S., M. Masselot, J. Guerdoux, A. Tillier. 1994. Monophyly of major gastropod taxa tested from partial 28S rRNA sequences, with emphasis on euthyneuran and hot–vent limpets Peltospiroidea. The Nautilus, Supplement 2 108:122–140.

Tillier, S., M. Masselot, H. Philippe, and A. Tillier. 1992. Phylogénie moléculaire des Gastropoda fondée sur le séquençage partiel de l'ARN ribosomique 28S. Comptes-Rendus de l'Académie des Sciences, Paris, série III, 314:79–85.

Tillier, S., M. Masselot, and A. Tillier. 1996. Phylogenetic relationships of the pulmonate gastropods from rRNA sequences, and tempo and age of the stylommatophoran radiation. Pp. 267–284 in Origin and Evolutionary Radiation of the Mollusca (J. D. Taylor, ed.). Oxford Science Publications, London.

Ueshima, R. 2001. Phylogeny of opisthobranch gastropods inferred from mitochondrial gene arrangement: A case study for comparing phylogenetic performance between gene sequence and gene order data. P. 361 in World Congress of Malacology (L. Salvini-Plawen, J. Woltzow, H. Sattman, and G. Steiner, eds.). Vienna, Austria, Unitas Malacologica.

Wade, C. M., and P. B. Mordan. 2000. Evolution within the gastropod molluscs using the ribosomal RNA gene–cluster as an indicator of phylogenetic relationships. Journal of Molluscan Studies 66:565–570.

Wade, C. M., P. B. Mordan, and B. Clarke. 2001. A phylogeny of the land snails (Gastropoda: Pulmonata). Proceedings of the Royal Society of London B 268:413–422.

Wägele, H., and R. C. Willan. 2000. Phylogeny of the Nudibranchia. Zoological Journal of the Linnean Society 130:83–181.

Willan, R. C. 1987. Phylogenetic systematics of the Notaspidea (Opisthobranchia) with reappraisal of families and genera. American Malacological Bulletin 5:215–241.

Wollscheid, E., and H. Wägele. 1999. Initial results on the molecular phylogeny of the Nudibranchia (Gastropoda, Opisthobranchia) based on 18S rDNA data. Molecular Phylogenetics and Evolution 13:215–226.

Wollscheid-Lengeling E., J. Boore, W. Brown, and H. Wägele. 2001. The phylogeny of Nudibranchia (Opisthobranchia, Gastropoda, Mollusca) reconstructed by three molecular markers. Organisms Diversity and Evolution 1:241–256.

HEIKE WÄGELE, VERENA VONNEMANN, AND WOLFGANG WÄGELE

8

TOWARD A PHYLOGENY OF
THE OPISTHOBRANCHIA

The Opisthobranchia can be easily identified by external morphological and/or ecological characters. The "Cephalaspidea" are usually characterized by having a cephalic shield, whereas the Sacoglossa feed on algae and have only one pair of rolled rhinophores. The Anaspidea forage on algae and have two pairs of oral tentacles. The Tylodinoidea are characterized by their china-hat shaped shell, whereas within the Pleurobranchoidea, the cuplike shell may be reduced in more elaborate forms. In both of these groups, the gill is lying on the right side, rather free, and not in any cavity. The Nudibranchia, the most species-rich group, can be identified by their digitiform rhinophores. Smaller groups are also included, but are less often addressed because of their specialized habitats (e.g., Gymnosomata and Thecosomata are pelagic forms), or because they are infrequently encountered (e.g., Acochlidiacea). Although the monophyly of the Opisthobranchia and its possible relationship to the Pulmonata are still in debate (Haszprunar 1988; Ponder and Lindberg 1997; Lindberg and Ponder 2001; Chapter 7, this volume), this should not hinder our efforts to begin to elucidate phylogenetic relationships among this diverse group of gastropods. With appropriate taxonomic sampling, we should be able to ascertain whether some opisthobranch lineages are more closely related to other opisthobranchs or pulmonates, as well as to test the monophyly of traditionally recognized opisthobranch subgroups.

Monophyly of several subgroups of Opisthobranchia has been shown by morphological datasets: Mikkelsen (1996) could demarcate a smaller group of the former Cephalaspidea, which she named the Cephalaspidea *sensu stricto* (in the strict sense, s.str.) based on several apomorphies (e.g., three gizzard plates). Mikkelsen (1996) also demonstrated monophyly of the Sacoglossa, which is characterized by a ventral pouch (ascus) retaining worn teeth in the buccal mass and two lateral pouches at the pharynx. Monophyly of the Anaspidea is supported by several apomorphies, including a secondary gizzard and a caecum extending from the stomach (Mikkelsen 1996). Willan (1987) provided evidence for the monophyly of the Pleurobranchoidea, but the described sister-taxon relationship of the Pleurobranchoidea to the Tylodinoidea (formerly united under the name Notaspidea) was rejected by Schmekel (1985), Salvini-Plawen (1991), Salvini-Plawen and Steiner (1996), and Wägele and Willan (2000). The latter authors named several apomorphies in favor of a Pleurobranchoidea/Nudibranchia relationship (e.g., loss of an osphradium, presence of an androdiaulic genital system), and also gave evidence for the monophyly of the Nudibranchia. The Nudibranchia possess several unique features, such as the special vacuolated cells in the epidermis and the horizontal orientation of the heart complex.

Molecular studies of opisthobranch taxa have supported aspects of these morphological studies; however, most molecular works include few members of the diverse subgroups. Monophyly of the Anthobranchia, Cladobranchia (both belonging to the Nudibranchia), and Anaspidea has been supported by Thollesson (1999a), Wollscheid and Wägele (1999), and Wollscheid-Lengeling et al. (2001). Monophyly of the Sacoglossa (based on two species), Pleurobranchoidea (based on two species), and the sister-taxon relationship of the Pleurobranchoidea with the Nudibranchia was revealed by Wollscheid-Lengeling et al. (2001). In contrast, several analyses of molecular markers contradict results based on morphological or histological data. Thollesson (1999a), using 16S rDNA sequences, considered the Nudibranchia to be paraphyletic with the single representative pleurobranchoid species (*Berthella sideralis*), sister to the nudibranch group Cladobranchia.

Wollscheid-Lengeling et al. (2001) were the first to compare three different molecular markers (18S rDNA, 16S rDNA, and cytochrome oxidase I [COI] gene) and found congruence among the gene trees for the monophyly of major taxa within the Opisthobranchia. These authors did not address the problem of monophyly/paraphyly of the Opisthobranchia, because their main aim was to elucidate the phylogenetic relationships within the Nudibranchia. They argued that the phylogeny estimated with the 18S rRNA gene showed higher congruence with phylogenetic analyses based on morphology, especially when looking at older divergences. They also mentioned that the 16S rDNA did not resolve the relationships on the family or genus level, but showed coincident results with the 18S rRNA gene in deeper nodes, whereas the COI gene resolved relationships on the genus level.

Many more sequences are now available for all three of the molecular markers mentioned. In this chapter, we will discuss our present knowledge of available molecular data in the major subgroups within the Opisthobranchia and compare the molecular phylogenetic hypotheses with those based on morphological characters. The number of species, within the different opisthobranch taxa differs considerably: about 2,000 cephalaspid species are described, compared with about 3,000 nudibranch species, 250 Sacoglossa species, 100 Anaspidea species, 55 Pleurobranchoidea species, and 10 Tylodinoidea species. Ideally, one should attempt to adequately represent each lineage for a thorough phylogenetic analysis; however, at this time, only the Nudibranchia is well represented. Therefore, our discussion will be biased in favor of the Nudibranchia.

TAXON SAMPLING

We drew sequences of mitochondrial 16S rDNA and COI genes from GenBank (Table 8.1). Unfortunately, we could not use the same taxa for all genes, although many more species are listed in GenBank with shorter sequences. Nuclear 18S rDNA sequences are partly drawn from GenBank, but 17 new sequences are included in this analysis (Table 8.1). Voucher specimens are available from the first author (H.W.). To investigate monophyly/paraphyly of the Opisthobranchia, we included several sequences of the Basommatophora (Pulmonata) and used species belonging to the caenogastropod family Littorinidae as outgroups in the 18S rDNA and 16S rDNA analyses. In choosing these taxa, we were guided by the results of Haszprunar (1985, 1988), Ponder and Lindberg (1997), and Winnepenninckx et al. (1998). The Littorinidae belongs to the sister taxon to the Heterobranchia. For the COI alignment, littorinids were not used as outgroup, because the distance of the COI sequence of *Littorina obtusata* was deemed too high compared with the ingroup taxa. Instead, we used the sequence of the more closely related stylommatophoran, *Cepaea nemoralis*.

The dataset consists of 80 species (81 sequences) with 2,668 bp for 18S rDNA, 108 species (123 sequences) with 516 bp for 16S rDNA, and 77 species (77 sequences) with 605 nucleotides and 165 amino acid characters for the COI gene. In some cases, several available sequences of the same species were included, because analyses showed that they were not identical.

DNA EXTRACTION AND SEQUENCING

Genomic DNA was extracted from alcohol-preserved specimens with the Blood and Tissue-Kit (Qiagen). The 18S rDNA region was amplified by polymerase chain reaction (PCR) with primers developed by Wollscheid and Wägele

Table 8.1

Species investigated, with collection site of those species whose sequences have not yet been published*, and GenBank accession numbers for all three genes

Taxon	Higher Category	18S	16S	COI
		\multicolumn{3}{c}{Genes Investigated}		

Taxon	Higher Category	18S	16S	COI
"PROSOBRANCHIA"				
Littorina littorea (Linné 1758)	Littorinidae	X91970	—	—
Tectarius viviparus (Rosewater 1982)		—	U66352	—
PULMONATA				
Stylommatophora				
Cepaea nemoralis Linné 1758	Helicidae	—	—	U23045
"Basommatophora"				
Amerianna carinata (H. Adams 1861)	Planorbidae	—	U82065	—
Biomphalaria alexandrina Ehrenberg		U65225	—	—
Biomphalaria peregrina (d'Orbigny 1835)		—	AY030231	—
Biomphalaria schrammi (Crosse 1864)		AY030402	AY030233	—
Helisoma trivolvis (Say 1816)		—	AY030234	—
Ancylus fluviatilis O. F. Müller 1774	Ancylidae	—	AY326926	—
Austropeplea lessoni (Deshayes 1830)	Lymnaeidae	—	U82066	—
Austropeplea ollula (Gould 1859)		—	U82067	—
Bullastra cumingiana (Pfeiffer 1845)		—	U82068	—
Bulimnea megasoma (Say 1824)		—	U82069	—
Fossaria truncatula (O. F. Müller 1774)		Z73985	—	—
Lymnaea stagnalis (Linné 1758)		Z73984	U82071	—
Radix peregra (O. F. Müller 1774)		Z73981	—	—
Radix quadrasi (von Möllendorff 1898		—	U82075	—
Radix rubiginosa (Michelin 1831)		—	U82076	—
Stagnicola emarginata (Say 1821)		—	U82081	—
Stagnicola palustris (O. F. Müller 1774)		Z73983	U82082	—
Bulinus forskalii Ehrenberg 1831	Bullinidae	—	AY029550	—
Bulinus wrighti Mandahl Barth 1965		—	AY029552	—
Siphonaria algesirae Quoy & Gaimard 1834	Siphonariidae	X91973	—	—
Opisthobranchia				
GYMNOSOMATA				
Clione limacina Martens 1675	Clionidae	—	AJ223406	—
"CEPHALASPIDEA" s.l.	Incertae sedis			
Acteon tornatilis (Linné 1767)	Acteonidae	—	AJ223405	—
Pupa strigosa (Gould 1859)		—	—	AB028237
	Cephalaspidea *s.str.*			
Bulla gouldiana Pilsbry 1895	Bullidae	—	AF156125	AF156141
Bullacta exarata (Philippi 1848)	Haminoeidae	AF18867	—	—
Haminoea cymbalum (Quoy & Gaimard 1933)		AF249221	AF249258	—
Haminoea virescens Sowerby 1833		—	AF156126	AF156142
Smaragdinella sp.		AJ224789	AF249257	AF249806

Table 8.1 continued

Taxon	Higher Category	Genes Investigated		
		18S	16S	COI
Chelidonura inornata Baba 1949	Aglajidae	*Lizard Isl., Australia	—	—
Diaphana minuta Brown 1827 Diaphanidae		—	AJ223404	—
Philine aperta (Linné 1767) Philinidae		—	AJ223402	—
Scaphander punctostriatus (Mighels & Adams 1841)	Cylichnidae	—	AJ223403	—
SACOGLOSSA				
Placobranchoidea				
Elysia timida Risso 1818		—	—	AF249818
Elysia viridis (Montagu 1804)		—	AJ223398	—
Thuridilla bayeri Marcus 1965		AF249220	—	—
Thuridilla hopei (Verany 1853)		—	—	AF249810
Thuridilla ratna Marcus 1965		—	AF249256	—
Limapontia capitata (O. F. Müller 1773)	Limapontioidea	AJ224920	—	—
Placida dendritica (Alder & Hancock 1843)		—	AJ223399	—
ANASPIDEA				
Akera bullata O. F. Müller 1776	Akeridae	—	AF156127 AJ223401	AF156143
Aplysia brasiliana Rang 1828	Aplysiidae	—	AF192296	—
Aplysia californica Cooper 1863		AY039804	AF192295	—
Aplysia cervina (Dall & Stimpson 1901)		—	AF156128	AF156144
Aplysia dactylomela Rang 1828		—	AF192297	—
Aplysia depilans Gmelin 1791		AJ224918	AF192294	AF249824
Aplysia extraordinaria Allan 1932		AF249193	AF249255	AF249823
Aplysia fasciata Poire 1789		—	AF192298	—
Aplysia gigantea Sowerby 1869		—	AF192299	—
Aplysia Juliana Quoy & Gaimard 1832		—	AF192292	—
Aplysia kurodai Baba 1937		—	AF192300	—
Aplysia morio Verrill 1901		—	AF192301	—
Aplysia oculifera Adams & Reeve 1850		—	AF192302	—
Aplysia parvula Mörch 1863		—	AF192291	AF2449822
Aplysia punctata Cuvier 1803		AJ224919	AF156129 AJ223400 AF249253	AF156145
Aplysia sp.		AF249192 X94268	AF249254	—
Aplysia vaccaria Winkler 1955		—	AF192293	—
Bursatella leachii de Blainville 1817		—	AF156130	AF156146
Dolabella auricularia (Lightfoot 1786)		—	AF156131 AF156132	AF156147
Dolabrifera dolabrifera (Cuvier 1817)		—	AF156133	AF156149
Notarchus indicus Schweigger 1820		—	AF156134 AF156135	AF156150 AF156151

Continued on next page

Table 8.1 continued

Taxon	Higher Category	Genes Investigated		
		18S	16S	COI
Petalifera ramosa Baba 1959		—	AF156136	AF156152
			AF156137	
Phyllaplysia sp.		—	AF156138	AF156155
Phyllaplysia taylori Dall 1900		—	AF156139	AF156155
Stylocheilus longicauda Quoy & Gaimard 1832		—	AF156140	AF156156
TYLODINOIDEA				
Tylodina perversa (Gmelin 1790) Tylodinidae		—	—	AF249809
Umbraculum mediterraneum Lamarck 1812	Umbraculidae	*Meteor Bank, Atlantic	—	—
PLEUROBRANCHOIDEA				
Bathyberthella antarctica Willan & Bertsch 1987	Pleurobranchidae	AF249219	—	—
Berthella sideralis (Lovén 1846)		—	AJ225181	AJ223257
Berthellina citrina (Rüppell & Leuckart 1828)		—	—	AF249785
Euselenops luniceps (Cuvier 1817)	Pleurobranchaeidae	AF249218	—	—
NUDIBRANCHIA				
Bathydoridoidea				
Bathydoris clavigera Thiele 1912	Bathydorididae	*Weddell Sea, Antarctica	AF249222	AF249808
Doridoidea				
	Phanerobranchia			
Acanthodoris pilosa (O. F. Müller 1776)	Ochidorididae	AF249770	AF249236	AJ223254
			AJ225177	
Adalaria proxima (Alder & Hancock 1854)		—	AF249225	—
Diaphorodoris luteocincta (Sars 1870)		AJ224775	AF249230	AF249796
Diaphorodoris papillata Portmann & Sandmeier 1960		—	—	AF249819
Onchidoris bilamellata (Linné 1767)		AJ224776	AJ225195	—
			AF249235	
Onchidoris muricata (O. F. Müller 1776)		—	AJ225196	AJ223271
Ancula gibbosa (Risso 1818) Goniodorididae		—	AJ225179	AJ223255
Goniodoris castanea Alder & Hancock 1845		—	AJ225187	AJ223263
Goniodoris nodosa (Montagu 1808)		AJ224783	AF249226	AJ223264
			AJ225188	
Okenia aspersa Alder & Hancock 1845		—	AJ225194	AJ223270
Crimora papillata Alder & Hancock 1862	Triophidae	—	—	AF249821
Plocamopherus ceylonicus (Kelaart 1885)		AF249207	—	—
Triopha catalinae (Cooper 1863)		AJ224782	AF249227	—
Limacia clavigera (Müller 1776)	Polyceridae	AJ224778	AJ225192	AJ223268
Palio dubia Sars 1829		—	AJ225197	AJ223272
Polycera aurantiomarginata García & Bobo 1984		—	AJ225199	AJ223274

Table 8.1 continued

Taxon	Higher Category	Genes Investigated		
		18S	16S	COI
Polycera quadrilineata (Müller 1776)		AJ224777	AJ225200 AF24922	AJ223275
Polycerella emertoni Verrill 1881		—	AJ225198	AJ223273
Thecacera pennigera Montagu 1815		—	AJ225202	AJ223277
Aegires punctilucens d'Orbigny 1837	Aegiridae Cryptobranchia	—	AJ225178	—
Archidoris pseudoargus (Rapp 1827)	Dorididae	AF249217	AJ225180 AF249224	AJ223256
Austrodoris kerguelenensis (Bergh 1884)		AF249771	AF249234 AF249233	—
Discodoris concinna (Alder & Hancock 1864)		AJ224781 AF249213	AF249228	AF249801
Doriopsis granulosa Pease 1860		AF249212	AF249223	AF249798
Jorunna tomentosa (Cuvier 1804)		AF249210	AJ225191	AJ223267
Platydoris argo (Quoy & Gaimard 1832)		—	—	AF249811
Cadlina laevis (Linné 1767)	Chromodorididae	—	AJ225182	AJ223258
Cadlina luteomarginata (MacFarland 1966)		AJ224772	AF249231	AF249803
Chromodoris krohni (Verrany 1846)		AJ224774	AF249239	AF249805
Chromodoris kuiteri (Rudman 1982)		AF249214	AF249240	AF249804
Chromodoris luteorosa (Rapp 1827)		—	AJ225183	AF249815
Chromodoris purpurea (Laurillard 1831)		—	AJ225184	AJ223260
Chromodoris quadricolor (Rüppel & Leuckart 1828)		AJ224773	AF249241	AF249802
Chromodoris tinctoria (Rüppell & Leuckart 1828)		AF188676	—	—
Durvilledoris pusilla (Bergh 1874)		—	AJ225193	AJ223269
Glossodoris atromarginata (Cuvier 1804)		AF249211	—	AF249789
Hypselodoris elegans (Cantraine 1834)		AJ224779	AF249238	AF249787
Hypselodoris orsinii (Verany 1846)		—	AJ225189	AJ223265
Hypselodoris villafranca (Risso 1818)		AJ224780	AF249237 AJ225190	AJ223266
Porostomata				
Dendrodoris fumata (Rüppell & Leuckart 1828)	Dendrodorididae	AF249216	—	AF249799
Dendrodoris nigra (Stimpson 1855)		AF249215	AF249242	AF249795
Doriopsilla areolata Bergh 1880		—	AJ225186	AJ223262
Phyllidia coelestis Bergh 1905	Phyllidiidae	AF249209	—	—
Phyllidia elegans Bergh 1869		—	AJ225201	AJ223276
Phyllidiella pustulosa (Cuvier 1804)		AF249208	AF249232	—
Dendronotoidea				
Bornella stellifer (Adams & Reeve in Adams 1848)	Bornellidae	*Magnetic Isl. Australia	—	—
Dendronotus dalli Bergh 1879	Dendronotidae	*USA, Atlantic	AF249252	AF249800
Dendronotus frondosus (Ascanius 1774)		AF249206	AF249251 AJ22518	AJ223261

Continued on next page

Table 8.1 continued

Taxon	Higher Category	Genes Investigated		
		18S	16S	COI
Dendronotus iris Cooper 1863		*Ross Isl. Canada	—	—
Doto coronata (Gmelin 1791)	Dotidae	AF249203	—	AF24979
Doto eireana Lemche 1976		AF249204	AF249248	—
Doto floridicola Simroth 1888		*Spain, Mediterranean	—	AF249820
Doto fragilis (Forbes 1838)		—	AJ223392	—
Doto koenneckeri Lemche 1976		AF249205	AF249249	AF249797
Doto pinnatifida (Montagu 1804)		AF249202	AF249250	AF249793
Hero formosa (Lovén 1841)	Heroidae	—	AJ223395	—
Melibe leonina (Gould 1852)	Tethydidae	AJ224784	—	—
Marionia blainvillea Risso 1828	Tritoniidae	—	—	AF249812
Tritonia nilsodhneri Marcus 1983		AF249200	—	—
Tritonia plebeia Johnston 1828		—	AJ223393	—
Tritoniella belli Eliot 1907		AF249201	—	—
"Arminoidea"				
Armina loveni (Bergh 1860)	Arminidae	AF249196	AJ223394 AF249243	—
Dermatobranchus semistriatus Baba 1949		AF249195	AF249244	—
Janolus cristatus Delle Chiaje 1841	Zephyrinidae	AF249194	—	AF249813
Aeolidoidea				
Cuthona caerulea (Montagu 1804)	Tergipedidae	AF249199	—	AF249807
Cuthona nana (Alder & Hancock 1842)		*Helgoland, North Sea	—	—
Cuthona sibogae Bergh 1905		*Lizard Isl., Australia	—	—
Tergipes tergipes (Forskal 1775)		AF249197	—	—
Eubranchus exiguus (Alder & Hancock 1848)	Eubranchidae	AJ224787	AF249246	AF249792
Eubranchus farrani (Alder & Hancock 1844)		—	AJ223396	—
Eubranchus sp.		AJ224786	—	AF249791
Cratena peregrina Gmelin 1791		—	—	AF249786
Facelina auriculata (Müller 1776)	Facelinidae	*Helgoland, North Sea	—	—
Facelina punctata Alder & Hancock 1845		—	—	AF249816
Facelina bostoniensis (Couthouy 1838)		*Kieler Förde, Baltic Sea	—	—
Godiva banyulensis (García & García 1985)		*Spain, Mediterranean	—	AF249782
Phidiana lynceus Bergh 1867		*Curaçao, Caribbean Sea	—	—
Phyllodesmium briareum (Bergh 1896)		*Lizard Isl., Australia	—	—
Flabellina affinis (Gmelin 1791) Flabellinidae		*Spain, Mediterranean	—	—
Flabellina babai Schmekel 1972		*Spain, Mediterranean	—	—

Table 8.1 continued

Taxon	Higher Category	Genes Investigated		
		18S	16S	COI
Flabellina ischitana Hirano &Thompson 1990		—	—	AF249814
Flabellina lineata (Lovén 1846)		—	AJ223397	—
Flabellina pedata (Montagu 1814)		AJ224788	AF249247	AF249817
Flabellina verrucosa (Sars 1829)		AF249198	AF249245	AF249790
Flabellina sp.		*Balgal Beach Australia	—	—
Notaeolidia depressa Elliot 1905	Notaeolidiidae	*Weddell Sea, Antarctica	—	—

(1999), under the following conditions: 95°C for 4 minutes, followed by 38 cycles of 30 seconds at 94°C, 30 seconds at 52.5°C, 2.5 minutes at 72°C, and a final extension at 72°C for 10 minutes. Each PCR reaction mix (50 µl) contained 5 µl of 10×PCR buffer (Qiagen), 10 µl of Q-Solution (Qiagen), 5 µl of dNTP-mix (2 mM per dNTP), 0.5 µl of each primer (18A1, 1800), 0.3 µl of *Taq* polymerase (Qiagen), 0.25 to 3.0 µl of genomic DNA, and 21.55 to 25.7 µl H_2O. After purification of the PCR products with the QIAquick PCR Purification Kit (Qiagen), the 18S rDNA was ligated into the pCR2.1 Vector (InVitrogen) and transformed in *E. coli* cells by heat shock. Plasmid purification was performed using the S.N.A.P Miniprep Kit (InVitrogen).

Sequencing of the Plasmid-DNA was realized by the chain-termination method (Sanger et al. 1977) with fluorescent-labeled primer on the automated sequencers *4000* and *4000IR²* (Li-Cor). Both strands of the 18S rDNA were sequenced using the primers *Universal* and *Reverse* that were binding at the polylinker vector sites and several internal primers (see Wollscheid and Wägele 1999). Primer fitting was not sufficient in one case; therefore, the 18S rRNA of *Bornella stellifer* could not be completely sequenced.

DNA ALIGNMENT

Sequence alignment produces homology hypotheses on which all of the following phylogenetic reconstructions, as well as statistical analyses, are based. Therefore, we gave special attention to the alignment. During subsequent addition of taxa and realignment of these additional sequences to the existing alignments, the program CLUSTAL X's (Thompson et al. 1997) mistakes were obvious in unambiguous areas of the alignment, especially when the same sequence was added several times. In these cases, we realigned by eye. For more difficult alignment regions, we searched for similar patterns (three or more nucleotides in sequence) and then aligned by eye. When a decision had to be made about transition versus transversion, we selected transitions. This resulted in an alignment with more gaps than in the beginning, after the first alignment with CLUSTAL X. Alignment by eye was performed with Genedoc (Nicholas and Nicholas 1997). Some areas remained ambiguous, but we chose to include these areas in the analyses, following Wenzel and Sidall (1999).

Larger gaps in the alignment in the 18S and 16S rRNA genes were treated as a 5th nucleotide in all parsimony analyses (Simmons and Ochoterena 2000). Two gaps in the COI gene were caused by an additional base at position 68 in the sequence of *Diaphorodoris luteocincta,* and an additional base at position 494 in *Cepaea nemoralis.* We presume that these represent amplification or sequencing errors and subsequently deleted them for translation of the DNA sequences into amino acid sequences. Any additional codons observed in other species were treated as missing data in all other sequences.

PHYLOGENETIC ANALYSIS

A relative rate test (Wu and Li 1985) was used to identify long-branch taxa (Table 8.2). The outgroup in the phylogenetic analyses was also chosen as the reference outgroup taxon. Because of the extremely high number of taxa, all ingroup taxa were tested against one species of the ingroup that did not belong to the Opisthobranchia. For the 18S rDNA analysis, all taxa were tested against the basommatophoran *Siphonaria algesirae,* with *Littorina littorea* as outgroup. *Siphonaria* was chosen because it seems to represent the most basal basommatophoran (Dayrat et al. 2001). For the 16S rDNA analysis, the taxa were tested against the basommatophoran *Stagnicola palustris* (a *Siphonaria* sequence was not available for this gene), with the caenogastropod *Tectarius viviparus* as outgroup. For the COI gene, all taxa were tested against the acteonid *Pupa strigosa* (Mikkelsen 1996 excluded the whole family Acteonidae from the Opisthobranchia) with *Cepaea nemoralis* as outgroup. An analysis of base composition by means of the χ^2 test (PAUP* 4.0; Swofford 1998) served for recognition of sequences with a significantly deviating distribution of bases. Saturation of the sequences by multiple substitutions was estimated for transitions and transversions separately by plotting the absolute number of substitutions against uncorrected p-distances (results not shown).

Maximum parsimony and distance methods were used for phylogenetic analyses using PAUP* 4.0 (Swofford 1998). The following settings were used

Table 8.2

Results of the relative rate test (Z-score values) comparing major taxonomic groups

Taxon	Z-score Values (Range)		
	18S	16S	COI
Cladobranchia	11.13–15.92	0.28–1.73	0.16–1.8
Aeolidoidea	11.80–15.92	0.28–0.92	0.38–1.58
Facelina bostoniensis	**15.84**	—	—
Phidiana lynceus	**15.92**	—	—
Godiva banyulensis	**15.92**	—	—
Aeolidoidea without the 3 taxa above	11.80–12.81	—	—
Phyllodesmium briareum	12.25	—	—
Facelina auriculata	12.14	—	—
Dendronotoidea	11.13–14.26	0.62–1.73	0.16–1.81
Arminoidea	12.91–14.27	0.49–1.45	0.89
Anthobranchia (complete with *Bathydoris* and all doridoidean species)	5.40–9.55	0.0044–1.71	0.06–3.57
Bathydoris	6.16 0	38	2.14
Dendrodoris	**5.40–5.44**	**1.48**	**3.24–3.57**
Anthobranchia without *Dendrodoris*	see Anthobranchia	see Anthobranchia	0.06–2.78
Jorunna tomentosa	**9.55**	**0.02**	**0.76**
Anthobranchia without *J. tomentosa*	5.40–6.24	see Anthobranchia	see Anthobranchia
Nudibranchia	5.40–15.92	0.0044–1.73	0.06–3.57
Pleurobranchoidea	5.90–6.29	1.79	0.34–0.63
Tylodinoidea	1.54	—	1.73
Anaspidea	0.86–2.26	0.47–1.79	0.25–1.98
Sacoglossa	2.27–2.47	0.25–1.14	0.90–0.99
Cephalaspidea	0.86–1.32	0.91–2.1 0	034–0.41
Acteon tornatilis	—	**3.28**	—
"Basommatophora"	0.82–1.75	0.015–2.24	—

Note: Species with deviating values are shown in bold font.

most often for parsimony analyses: character state optimization, ACCTRAN; gaps coded as 5th nucleotide; heuristic search, starting trees stepwise-addition (random), TBR branch-swapping. The same settings were used for the bootstrap analyses with 100 (18S, 16S, COI genes) or 1,000 pseudoreplicates (combined analysis). Extreme computation times impeded the use of a larger number of replications. For distance analyses, neighbor-joining (Saitou and Nei 1987) was used with the following evolutionary models tested for congruence: uncorrected *p*-distances, Jukes Cantor, Kimura 2-parameter, HKY 85, logdet transformation; heuristic search: starting-trees obtained via neighbor-joining. We used logdet transformations (Lockhart et al. 1994) for distance analyses to avoid groupings based on nucleotide composition and asymmetric rates of nucleotide change. Topologies based on logdet transformation differed markedly from results obtained with other models combined with neighbor-joining (Saitou and Nei 1987) as implemented in PAUP* 4.0. Bootstrap values of distance analyses with 100 replicates were obtained for the logdet transformation model. Length of alternative topologies were checked with the tree options implemented in MacClade v3. (Maddison and Maddison 1992).

It was beyond the scope of this study to analyze secondary structures, and it was impossible to perform maximum likelihood analyses because of computational constraints associated with the analysis of large datasets. A combined analysis was performed with those taxa for which data were available for the 18S and 16S rRNA genes. This guaranteed that no assumptions were made by the algorithms for the missing data in those taxa where not all genes were available. For the COI data, there was little taxonomic overlap with the other gene sequences, so it was not included in the combined analysis.

NOMENCLATURE

Because paraphyly is demonstrated convincingly for Basommatophora (Tillier et al. 1996) and Arminoidea (Wägele and Willan 2000), we show these names in quotation marks. Cephalaspidea is used for the monophyletic group identified by Mikkelsen (1996) as Cephalaspidea *sensu stricto* (s.str.). Thus, the species belonging to the family Acteonidae (*Acteon* and *Pupa*) are not considered as cephalaspid species. The Acteonidae is not assigned to a superordinated taxon.

18S rRNA GENE TREE

The alignment of 81 sequences (80 species) resulted in 2,668 positions: 547 characters are constant, 512 positions are variable and parsimony-uninformative, and 1,609 characters are parsimony-informative.

Relative Rate Test

When estimating relative distances from all taxa to the basommatophoran *Siphonaria algesirae,* with *Littorina littorea* as outgroup, branch-length differences, as measured by Z-scores, range from lowest in "Basommatophora"

(0.82) to 15.92 in members of the Aeolidoidea (Table 8.2). Highest Z-scores, and therefore highest distances, showed the aeolidoidean species *Facelina bostoniensis, Phydiana lynceus,* and *Godiva banyulensis* with Z-scores of 15.84 to 15.92. The effect is a misplacement of these taxa, belonging to the family Facelinidae, in the reconstructed tree (see below and Figure 8.1). These Z-scores are considerably higher than those of all other Aeolidoidea (highest score 12.81). In general, values for the Cladobranchia were twice as high as for the Anthobranchia. Z-scores similar to Anthobranchia also occur in the Pleurobranchoidea. All other investigated taxa have considerably lower Z-scores (Table 8.2).

Substitution Analysis

Plotting the pairwise transitions and transversions against the uncorrected pairwise genetic distances revealed no evidence of saturation for either transitions or transversions (not shown).

Base Frequencies

Observed base frequencies within the different Opisthobranchia clades are reported in Table 8.3. Base composition is similar throughout the different taxa, although the Nudibranchia generally tended to have a lower AT content than the other opisthobranch taxa. Considerably different base frequencies were observed in two facelinid species (*Phyllodesmium briareum* and *Facelina auriculata*), which have a much higher AT content. Base composition of these sequences differs significantly from the average (χ^2 test, $p < 0.001$). The phylogenetic effect of this base pair compositional bias is an erroneous placement of these taxa outside the Euthyneura (Figure 8.1). The lowest AT and highest CG contents were measured for the dendronotoid *Bornella stellifer*. Base composition of this sequence differs significantly from the average (χ^2 test, $p < 0.05$).

Phylogenetic Analysis

A 50% majority-rule consensus tree of 48 most parsimonious trees (9,405 steps, consistency index [CI] = 0.469, retention index [RI] = 0.789, homoplasy index [HI] = 0.531) from a maximum parsimony analysis of the complete 18S dataset is shown in Figure 8.1. The following monophyletic groups were observed: Cladobranchia (bootstrap value 99), Doridoidea, Anthobranchia (not recovered in bootstrap analysis), Nudibranchia (62), Nudipleura (58), Anaspidea (99), Sacoglossa (100), Cephalaspidea s.str. (100), and "Basommatophora" (92), with *Siphonaria algesirae* being the most basal species within the "basommatophoran" group. The Opisthobranchia are not monophyletic, since the "ba-

Table 8.3

Base frequencies for 18S rDNA, 16S rDNA, and COI genes, summarizing major taxa

	Nucleotides			
	A	C	G	T
18S rDNA				
Mean value	0.21423	0.26770	0.30665	0.21142
Nudibranchia (*P. briareum, F. auricularia* and *B. stellifer* excluded)	0.18406–0.21938	0.26355–0.30773	0.30074–0.34293	0.16658–0.22190
Phyllodesmium briareum	**0.25955**	**0.19888**	**0.26461**	**0.27697**
Facelina auriculata	**0.25774**	**0.20540**	**0.26730**	**0.26956**
Bornella stellifer	**0.16863**	**0.29893**	**0.34798**	**0.18447**
Pleurobranchoidea	0.21484–0.21545	0.26576–2.7159	0.29680–0.29463	0.22260–0.21833
Tylodinoidea (*Umbraculum*)	0.23739	0.23958	0.28015	0.24287
Anaspidea	0.23600–0.23696	0.23640–0.24616	0.27906–0.28618	0.24232–0.24584
Sacoglossa	0.23642–0.23862	0.23587–0.24191	0.28195–0.28415	0.23752–0.24355
Cephalaspidea	0.23998–0.24466	0.23426–0.23828	0.27750–0.28081	0.23937–0.24357
"Basommatophora"	0.23885–0.24125	0.23417–0.23958	0.27992–0.28344	0.24205–0.24661
Littorina littorea (outgroup)	0.24485	0.23273	0.27445	0.24797
16S rDNA				
Mean value	0.32389	0.14034	0.21639	0.31938
Nudibranchia	0.28883–0.36508	0.11622–0.16442	0.19211–0.24590	0.27913–0.34865
Pleurobranchoidea	0.33528	0.12011	0.22565	0.31896
Anaspidea	0.29775–0.32773	0.13202–0.16338	0.21289–0.23513	0.29745–0.33521

Table 8.3 continued

	Nucleotides			
	A	C	G	T
Sacoglossa	0.28771–0.30137	0.14795–0.16760	0.21096–0.23880	0.30423–0.33973
Cephalaspidea (*A. tornatilis* excluded)	0.29942–0.32362	0.12968–0.16422	0.21387–0.24927	0.28152–0.31988
Acteon tornatilis	0.29984	**0.11205**	0.20771	**0.38040**
"Basommatophora" (*C. limacina* excluded)	0.32558–0.38746	0.10894–0.12209	0.16024–0.18421	0.32764–0.37243
Clione limacina	0.34560	**0.14319**	0.19020	**0.32102**
Tectarius viviparus (outgroup)	0.35252	0.18465	0.18945	0.27338
COI				
Mean value	0.23437	0.16224	0.19785	0.40554
Nudibranchia	0.19629–0.26599	0.12860–0.20981	0.17677–0.24704	0.35533–0.44162
Pleurobranchoidea	0.23559–0.23689	0.13559–0.15771	0.20858–0.21017	0.39682–0.41864
Tylodinoidea (Tylodina)	0.22843	0.17259	0.20643	0.39255
Anaspidea	0.20474–0.24365	0.15567–0.19120	0.17597–0.22504	0.37902–0.43824
Sacoglossa	0.23898–0.24704	0.16610–0.18443	0.19459–0.20000	0.37394–0.39492
Cephalaspidea	0.21320–0.23858	0.17090–0.19145	0.18274–0.22335	0.37563–0.39594
Cepaea nemoralis (outgroup)	0.22735	0.19145	0.20513	0.37607

Note: Species with deviating values and the deviating values are shown in bold font.

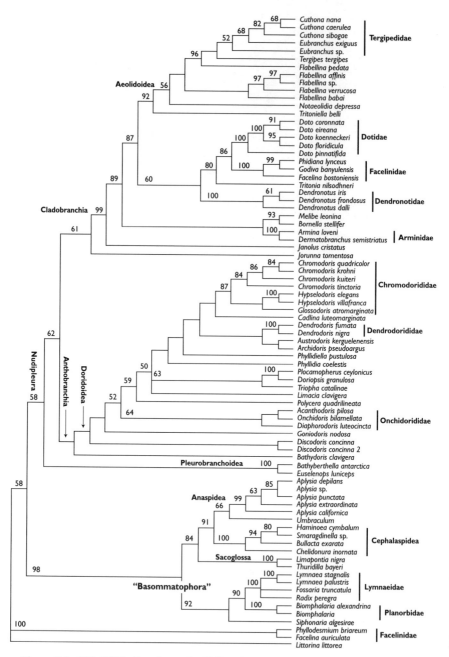

Figure 8.1. 18S rDNA: Topology of a 50% majority-rule consensus tree from a maximum parsimony analysis. Numbers on branches are bootstrap values (above 50%); branches without numbers were not recovered in the maximum parsimony bootstrap analysis, or the value is lower than 50%. Family names are indicated on the right, especially when arranged as monophyletic entities. Higher categories are usually indicated above (or with an arrow toward) the stem line.

sommatophoran" species are grouped with all opisthobranch taxa except for the Nudipleura (Nudibranchia and Pleurobranchoidea). The bootstrap value is rather high (98) for this grouping, but the placement as sister taxon to the Nudipleura (58) has weak support (58). The tylodinoid species *Umbraculum mediterraneum* never groups with the Pleurobranchoidea, but its position varies as sister taxon to the Cephalaspidea or to the Anaspidea.

Within the major taxa, several subordinate monophyletic groups can also be recognized. Within the Cladobranchia, the Aeolidoidea are monophyletic, but the species of the aeolid family Facelinidae (*Phidiana lynceus, Godiva banyulensis, Facelina bostoniensis, Facelina auriculata,* and *Phyllodesmium briareum*) are not included, possibly because of long-branch problems and base composition. The Dendronotoidea is rendered paraphyletic, because members of the Facelinidae are grouped within the Dendronotoidea. The "Arminoidea," with *Armina, Dermatobranchus,* and *Janolus,* is paraphyletic. On the family level, the following monophyletic taxa are recognized: Tergipedidae, Dotidae (100), Dendronotidae (100), Arminidae (99), Chromodorididae without *Cadlina* (87), Haminoeidae (94), and Lymnaeidae (100). We observed some notable incongruencies to traditional systematics. There is no clear resolution concerning the position of the genus *Cadlina,* which was traditionally considered to be a "chromodorid." The Triophidae is not monophyletic, because the dorid species *Doriopsis granulosa* groups together with *Triopha catalinae* and *Plocamopherus ceylonicus.* The Polyceridae (*Polycera, Limacia*), as well as the Phyllididae (*Phyllidia, Phyllidiella*), are not monophyletic. The dorid species *Jorunna tomentosa* appears as the sister taxon to the Cladobranchia. *Tritoniella belli* is the sister taxon to the Aeolidoidea; therefore the Tritoniidae (*Tritonia* and *Tritoniella*) is paraphyletic. The Facelinidae are paraphyletic, with *Phyllodesmium briareum* and *Facelina auriculata* being the sister taxon to all investigated euthyneuran sequences, possibly an artifact caused by a bias in base frequencies (high AT content).

The neighbor-joining tree based on distance criteria (Figure 8.2) is similar to the topology obtained in the maximum parsimony analysis. The Nudipleura (bootstrap value 95) is the sister taxon to all other opisthobranch and "basommatophoran" species, a clade supported by a bootstrap value of 100. *Umbraculum mediterraneum* (Tylodinoidea) is the sister taxon to the monophyletic Cephalaspidea. There is no support in the bootstrap analysis for the monophyly of "Basommatophora." The sequences of *Phyllodesmium briareum* and *Facelina auriculata* group within the Cladobranchia (bootstrap value 100). These sequences were placed outside the Opisthobranchia in the parsimony analysis. Analyzing the dataset with the logdet transformation, the Facelinidae is monophyletic (54), but branches off as the most basal group within the Clado-

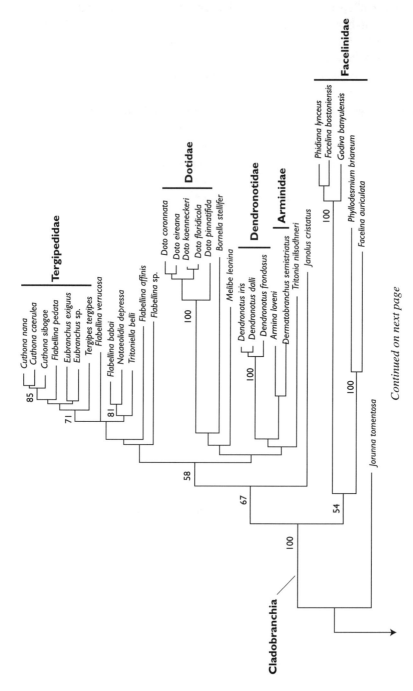

Figure 8.2. 18S rDNA: Distance analysis: Neighbor joining, using logdet transformation as the evolutionary model. Numbers on branches are bootstrap values (above 50%); branches without numbers were not recovered in the maximum parsimony bootstrap analysis, or the value is lower than 50%.

Continued on next page

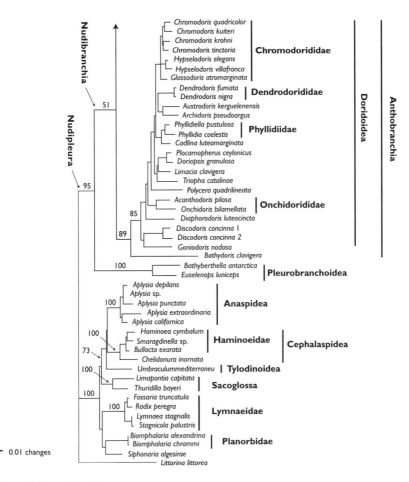

Figure 8.2 continued

branchia (Figure 8.2). Therefore, the Aeolidoidea is still paraphyletic. This is possibly the consequence of consideration of asymmetric rates of nucleotide substitutions that led to a high AT content in some Facelinidae. Using uncorrected distances or applying different evolutionary models, Pleurobranchoidea are no longer sister taxon to Nudibranchia, but to Anthobranchia—thus rendering the Nudibranchia paraphyletic (not shown). This is not the case when applying the logdet function, in which the Nudibranchia are still monophyletic (51), with the Pleurobranchoidea as the sister taxon (Nudipleura: bootstrap value, 95). Long distances are evident for the five facelinid species, whereas distances within the Doridoidea and all other investigated species (excluded Cladobranchia) are very short. Except for the position of the Pleurobranchoidea and the relationship of the facelinid species in the logdet analysis, there is a high congruency of results when applying different evolutionary models.

16S rRNA GENE TREES

The alignment of 123 sequences (108 species) resulted in 516 positions: 78 characters are constant, 92 are variable and parsimony-uninformative, and 346 parsimony-informative.

Relative Rate Test

The Z-score values range from 0.0044 (*Chromodoris luteorosea*) to 2.24 for the "basommatophoran" species *Biomphalaria schrammi*. All Z-scores are evenly distributed, and none are considerably higher in any particular group (Table 8.2).

Substitution Analysis

When comparing sequences within the Opisthobranchia, for distances less than 0.3, transition and transversion rates are similar and follow a linear pattern. However, at distances of more than 0.3, the 16S rRNA gene has more transversions than transitions and the absolute number of transitions does not increase in proportion to the p-distances, therefore a saturation is evident.

Base Frequencies

Base frequencies were regular for all major groups (Table 8.3). CG content was low within the "Basommatophora," and AT content was higher than the mean values. *Acteon tornatilis* showed a very low C content, and the T value was the highest observed.

Phylogenetic Analysis

Figure 8.2 shows the phylogenetic hypotheses obtained from the parsimony analysis of the 16S rDNA dataset (596 shortest trees with 4,102 steps; CI = 0.270, RI = 0.725, HI = 0.730). The following major taxa are recognized with bootstrap values higher than 50: Cladobranchia (95), Anthobranchia (position of Bathydoridoidea not resolved in the bootstrap analysis), Doridoidea (62), Nudibranchia (67), Nudipleura (59), Sacoglossa (100), Anaspidea (99), Cephalaspidea (56), Opisthobranchia (95), and "Basommatophora" (93). *Acteon* and *Clione* are the most basal taxa (united and supported by a bootstrap value of

69), followed by the "Basommatophora." The Sacoglossa branch off next, leaving the Nudipleura as sister taxon to the united Anaspidea and Cephalaspidea s.str. In the bootstrap analysis, the Sacoglossa is sister taxon to the Nudipleura (both groups united with a value of 59).

Within the major taxa, the following groups and families were monophyletic: *Chromodoris/Hypselodoris* (62), Goniodorididae, Onchidorididae, Dendrodorididae (81), Arminidae (87), Dotidae (100), Philinoidea (96), Haminoeidae (90), Lymnaeidae (87), Planorbidoidea (87), and Planorbidae (91). Within the Doridoidea, the Dendrodorididae are the most basal family. The Polyceridae (82) is clearly not monophyletic, since *Triopha catalinae,* usually assigned to a separate family (Triophidae), is an offspring of this clade. *Cadlina* and *Durvilledoris pusilla* do not group with the other chromodorid species, rendering the Chromodorididae paraphyletic. The Phyllidiidae is not monophyletic. Within the Cladobranchia, *Doto* (100) is the sister taxon to all other cladobranch taxa, but the latter grouping is supported only by a low bootstrap value (51). The Aeolidoidea, as well as the Dendronotoidea, appear as polyphyletic groups. Within the Anaspidea, the genus *Aplysia* is the sister taxon to all other anaspidean genera. Within the Cephalaspidea, *Diaphana* is the most basal genus and the family Haminoeidae is sister taxon to the Philinoidea.

Resolution at the genus and species level is high (Figure 8.3). When including several different sequences of the same species, the species group generally together with two exceptions: one *G. nodosa*2 groups with *Okenia aspersa,* whereas the other *G. nodosa*1 groups with *Goniodoris castanea.* The two included sequences of *Hypselodoris villafranca* also do not group together; one is more closely related to *H. elegans.*

Results of the distance analyses are similar for the major groups (Figure 8.4, A and B; Kimura two-parameter evolutionary model). The most basal species are *Acteon tornatilis* and *Clione limacina.* The Opisthobranchia (*Dendrodoris* excluded) are monophyletic, but not supported in the bootstrap analysis. The monophyletic Cephalaspidea is the most basal group within the Opisthobranchia. The Sacoglossa is the sister taxon to the Anaspidea. All of the aforementioned nodes, however, are not supported in the bootstrap analysis (logdet transformation). *Bathydoris clavigera* becomes the sister taxon to the Cladobranchia (93), a relationship that is supported by a bootstrap value of 69. In contrast to maximum parsimony analyses, the dorid family Dendrodorididae is paraphyletic, with *Dendrodoris nigra* the sister taxon to all opisthobranchiate species, whereas *Doriopsilla areolata* represents the most basal species of the Doridoidea. No different results were obtained by applying different evolutionary models, except for the logdet transformation. Contrary to the other models, the logdet correction indicates that the Chromodorididae are monophyletic, including *Durvilledoris pusilla* at the base of the clade, but the genus *Cadlina*

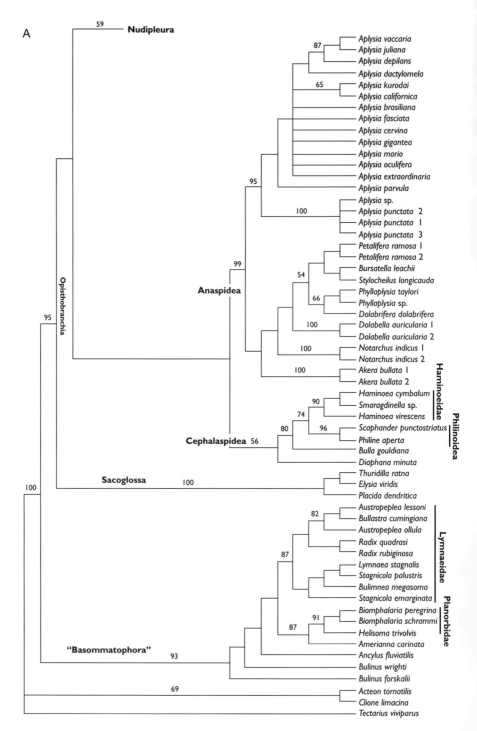

A

59 —— **Nudipleura**

87 ┌ *Aplysia vaccaria*
   ├ *Aplysia juliana*
   ├ *Aplysia depilans*
   └ *Aplysia dactylomela*
65 ┌ *Aplysia kurodai*
   └ *Aplysia californica*
     Aplysia brasiliana
     Aplysia fasciata
     Aplysia cervina
     Aplysia gigantea
     Aplysia morio
     Aplysia oculifera
     Aplysia extraordinaria
     Aplysia parvula
     Aplysia sp.
100 ┌ *Aplysia punctata* 2
    ├ *Aplysia punctata* 1
    └ *Aplysia punctata* 3

95

Anaspidea

99

54 ┌ *Petalifera ramosa* 1
   ├ *Petalifera ramosa* 2
   ├ *Bursatella leachii*
   └ *Stylocheilus longicauda*
66 ┌ *Phyllaplysia taylori*
   ├ *Phyllaplysia* sp.
   └ *Dolabrifera dolabrifera*
100 ┌ *Dolabella auricularia* 1
    └ *Dolabella auricularia* 2
100 ┌ *Notarchus indicus* 1
    └ *Notarchus indicus* 2
100 ┌ *Akera bullata* 1
    └ *Akera bullata* 2

Haminoeidae

90 ┌ *Haminoea cymbalum*
   ├ *Smaragdinella* sp.
74 └ *Haminoea virescens*
80   96 ┌ *Scaphander punctostriatus*
         ├ *Philine aperta*
Cephalaspidea 56   └ *Bulla gouldiana*
            Diaphana minuta

Philinoidea

Sacoglossa 100 ┌ *Thuridilla ratna*
                   ├ *Elysia viridis*
                   └ *Placida dendritica*

95

82 ┌ *Austropeplea lessoni*
   ├ *Bullastra cumingiana*
   └ *Austropeplea ollula*
     ┌ *Radix quadrasi*
87   └ *Radix rubiginosa*
       ┌ *Lymnaea stagnalis*
       ├ *Stagnicola palustris*
       ├ *Bulimnea megasoma*
       └ *Stagnicola emarginata*

Lymnaeidae

91 ┌ *Biomphalaria peregrina*
   ├ *Biomphalaria schrammi*
87 ├ *Helisoma trivolvis*
   └ *Amerianna carinata*

Planorbidae

"Basommatophora"
93   *Ancylus fluviatilis*
      Bulinus wrighti
      Bulinus forskalii

100

69 ┌ *Acteon tornatilis*
   ├ *Clione limacina*
   └ *Tectarius viviparus*

Opisthobranchia

Figure 8.3. 16S rDNA: Topology of a 50% majority-rule consensus tree from a maximum parsimony analysis. Numbers on branches are bootstrap values (above 50%);
branches without numbers were either not recovered in the analysis or were lower

B

than 50%. (A) Basal divergences of the maximum parsimony tree; the Nudipleura are shown in B.

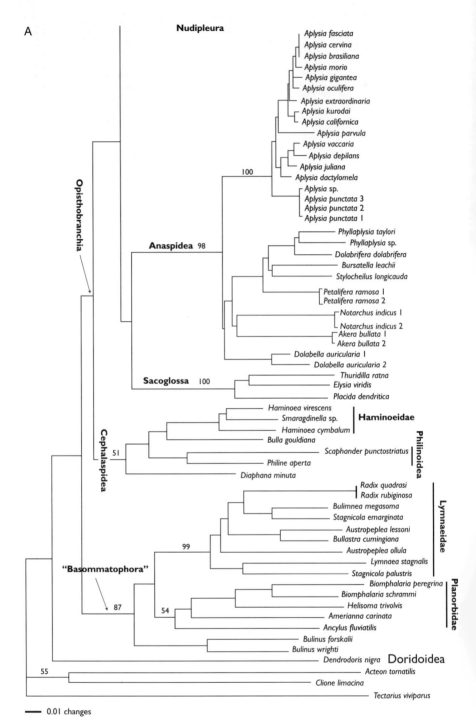

Figure 8.4. 16SrDNA: Distance analysis: Neighbor joining, Kimura two-parameter evolutionary model. Numbers on branches are bootstrap values (above 50%);

B

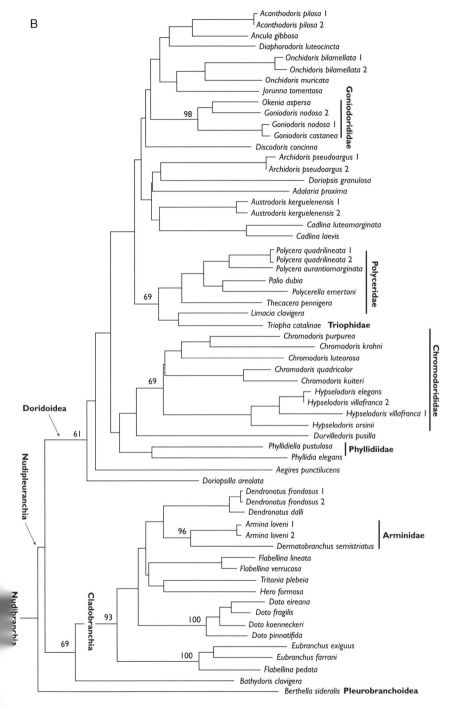

branches without numbers were not recovered in the maximum parsimony bootstrap analysis, or the value is lower than 50%.

still does not group within the Chromodorididae. The Phyllidiidae become monophyletic (no support in bootstrap analysis), but the Cephalaspidean species group with the "basommatophoran" species, rendering the Opisthobranchia paraphyletic (but this is without bootstrap support).

COI GENE TREE

The alignment of the DNA sequences of 77 taxa resulted in 605 sites, with 225 constant characters, and 338 parsimony-informative and 42 uninformative variable characters.

Relative Rate Test

Z-scores (Table 8.2) are evenly distributed among the major taxa and range from 0.034 (*Smaragdinella* sp.) to 3.37 (*Dendrodoris fumata*). The two *Dendrodoris* species showed the highest Z-scores, followed by *Chromodoris luteorosea* with 2.78. Z-scores within genera were highly variable (e.g., Z-scores of the three *Chromodoris* species were 0.82, 1.37, and 2.78, and of the two *Cadlina* species, 0.24 and 1.03).

Substitution Analysis

We observed a higher number of transitions when the absolute number of transversions and transitions were plotted against uncorrected distances (not shown). However, the data points were scattered, with two separate groups of points at greater distances. These two "clouds" represent pairwise comparisons of the two *Dendrodoris* species with other investigated species. Both transitions and transversions seem to indicate saturation to a lower degree.

Base Frequencies

Base frequencies (Table 8.3) were roughly similar throughout the major groups, but showed a high range of variation, especially within the Nudibranchia. No taxa deviated substantially from the observed range.

Phylogenetic Analysis

Parsimony analysis of the COI nucleotide sequences resulted in four short trees (4,867 steps, CI = 0.149, RI = 0.379, HI = 0.851). Confirmation of former results in deeper nodes is low, sometimes even on genus and species levels (tree not shown). A distance analysis based on logdet transformation is shown in Figure

8.5. Here some major groups are recognizable, although one or the other species are placed in groups incongruent with traditionally recognized taxa. For example, *Pupa strigosa* is certainly erroneously grouped with two aeolids. Because of this grouping, the Cladobranchia and again the Nudibranchia are not monophyletic. The pleurobranchoid species *Berthella sideralis* and *Berthellina citrina* group separately with members of the Doridoidea, rendering the Pleurobranchoidea polyphyletic and the Nudibranchia paraphyletic. As in the 16S rDNA topology, *Goniodoris nodosa* (sequence of Wollscheid and Wägele 1999) is more closely related with *Okenia aspersa* than with *Goniodoris castanea*. The sister-taxon relationship of *G. nodosa* with *O. aspersa* is supported by a bootstrap value of 98, whereas the sister-taxon relationship (*G. castanea* (*G. nodosa*/*O. aspersa*)) is supported by a bootstrap value of only 58 (maximum parsimony) and without support in the distance analyses. These results indicate a possible mistake in species identification. *Dendrodoris* is the most basal part of the opisthobranch clade.

In analyzing the amino acid sequences, we found almost no congruence with earlier results in the deeper nodes. Due to the lack of phylogenetic signal, it was not possible to run even a heuristic search for a maximum parsimony analysis because the tree buffer memory was full. In the distance analyses (Figure 8.6), several taxa are recognized, although the relationship among them is considerably different from all other analyses. The following groups are recognized: Cephalaspidea (2 species of Haminoeidae and 1 from the Bullidae), Sacoglossa (2 species of the family Placobranchidae), Anaspidea, Pleurobranchoidea, and Cladobranchia. On a family level, confirmation is evident for several lineages (e.g., Goniodorididae, Flabellinidae, Tergipedidae, Haminoeidae), but sometimes one or the other genus or species is not included, or a taxon of another family is included. For example, the Chromodorididae are recognized, with *Durvilledoris* included, but the onchidoridid genus *Diaphorodoris* is grouped within this family. The Polyceridae are monophyletic, but *Limacia* is not included. All genera are monophyletic, except *Polycera, Doto,* and *Goniodoris* (the same situation as described for the COI nucleotide analysis). The following higher groups are not recognized: Nudipleura, Nudibranchia, Doridoidea, Dendronotoidea, and Aeolidoidea. The most basal branch is the genus *Dendrodoris* with a very long branch. The next branch is the acteonid species *Pupa strigosa,* which represents the sister taxon to the rest of the investigated species.

COMBINED 18S AND 16S rDNA ANALYSIS

A combined 18S rDNA and 16S rDNA dataset was created using 39 species, resulting in 2,921 positions, with 1,608 constant characters, 970 parsimony-informative characters, and 343 parsimony-uninformative characters.

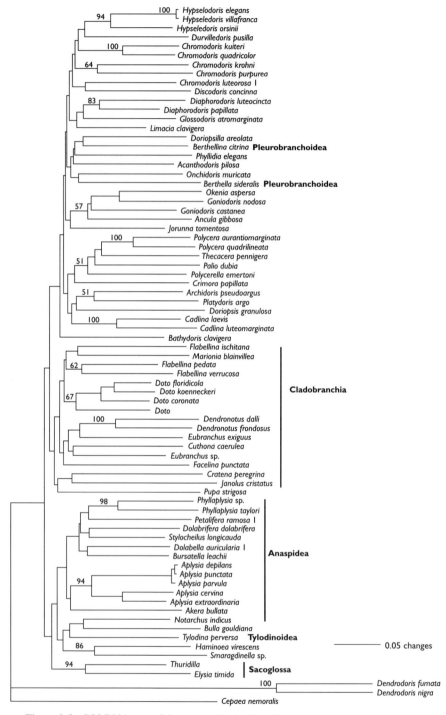

Figure 8.5. COI DNA gene: Distance analysis: Neighbor joining, using logdet transformation as evolutionary model. The low support for clades is visualized with the very short internal branches, the topology is not reliable. Numbers on branches are bootstrap values (above 50%); branches without numbers were not recovered in the maximum parsimony bootstrap analysis, or the value is lower than 50%.

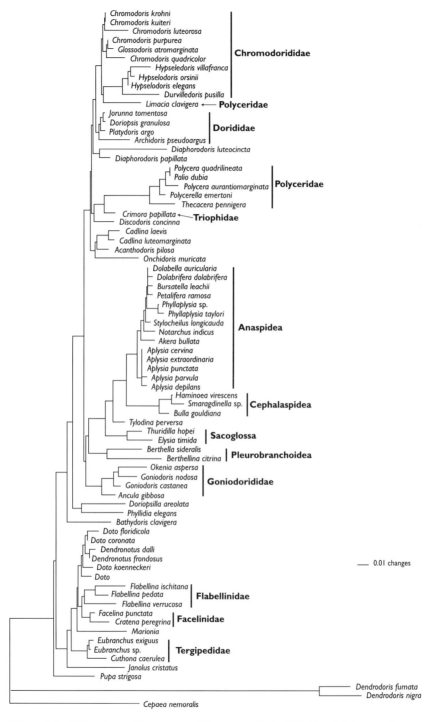

Figure 8.6. COI amino acid sequences: Distance analysis: Neighbor-joining topology. Resolution is better than in Figure 8.5, but larger taxa are not recovered (e.g., Doridoidea, Nudibranchia, Nudipleura).

Due to the smaller taxon sample with fewer taxa of higher order, the number of constant characters increased considerably to the detriment of informative positions (compare with results of 18S rDNA analyses). Figure 8.7 shows the single most parsimonious tree obtained from the maximum parsimony analysis (3,216 steps, CI = 0.581, RI = 0.761, HI = 0.419). The monophyletic Cephalaspidea (two species: 100) is the sister taxon to the monophyletic Anaspidea (represented here only by the genus *Aplysia:* 100), united with a bootstrap value of 95. The Nudibranchia is monophyletic (100). *Bathydoris clavigera* is recognized as sister taxon to the Doridoidea (97), therefore the Anthobranchia (85) are monophyletic. Within the Doridoidea, the Chromodorididae (94, *Cadlina* excluded) are supported. The dorid species *Jorunna tomentosa* is included in the Doridoidea and is supported as sister taxon to *Discodoris concinna* with a bootstrap value of 89. Contrary to the shortest tree of the maximum parsimony analysis, the consensus topology of the bootstrap analysis shows a polytomy concerning the relationship of families within the Doridoidea, but recovered the following groups, although only supported with low values: Polyceridae (66), Onchidorididae (63), and Porostomata (51), the latter uniting *Dendrodoris nigra* and *Phyllidiella pustulosa,* an old but often discussed and denied group (Valdes and Gosliner 1999). The Cladobranchia (100) is the sister taxon to the Anthobranchia. Within the Cladobranchia, the Aeolidoidea (98), the family Arminidae (100), and the genera *Dendronotus* (100) and *Doto* (100) are monophyletic. When comparing the results of maximum parsimony and distance analyses, the branching pattern is the same under all different evolutionary models, with the exception of *Jorunna tomentosa* as the sister taxon to the Cladobranchia, and *Bathydoris clavigera* as the most basal group within the Nudibranchia in all distance analyses.

PHYLOGENETIC IMPLICATIONS

Wollscheid-Lengeling et al. (2001) used the three different genes to reconstruct the phylogeny of the nudibranch taxa. They used 54, 38, and 45 species for the 18S rDNA, 16S rDNA, and COI genes, respectively. The final tally in this study is 73 (18S), 106 (16S), and 76 (COI) species. Although we included many more opisthobranch sequences, as well as additional "basommatophoran" sequences, the COI data are obviously not appropriate for reconstructing deeper nodes. Based on the substitution rates, the 18S rRNA gene seems to be appropriate for analyzing the deeper nodes of the Opisthobranchia, since no saturation effects are recognizable. More transversions than transitions in 16S rDNA sequences were also observed by Lydeard et al. (2000) in analyzing several gas-

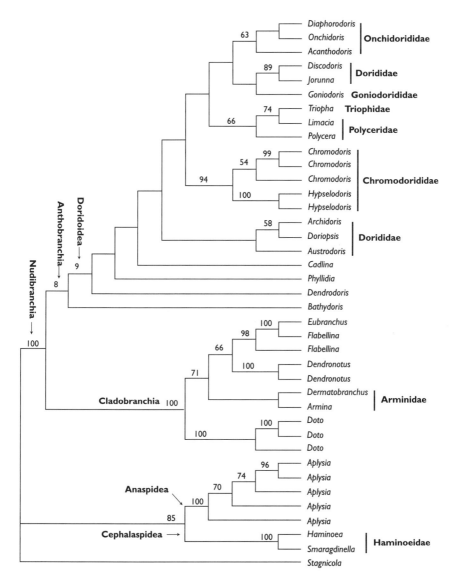

Figure 8.7. Combined analysis of 18SrDNA and 16SrDNA for taxa for which both sequences are available. Only one shortest tree was found in a maximum parsimony analysis. Numbers on branches are bootstrap values (higher than 50%).

tropods and other mollusks. These authors explained this as an effect of biased sampling and a possible site saturation of transitions. This seems to be the case for the deeper nodes, but not within the Opisthobranchia.

Many studies dealing with the phylogeny of gastropods based on different genes have been published (e.g., 28S rDNA-Tillier et al. 1996; histone H3-Colgan et al. 2000; Chapter 6, this volume), but usually only single species representing diverse lineages are included and compared (e.g., one species of the Anaspidea, one of the Cephalaspidea, one from the Onchidiidae; Colgan et al. 2000). In the next section, we will discuss only those analyses that included several members of each taxon.

OPISTHOBRANCHIA

According to Mikkelsen (1996), apomorphies for the Opisthobranchia include operculum in adults absent and parapodia present. Salvini-Plawen (1991) presented nerve characters as synapomorphies for the Opisthobranchia: head shield with bifurcated tentacle (= clypeo-capitis) nerves, Hancock's sense organs with external branch of labiotentacularis nerve. Synapomorphies for the sister taxon to the Opisthobranchia, the Pulmonata/Gymnomorpha clade, include a procerebrum with cerebral glands and dorsal bodies, anterior shift of female genital opening outside the mantle cavity, and restriction of the opening of the mantle cavity (Salvini-Plawen 1991). Previous molecular analyses that included several species of the representative taxa indicated paraphyly of the Opisthobranchia (Winnepenninckx et al. 1998; Thollesson 1999a; Wollscheid and Wägele 1999; Dayrat et al. 2001). Clarification of the monophyly/paraphyly of Opisthobranchia was not possible with the data used in this study (Chapter 7, this volume). Although we included "basommatophoran" species in this analysis (not in the COI gene analysis owing to lack of data), we analyzed additional sequences of the Stylommatophora, as well as of the basal Heterobranchia (e.g., Pyramidelloidea and Architectonicoidea), to fully assess opisthobranch monophyly. According to the preliminary results for the 18S rRNA gene, the Opisthobranchia are paraphyletic; according to the 16S rRNA gene, they are monophyletic. Analyzing the characters (positions) of the 18S rDNA, which are recognized by the computer program PAUP* in a heuristic search (maximum parsimony) as synapomorphic for the clade ("Basommatophora" / Cephalaspidea / Tylodinoidea / Sacoglossa / Anaspidea), the number is rather low (37 apomorphic positions of 1,609 parsimony-informative characters = 2%). In 3 of these 37 positions, the clade has the same base as the outgroup *Littorina littorea*. In 11 of the 37 characters, nearly all members of the Nudipleura possess the same nucleotide as the outgroup; therefore, the signal is not strong. Monophyly of Opisthobranchia is only 19 steps longer than the most parsimonious tree (total length = 9,405), with "Basommatophora" being the sister group. The Opisthobranchia is supported by 48 apomorphic characters of 349 parsimony-informative characters (14%) for the 16S rRNA sequence data, but the signal-to-noise ratio is low.

Mikkelsen (1996) recognized the basal position of *Acteon* in her unique study on the phylogeny of lower Opisthobranchia, based on morphological and histological characters. She excluded *Acteon* not only from the Cephalaspidea, but also from the Opisthobranchia. This exclusion is supported by the analyses of Thollesson (1999a) and by our current 16S rDNA data. Nevertheless, it should be mentioned that the Z-score of *Acteon* (3.28) is the highest measured in the 16S rDNA analysis. This indicates a long-branch sequence that may be attracted to other unrelated taxa because of multiple convergent substitutions (Felsenstein 1978; Hendy and Penny 1989; Wägele 2000). Additionally, this species has a much higher T and a considerably lower C nucleotide content. Nevertheless, results of the COI amino acid sequence of *Pupa strigosa,* a species also belonging to the Acteonidae, indicate the basal position of this family. The grouping of *Clione limacina* with *Acteon tornatilis* contradicts the assumption of Dayrat et al. (2001) and Dayrat and Tillier (Chapter 7, this volume) that Gymnosomata and Thecosomata are directly related to the Anaspidea. It cannot be denied that the outgroup (*Tectarius viviparus*, *A. tornatilis,* and *Clione limacina*) share some indels, but these may represent a "plesiomorphic trap" (see Wägele 2000). Since few species are included in these analyses, the results must be considered as preliminary, and giving new names to these "new clades" should be avoided at present.

ANASPIDEA

The inclusion of the Akeridae into the clade Anaspidea was suggested by Thompson and Seaward (1989) on the basis of unequal cleavage of the ovum, a character that is rarely seen in opisthobranchs. Mikkelsen (1996) listed two synapomorphic characters for this clade (i.e., filter chamber present and stomach caecum present). Thollesson (1999a) included *Aplysia* and *Akera* in a 16S rDNA study on euthyneurans; the results supported the monophyly of the Anaspidea. Medina and Walsh (2000) analyzed the phylogeny of the Anaspidea by using partial sequences of 16S rDNA, 12S rDNA, and COI. They considered the Anaspidea to be monophyletic, with *Akera bullata* being the most basal species. Our study revealed a monophyletic *Aplysia,* which was usually the sister taxon to all other anaspidean genera. In our analyses, the genus *Akera* is always grouped within the Anaspidea, but usually not as the most basal branch. Within the genus *Aplysia,* the branching pattern differed from that of Medina et al. (2001), whose reconstruction of the phylogeny of the genus was based on a combined analysis of partial 16S rDNA and 12S rDNA sequences, using *Akera* as outgroup.

SACOGLOSSA

Monophyly of the Sacoglossa was assumed by Jensen (1996), who mentioned the ascus as one possible synapomorphy, whereas Mikkelsen (1996) provided nine additional morphological apomorphies based on a cladistic analysis. Sacoglossa monophyly is supported in all analyses in our study, but the number of investigated species is still insufficient. Actually, only members of the Placobranchoidea and Limapontioidea were included, whereas the third-largest sacoglossan group, the Oxynoacea, was not considered. The position of the Sacoglossa varies considerably depending on the gene and method applied. Nevertheless, a grouping of the Sacoglossa as sister taxon to the "Basommatophora" (results obtained by Thollesson 1999a) was not found.

CEPHALASPIDEA S.STR.

Mikkelsen (1996) listed characters of the mantle cavity and digestive system (i.e., marginal flexure of the ciliated strips, the number of gizzard plates is changed from many to three), which support monophyly of the clade Cephalaspidea s.str. In Mikkelsen's analysis, this group comprises the families Bullidae, Haminoeidae, Retusidae, and Cylichnidae. Monophyly of the Cephalaspidea s.str. is supported in our study by high bootstrap values in the 18S rDNA analyses, and low values in the 16S rDNA topology. Thollesson (1999a) and Dayrat et al. (2001) presented similar results concerning the Cephalaspidea based on the 16S rRNA and 28S rRNA gene, respectively. It should be emphasized, however, that the sample of sequences is not representative of the entire diversity of the group.

TYLODINOIDEA

In the past, the Tylodinoidea (= Umbraculoidea) have often been united with the Pleurobranchoidea under the name "Notaspidea." Schmekel (1985) and

Salvini-Plawen (1991) considered the Notaspidea to be paraphyletic. The latter author grouped the Tylodinoidea with the Anaspidea, Sacoglossa, Cephalaspidea (*sensu lato,* s.l., in the former sense), Thecosomata, and Gymnosomata on the basis of the apomorphy: possession of an "anterior gizzard." But an anterior gizzard is also present in *Bathydoris* spp. (Wägele 1989). Salvini-Plawen and Steiner (1996) placed the Tylodinoidea as sister taxon to the Nudipleura and listed the following synapomophies: reduction of head-shield and clypeo-capitis nerves, and reduction of pallial cavity, but plicatidium retained. Paraphyly of the Notaspidea was also strengthened by Wägele and Willan (2000), who united the Nudibranchia with the Pleurobranchoidea without discussing the possible position of the Tylodinoidea. Further evidence of paraphyly is provided by our study, in which the two members of the Tylodinoidea never group with the members of the Pleurobranchoidea. Nevertheless, the position of the tylodinoid species varies to a great extent, depending on the gene and analytical method used. Analyzing the COI gene, *Tylodina perversa* groups with cephalaspid species, or is sister taxon to the Anaspidea + Cephalaspidea. In contrast, the 18S rRNA gene places the tylodinoid *Umbraculum mediterraneum* as a sister taxon to the Anaspidea (maximum parsimony, bootstrap value 66) or in a polytomy with Anaspidea and Cephalaspidea.

NUDIPLEURA

Salvini-Plawen and Steiner (1996) favored the Nudipleura concept, which was substantiated by Wägele and Willan (2000). The sister-taxon relationship of Nudibranchia and Pleurobranchoidea is supported in all maximum parsimony analyses of the 18S rRNA gene, although the bootstrap value is low (58). In the distance analyses, this clade was found only in the logdet transformation analysis, but with a high bootstrap value of 95. Although this clade was recognized in all analyses of the 16S rRNA gene, there is no support in bootstrap analyses.

PLEUROBRANCHOIDEA

Monophyly of this clade was thoroughly discussed by Willan (1987). Too few species are included in our study to discuss the results within this taxon. Nevertheless, the long indel of about 100 base pairs might be interpreted as an apomorphy for this taxon. Inclusion of many more species will show the possible phylogenetic information of this putative apomorphy. The lack of resolution concerning the position of the pleurobranchoid species *Pleurobranchus* sp. in the 28S rDNA analysis of Dayrat et al. (2001) certainly is due to the low number of Nudibranchia (only one species) included in the analysis.

NUDIBRANCHIA

Wägele and Willan (2000) discussed traditional hypotheses on paraphyly versus monophyly of Nudibranchia and gave evidence of monophyly by naming several apomorphies. Results of recent published molecular analyses (Thollesson 1999a; Wollscheid-Lengeling et al. 2001) are contradictory. The analyses of 18S rDNA sequences indicated monophyly, whereas the analyses of the 16S rRNA gene favored paraphyly, because of inclusion of a pleurobranchoid species. Although the number of included pleurobranchoid species is not higher in our study, the results for both genes now indicate monophyly of Nudibranchia, although bootstrap values are not high (18S: maximum parsimony, 62, neighbor-joining, 51; 16S: maximum parsimony, 67, neighbor-joining, 61).

Although the number of available sequences of nudibranchs have increased considerably, some major nudibranch groups are still underrepresented, which makes discussion of their position difficult. This is the case for the Bathydoridoidea, where only one species (*Bathydoris clavigera*) could be included in the present analyses. The Bathydoridoidea are monogeneric and fewer than 10 species are known, which are distributed in polar regions and the deep sea (Wägele 1989). The position of this taxon depends on the dataset and analytical method used. Since *B. clavigera* shows no bias in all three datasets concerning base composition and substitution rate, we can assume that the "switching" of the relationship between Cladobranchia or the Doridoidea in different topologies presented here is due to lack of a strong phylogenetic signal in favor of one or the other position. In the combined analysis, the signal is stronger in favor of a sister-taxon relationship to the Doridoidea. This relationship is present in the maximum parsimony analysis (bootstrap value, 85) and also in distance analyses using uncorrected distances, but *Bathydoris clavigera* is the sister taxon to the Cladobranchia in many distance analyses (bootstrap value for neighbor-joining, logdet transformation: 69) or to all Nudibranchia, when analyzing the combined data with evolutionary model-dependent methods.

DORIDOIDEA

Monophyly of Cryptobranchia and Phanerobranchia, which are traditionally recognized within the Doridoidea (Wägele and Willan 2000), is not documented in any analysis based on molecular data (Wollscheid and Wägele 1999; Thollesson 1999b, 2000; Wollscheid-Lengeling et al. 2001, and this study). Although Wägele and Willan (2000) discussed the monophyly of both groups (Cryptobranchia with gills retractile into branchial pockets; Phanerobranchia with second lateral larger and relatively broader than succeeding laterals, and outer laterals platelike), this question is still open to debate. The aberrant position of *Jorunna tomentosa* in 18S rRNA gene analyses is discussed in Wollscheid-Lengeling et al. (2001) and still remains to be elucidated by including more species of this genus. The Z-score value is the highest within the Anthobranchia and the sequence of this species is possibly misplaced because of long-branch effects. A basal position is also recognized by Thollesson (2000), who performed a combined analysis of 16S rDNA and COI sequences of doridoidean taxa. Thollesson (2000) used *Berthella sideralis* and *Dendronotus frondosus* as outgroups. When using different weighting schemes, *Jorunna* becomes the sister taxon to the Goniodorididae. This position was found only in our study of the COI nucleotide dataset.

The family Chromodorididae comprises about 15 genera, with *Chromodoris* being the most species-rich genus (probably up to 100 species). Rudman (1984) considered *Cadlina*, next to *Cadlinella*, as being a basal genus within the family. Usually the presence of so-called mantle dermal formations, a kind of repugnatorial gland, is considered to be the uniting character for all of these genera. Since mantle dermal formations are also known from other taxa that are not closely related even to the Nudibranchia (e.g., the sacoglossan *Placobranchus ocellatus;* H.W., unpublished data), this character is a weak synapomorphy for the Chromodorididae. Neither Thollesson (1999b, 2000) nor Wollscheid-Lengeling et al. (2001) found evidence for the inclusion of *Cadlina* in the monophyletic family Chromodorididae. Results in the 18S rDNA dataset are incongruent, but the 16S rDNA and the COI data strengthen the hypothesis that *Cadlina* should be excluded from this family.

The members of the Polyceridae (with *Polycera, Polycerella, Thecacera, Limacia*) usually group together (not in maximum parsimony analysis of 18S rDNA) (see also Thollesson 2000), but members of the family Triophidae (*Triopha, Plocamopherus, Crimora*) seem to be an offshoot of the former. Since only a few members of the Triophidae and Polyceridae could be considered until now, it seems to be premature to amalgamate both families.

The family Phyllidiidae (here represented with two genera and one species each) was not recovered in the maximum parsimony analyses (18S and 16S rRNA genes), but always in the distance analyses. The family is well studied morphologically and histologically (Wägele 1985; Brunckhorst 1993) and there is no doubt about its monophyly, which is based mainly on the unique forma-

222   ·   H. WÄGELE, V. VONNEMANN, AND W. WÄGELE

tion of secondary leaflets on the ventral side of the notum, and on the aberrant circulatory system (Wägele 1984). Z-scores and base frequencies do not show any particular bias. At the moment, it is not evident why maximum parsimony analyses do not recognize a closer relationship of these two taxa.

The position of the family Dendrodorididae is one of the most variable when considering the three different genes. This has already been observed by Wollscheid-Lengeling et al. (2001). The 18S rDNA sequence of *Dendrodoris* is placed within the Doridoidea. The maximum parsimony analysis of the 16S rRNA gene renders the family (with *Dendrodoris* and *Doriopsilla*) mono-phyletic, as the most basal offshoot of the Doridoidea, but the distance between the two genera is so divergent that, in distance analyses, *Doriopsilla* is the sister taxon to the Doridoidea, and *Dendrodoris* to all Opisthobranchia. Rate analy-ses of the COI gene indicate that the sequences of *Dendrodoris* species are satu-rated by multiple substitutions and the distance is the highest within the ana-lyzed species. Comparing these results, it seems obvious that the 16S rRNA gene evolved more quickly in *Dendrodoris* than in other gastropods. The Poro-stomata concept, uniting the Dendrodorididae and Phyllidiidae into one taxon on the basis of reduction of radula and jaws, has been discussed for many years. Recently, Valdes and Gosliner (1999) gave strong evidence in favor of this hy-pothesis: they analyzed many members of both families, as well as members of a new family that also lack cuticular structures in the buccal bulb, the Man-deliidae. They included several other doridoidean taxa. According to their mor-phological data, the taxon Porostomata is supported by nine autapomorphies. Similar results have been presented by Brodie (2001) using histological char-acters. Monophyly is also indicated by the molecular analyses of Thollesson (2000). The molecular data presented here do not support this clade, except for the bootstrap analysis of the combined dataset (not shown). But it is notewor-thy that the dendrodoridean species *Doriopsilla areolata* (but not *Dendrodoris*) and the phyllidiid species *Phyllidia elegans* are united in the distance analyses based on the amino acid sequences (Figure 8.6).

Thollesson (1999b) investigated the 16S rRNA sequences of two *Goniodoris* species from Kristineberg (Sweden, North Atlantic), *G. castanea* and *G. no-dosa*. Wollscheid-Lengeling et al. (2001) also included a species from Rosas (Spain, Mediterranean Sea), which they identified as *G. nodosa*. In our 16S rRNA gene analysis, the two *G. nodosa* sequences never group together, but the *G. nodosa* investigated by Wollscheid-Lengeling et al. (2001) is always the sis-ter taxon to *Okenia aspersa*. We assume the following possibilities: the sequence analyzed by Wollscheid-Lengeling et al. (2001) is not a *G. nodosa* sequence, but perhaps an *Okenia* species, or the two sequences of *G. nodosa* differ so much from each other because of the large distance between their localities. Only further investigation of new material can elucidate this problem. The same applies for the two *Hypselodoris villafranca* sequences (Figure 8.3). One sequence of *H. villafranca* comes from a specimen collected in Northern Spain, Atlantic (same locality as *H. elegans;* Wollscheid-Lengeling et al. 2001), whereas the other sequence of *H. villafranca,* described by Thollesson (1999b) was taken from an animal collected in Cadiz, Southern Spain, Mediterranean.

CLADOBRANCHIA

The Cladobranchia, supported by several morphology-based apomorphies (Wägele and Willan 2000), are usually supported by the molecular data. Although better resolution in favor of a traditional Aeolidoidea and Dendronotoidea within the Cladobranchia was expected by including many more sequences of this clade, this was not the case for the different datasets used in our study (not enough species included in the combined analysis). The Aeolidoidea are considered to be monophyletic, with members sharing the presence of cnidosacs and elaboration of the oral veil into oral tentacles (Wägele and Willan 2000). In the 18S rDNA analysis, all members of the included aeolidoidean species are grouped within one monophylum, except for members of the family Facelinidae. Because a thorough phylogenetic analysis of the taxon Aeolidoidea based on morphological characters is still missing, apomorphic characters cannot be listed. The Facelinidae is a highly diverse group with about 40 genera and these are united by diagnostic characters that are mainly apomorphic within the Aeolidoidea (e.g., radula formula 0.1.0, anus cleioproct, and simple oral glands present). But some of these also occur in other aeolidoidean families (e.g., radula 0.1.0 also in Aeolidiidae, Tergipedidae, etc.). Some of the facelinid genera show a high species number probably due to development of characters that allowed an adaptive radiation. This is certainly true for the genus *Phyllodesmium,* of which about 20 species are known. Many of these are able to house zooxanthellae and have a symbiotic relationship with these single-celled algae (e.g., *P. briareum,* the species investigated here, Rudman 1991, Wägele and Johnson 2001). A high deviation of the typical aeolid, or even nudibranch, sequence can be seen in the 18S rRNA gene of the facelinids, where even many areas of the so-called conservative regions differ considerably from those of all other included gastropod species (Figure 8.8). Furthermore, the Z-scores are considerably higher in three members of the facelinids (*Facelina bostoniensis, Phidiana lynceus, Godiva banyulensis*), and in the other two members (*Phyllodesmium briareum* and *Facelina auriculata*), the base frequencies differ considerably. Nevertheless, monophyly of the Facelinidae was recovered

```
                         1580           *         1600          *         1620
Phyllodesmium briareum  -GATT--GG-AGGT-CGTTAC TTGC AT-GA---CT CTTT TCAGCACC TTAT GAGAAATC
Facelina auriculata     -GATT--GGCAG--ACGTTTT TTTG AT-GA---CT CTG CCAGCACC TTAT GAGAAATC
Phidiana lynceus        ---TT AC- GCGGGGACGTCCCCAGT GG GGA AGC CT-CGCGAG--GCCCCGGAGAAATC
Godiva banylensis       ---TT AC- GCGGGAACGTCCCCAGT GG GGA AGT CT-CGCGAG--GCCCCGGAGAAATC
Facelina bostoniensis   ---TT AC- GCGGGGACGTCCCCAGT GG GGA AGC CT-CGCGAG--GCCCCGAAGAAATC
Cuthona nana            CGAT--CCGCGG-AA-GTCTCGC--ACGGA---CTGCGCGGGCGGCCCCCGGGAAACC
Cuthona caerulea        CGAT--CCGCGG-AA-GTCTCGC--ACGGA---CTGCGCGGGCGGCCCCCGGGAAACC
Cuthona sibogae         CGAT--CCGCGG-AA-GTCTCGC--ACGGA---CTGCGCGGGCGGCCCCCGGGAAACC
Eubranchus exiguus      CGAT--CCGCGG-AA-GTCTCGC--ACGGA---CTGCGCGGGCGGCCCCCGGGAAACC
Eubranchus sp.          CGAT--CCGCGG-AA-GTCTCGC--ACGGA---CTGCGCGGGCGGCCCCCGG-AAACC
Tergipes tergipes       CGAT--CCGCGG-AA-GTCTCGT--ACGGA---CTGCGCGGGCGGCCCCCGGGAAACC
Flabellina affinis      CGAT--CCGCGG-AA-GTCTTTAA-ACGGA---CTGCGCGGGCGGCCCCCGGGAAACC
Flabellina sp.          CGAT--CCGCGG-AA-GTCTCGC--ACGGA---CTGCGCGGGCGGCCCCTGGGAAACC
Flabellina babai        CGAT--CCGCGG-AA-GTCTCGC--ACGGA---CTGCGCGGGCGGCCCCCGGGAAACC
Flabellina pedata       CGAT--CCGCGG-AA-GTCTCGC--ACGGA---CTGCGCGGGCGGCCCCCGGGAAACC
Flabellina verrucosa    CGAT--CCGCGG-AA-GTCTCGC--ACGGA---CTGCGCGGGCGGCCCCCGGGAAACC
Notaeolidia depressa    CGAT--CCG GG-AA-GTCTCGC--ACGGA---CTGCGCGGGCGGCCCCCGGGAAACC
```

Figure 8.8. Conservative part of the alignment of 18SrDNA from position 1,577 to 1,634 showing several synapomorphic patterns (boxed sections) for the species *Phyllodesmium briareum* and *Facelina auriculata*, as well as for the clade *Phidiana lynceus*, *Godiva banyulensis*, and *Facelina bostoniensis*.

in distance analyses using logdet transformation. The paraphyly of the genus *Facelina* in this study is astonishing and has not been previously addressed. But it is premature to speculate about this genus before further analyses are performed on the whole family.

The paraphyly of the genus *Flabellina* is not unexpected. Wägele and Willan (2000) assumed that the genus, comprised of nearly 100 species, is paraphyletic. It represents an agglomerate of basal forms with mainly plesiomorphic aeolidoidean characters (e.g., *F. verrucosa* with a triseriate radula 1.1.1, pleuroproct anus, and a glandular stripe confined to the lateral right side of notum) and highly derived forms (e.g., *F. affinis* with flattened motile oral tentacles, oral glands, cerata on peduncles, rachis with retracted main cusp, and lateral glandular stripe reaching into cerata), and probably gave rise to several other aeolidoidean families. The suggested monophyly of the three *Flabellina* species in the COI amino acid analysis (distance methods) might be due to incomplete taxon sampling.

The position of the Antarctic species *Notaeolidia depressa*, which was considered to be the most basal aeolidoidean species in the study of Wägele and Willan (2000), cannot be clarified here. Although, according to the most parsimonious trees of the18S rDNA alignment, it seems to be a basal member of the Aeolidoidea, this is not supported in the distance analysis, and not resolved in the bootstrap analysis. The lack of a phylogenetic signal supporting the Dendronotoidea contradicts morphological studies (Wägele and Willan 2000), whereas the paraphyly of "Arminoidea" is strengthened.

FINAL REMARKS

What does this study show about our knowledge of the molecular phylogeny of opisthobranchs? First, looking at the alignment of the 18S rRNA gene, it is evident that this gene has great potential for use in further studies, especially for the Cladobranchia, when more species of this taxon are included. Substitution analyses showed that the mitochondrial genes (16S rDNA and COI) may be saturated in transitions and, to a certain extent, also in transversions, whereas no distinct saturation was observed for the 18S rDNA. Therefore, the 18S rDNA gene could be useful in the future for studying deeper nodes of the Opisthobranchia tree, whereas the 16S rDNA seems to be informative at the family level and, to a certain extent, also for deeper nodes. Second, many major taxa are still underrepresented. This study shows that inclusion of other taxonomic groups such as basal Heterobranchia, Pyramidelloidea, and Architectonicoidea is needed for clarifying the question of monophyly or paraphyly of the Opisthobranchia. We should be cautious before we cast the fatal blow to this taxon, as Thollesson (1999a: p. 81) recommends: "I think it is better to abandon the use of Opisthobranchia altogether."

ACKNOWLEDGMENTS

We would like to thank the following for providing material for this study: Nils Brenke (Bochum), Ulrike Englisch (Bochum), Tanya Fabricius (U.S.A.), Hermann Dreyer (Vienna), Christoph Held (Bochum), Annette Klussmann-Kolb (Mainz), Brian Penny (U.S.A.), Gabi Strieso (Bochum). We are grateful to Beate Hackethal (Bochum) for sequencing the *Umbraculum mediterraneum* species and for further help in the laboratories. Ulrike Englisch (Bochum) and Christoph Held (Bochum) were very supportive in analyzing the new sequences. This study is supported by the German Research Foundation to H. Wägele, V. Vonnemann (Wa 618/6-1 and Wa 618/7-1).

REFERENCES

Brodie, G. 2001. The systematics and phylogeny of the nudibranch Dendrodorididae (Opisthobranchia, Doridoidea). P. 40 in World Congress of Malacology (L. Salvini-

Plawen, Voltzow, J., Sattmann, H. and G. Steiner, eds.). Vienna, Austria, Unitas Malacologica.

Brunckhorst, D. J. 1993. The systematics and phylogeny of phyllidiid nudibranchs (Doridoidea). Records of the Australian Museum, Supplement 16:1–107.

Colgan, D. J., W. F. Ponder, and P. Eggler. 2000. Gastropod evolutionary rates and phylogenetic relationships assessed using partial 28S rDNA and histone H3 sequences. Zoologica Scripta 29:29–63.

Dayrat, B., A. Tillier, G. Lecointre, and S. Tillier. 2001. New clades of euthyneuran gastropods (Mollusca) from 28S rRNA sequences. Molecular Phylogenetics and Evolution 19:225–235.

Felsenstein, J. 1978. Cases in which parsimony and compatibility methods will be positively misleading. Systematic Zoology 27:401–410.

Haszprunar, G. 1985. On the anatomy and systematic position of the Mathildidae (Mollusca, Allogastropoda). Zoologica Scripta 14:201–213.

Haszprunar, G. 1988. On the origin and evolution of major gastropod groups, with special reference to the Streptoneura. Journal of Molluscan Studies 54:367–441.

Hendy, M. D., and D. Penny. 1989. A framework for the quantity study of evolutionary trees. Systematic Zoology 38:297–309.

Jensen, K. R. 1996. Phylogenetic systematics and classification of the Sacoglossa (Mollusca, Gastropoda, Opisthobranchia). Philosophical Transactions of the Royal Society, London B 351:91–122.

Lindberg, D. and W.F. Ponder. 2001. The influence of classification on the evolutionary interpretation of structure: A re-evaluation of the evolution of the pallial cavity of gastropod molluscs. Organisms, Diversity and Evolution 1:273–299.

Lockhart, L. J., M. A. Steel, M. D. Hendy, and D. Penny. 1994. Recovering evolutionary trees under a more general model of sequence evolution. Molecular Biology and Evolution 11:605–612.

Lydeard, C., Holznagel, W. E., Schnare, M. N. and R. R. Gutell. 2000. Phylogenetic analysis of molluscan mitochondrial LSU rDNA sequences and secondary structures. Molecular Phylogenetics and Evolution 15:83–102.

Maddison, W. P., and D. R. Maddison. 1992. MacClade ver. 3. Sinauer Associates, Sunderland MA.

Medina, M., T. M. Collins, and P. J. Walsh. 2001. mtDNA ribosomal gene phylogeny of sea hares in the genus Aplysia (Gastropoda, Opisthobranchia, Anaspidea): Implications for comparative neurobiology. Systematic Biology 50:676–688.

Medina, M., and P. J. Walsh. 2000. Systematics of Order Anaspidea based on mitochondrial DNA sequence (12S, 16S and COI). Molecular Phylogenetics and Evolution 15:41–58.

Mikkelsen, P. .M. 1996. The evolutionary relationships of Cephalaspidea s.l. (Gastropoda; Opisthobranchia): A phylogenetic analysis. Malacologia 37:375–442.

Nicholas, K. B., and H. B. G. Nicholas. 1997. Genedoc: A tool for editing and annotating multiple sequence alignment. Distributed by the author. Available online at www.eris.com/~ketchup/genedoc.shtml.

Ponder, W. F., and D. R. Lindberg. 1997. Towards a phylogeny of gastropod molluscs: An analysis using morphological characters. Zoological Journal of the Linnean Society 119:83–265.

Rudman, W. B. 1984. The Chromodorididae (Opisthobranchia: Mollusca) of the Indo–West Pacific: A review of the genera. Zoological Journal of the Linnean Society 81:115–273.

Rudman, W. B. 1991. Further studies on the taxonomy and biology of the octocoral feeding genus *Phyllodesmium* Ehrenberg, 1831 (Nudibranchia: Aeolidoidea). Journal of Molluscan Studies 57:167–203.

Salvini-Plawen, L. von. 1991. The status of the Rhodopidae (Gastropoda: Euthyneura). Malacologia 32:301–311.

Salvini-Plawen, L. von, and G. Steiner. 1996. Synapomorphies and plesiomorphies in higher classification of Mollusca. Pp. 29–51 in Origin and evolutionary radiation of the Mollusca (J. D. Taylor, ed.). Oxford University Press, Oxford.

Saitou, N., and M. Nei. 1987. The neighbor-joining method: A new method for reconstructing phylogenetic trees. Molecular Biology and Evolution 4:406–425.

Sanger, F., S. Nicklen, and A. R. Coulson. 1977. DNA sequencing with chain-terminating inhibitors. Proceedings of the National Academy of Sciences U.S.A. 74:5463–5467.

Schmekel, L. 1985. Aspects of evolution within the opisthobranchs. Pp. 221–267 in The Mollusca (K. M. Wilbur, ed.). Academic Press, London.

Simmons, M. P., and H. Ochoterena. 2000. Gaps as characters in sequence–based phylogenetic analyses. Systematic Biology 49:369–381.

Swofford, D. L. 1998. PAUP*. Phylogenetic Analysis Using Parsimony (*and other methods). Version 4. Sinauer Associates, Sunderland, MA. 128 pp.

Thollesson, M. 1999a. Phylogenetic analysis of Euthyneura (Gastropoda) by means of the 16SrRNA gene: Use of a "fast" gene for "higher-level" phylogenies. Proceedings of the Royal Society of London B 266:75–83.

Thollesson, M. 1999b. Phylogenetic analysis of dorid nudibranchs (Gastropoda: Doridacea) using the mitochondrial 16S rRNA gene. Journal of Molluscan Studies 65:335–353.

Thollesson, M. 2000. Increasing fidelity in parsimony analysis of dorid nudibranchs by differential weighting, or a tale of two genes. Molecular Phylogenetics and Evolution 16:161–172.

Thompson, J. D., T. J. Gibson, F. Plewniak, F. Jeanmougin, and D. G. Higgins. 1997. The ClustalX Windows interface: Flexible strategies for multiple sequence alignment aided by quality analysis tools. Nucleic Acids Research 24:4876–4882.

Thompson, T. E., and D. R. Seaward 1989. Ecology and taxonomic status of the aplysiomorph *Akera bullata* in the British Isles. Journal of Molluscan Studies 55:489–496.

Tillier, S., M. Masselot, and A. Tillier. 1996. Phylogenetic relationships of the pulmonate gastropods from rRNA sequences, and tempo and age of the stylommatophoran radiation. Pp. 267–284 in Origin and Evolutionary Radiation of the Mollusca (J. D. Taylor, ed.). Oxford University Press, Oxford.

Valdes, A., and T. M. Gosliner. 1999. Phylogeny of the radula-less dorids (Mollusca,

228 · H. WÄGELE, V. VONNEMANN, AND W. WÄGELE

Nudibranchia), with the description of a new genus and a new family. Zoologica Scripta 28:315–360.

Wägele, H. 1984. Kiemen und Hämolymphkreislauf von *Phyllidia pulitzeri* (Gastropoda, Opisthobranchia, Doridacea). Zoomorphology 104:246–251.

Wägele, H. 1985. The anatomy and histology of *Phyllidia pulitzeri* Pruvot–Fol, 1962, with remarks on the three Mediterranean species of *Phyllidia* (Nudibranchia, Doridacea). Veliger 28:63–79.

Wägele, H. 1989. A revision of the Antarctic species of *Bathydoris* Bergh, 1884 and comparison with other known bathydorids (Opisthobranchia, Nudibranchia). Journal of Molluscan Studies 55:343–364.

Wägele, H., and G. Johnson. 2001. Observations on the histology and photosynthetic performance of "solar-powered" opisthobranchs (Mollusca, Gastropoda, Opisthobranchia) containing symbiotic chloroplasts or zooxanthellae. Organisms, Diversity and Evolution 1:193–210.

Wägele, H., and Willan, R. C. 2000. On the phylogeny of the Nudibranchia. Zoological Journal of the Linnean Society 130:83–18.

Wägele, J. W. 2000. Grundlagen der Phylogenetischen Systematik. Verlag Dr. Friedrich Pfeil Munich, Germany. 315pp.

Wenzel, J. W., and M. E. Sidall. 1999. Noise. Cladistics 15:51–64.

Willan, R. C. 1987. Phylogenetic systematics of the Notaspidea (Opisthobranchia) with reappraisal of families and genera. American Malacological Bulletin 5:215–241.

Winnepenninckx, B., G. Steiner, T. Backeljau, and R. DeWachter. 1998. Details of gastropod phylogeny inferred from 18Sr RNA sequences. Molecular Phylogenetics and Evolution 9:55–63.

Wollscheid, E., and H. Wägele. 1999. Initial results on the molecular phylogeny of the Nudibranchia (Gastropoda, Opisthobranchia) based on 18S rDNA data. Molecular Phylogenetics and Evolution 13:215–226.

Wollscheid-Lengeling, E, J. Boore, W. Brown, and H. Wägele. 2001. The phylogeny of Nudibranchia (Opisthobranchia, Gastropoda, Mollusca) reconstructed by three molecular markers. Organisms, Diversity and Evolution 1:241–256.

Wu, C.-I., and W.-H. Li. 1985. Evidence for higher rates of nucleotide substitution in rodents than in man. Proceedings of the National Academy of Sciences U.S.A. 82:1741–1745.

JOHN P. WARES AND THOMAS F. TURNER

9
PHYLOGEOGRAPHY AND DIVERSIFICATION IN AQUATIC MOLLUSKS

A fundamental goal of ecology and evolutionary biology is to establish the factors that determine species distributions. These factors are both intrinsic to a species, such as traits that determine limits to physiological tolerance or reproductive capacity, and extrinsic, such as events that fragment continuously distributed species or populations. Direct analysis of the interactions among these forces is typically not possible because of the broad evolutionary time scale involved. However, the field of phylogeography (Avise 2000) provides a framework for inferring the relative importance of these factors in determining present-day species distributions. An inherent assumption in phylogeographic studies is that intraspecific descriptions of genetic and phenotypic variation can illuminate the same determinant factors that have shaped biotic diversity at higher levels of organization (Burton 1998; Wares et al. 2001; Wares 2002; Meyer and Paulay, in review).

The field of phylogeography, together with related population genetic methods, has been extensively reviewed elsewhere (e.g., Avise 2000, Grosberg and Cunningham 2001). These methods are used primarily to qualitatively discriminate among patterns of genealogical and geographic structuring (e.g., Avise 1992; Templeton et al. 1995), as well as to analyze the boundaries between recently diverged species (Knowlton 2000). New techniques allow hypothesis testing to be more quantitative, focusing on the resolution of temporal hypotheses (e.g., Edwards and Beerli 2000; Nielsen and Wakeley 2001). Comparative phylogeography applies these same techniques to a set of codistributed species. When phylogeographic patterns are concordant among a set of species, this implies the uniform action of extrinsic forces in shaping genetic diversity (Avise 1992; Bermingham and Moritz 1998; Wares 2002).

229

Understanding the broad interactions among intrinsic traits and extrinsic forces acting on species requires the comparison of a large number of taxa with variation in the traits that may influence the phylogeographic patterns for that species or species group. Aquatic mollusks are an exemplary taxon for this purpose. Nearly 100,000 molluscan species exist (Barnes et al. 2001), with broad variation in dispersal mode, development, physiological tolerances, sexual systems, and metabolic needs. We also have a detailed fossil record for mollusks, particularly among marine species, that can be used to clarify the temporal concordance among competing historical hypotheses (Cunningham and Collins 1994; Marko 2002). In this chapter, we review how phylogeographers use these attributes of mollusks in marine and freshwater environments to illuminate the evolutionary processes responsible for the diversification and survival of species in the face of climatic and environmental change. This is not intended to be an exhaustive review of the literature, but rather an exhibition of the promise that molluscan models hold for generating truly integrative studies of phylogeography and speciation.

FROM GENES TO SPECIES

Our review of the phylogeography of freshwater and marine mollusks is limited primarily to studies that use data that can be explicitly ordered into a genealogy using standard phylogenetic methods (e.g., DNA sequences). These data allow the simultaneous evaluation of spatial and temporal genetic variation. Placing phylogeographic studies in a temporal context is crucial (Cunningham and Collins 1994; Templeton et al. 1995; Avise 2000; Turner et al. 2000), and studies that rely on more traditional analysis of unordered gene frequencies (e.g., F-statistics based on allozyme data) do not offer this temporal context (see Grosberg and Cunningham 2001).

Many inferences made in phylogeographic studies are based on an elegant connection between the statistical expectation of genealogical patterns and the demography and ecology of particular organisms (coalescent theory; reviewed in Hudson 1990). Given a randomly mating population with genetic effective population size Ne, the genetic divergence between any pair or group of alleles sampled from this population is proportional to Ne. Any process that affects Ne (including mating system, variance in reproductive success, life span, abundance, and dispersal mechanisms) can alter the overall shape of a gene tree.

Thus, a number of particular features of the species themselves must be considered when performing phylogeographic analyses. These features contribute to the range of responses a species may have when faced with environmental change. Intrinsic features, such as mating system or body size, constrain the effects of extrinsic forces, such as glaciation or fragmentation events. In this chapter, we review the different patterns predicted by these extrinsic and intrinsic forces and conclude with a discussion of the genealogical signals generated by interactions between the two.

EXTRINSIC FORCES IN MARINE MOLLUSCAN PHYLOGEOGRAPHY

From our review of the literature, it is evident that marine species respond to very different forces than freshwater molluscan species. Extrinsic events, such as oceanographic or geological forces, have rarely led to vicariant speciation in marine species (Palumbi 1992, 1994), whereas vicariant forces (the isolation of aquatic habitats) dominate the discussion of phylogeographic patterns in freshwater species. In essence, marine species experience a geographically homogeneous landscape, whereas that of freshwater species is more fractal-like. The difference between these two environments has led researchers to look for different forces to play a role.

Geographic range expansion is one important phylogeographic process that may be captured in marine populations. The range fluctuations experienced by molluscan species during Pleistocene climatic changes are clearly important for genetic and phenotypic diversification of populations. A genetic analysis of populations of the marine gastropod *Acanthinucella spirata* to the north and south of Point Conception, California, indicate a northward range expansion from refugial populations south of Point Conception (Figure 9.1; Hellberg et al. 2001). However, although it is typical for expansion populations to exhibit lower genetic diversity, it is also possible for such expansion populations to experience a significant "founder effect." In the case of *Acanthinucella,* the expansion from refugial populations was accompanied by dramatic evolution of shell shape and size. Pleistocene populations of *Acanthinucella,* as evidenced by fossil data, do not differ significantly from modern populations of *A. spirata* in the refugial range; but the genetically depauperate expansion populations contain morphotypes not found in either ancestral populations or modern refugial populations. This suggests that the altered shell morphology is in some way favored in the northern populations and has become newly evolved or increased in frequency after range expansion.

These range expansion events typically generate patterns of lower genetic diversity in the expansion populations; the inference of such an event is stronger when multiple species share this pattern (Dillon and Manzi 1992; Marko 1998;

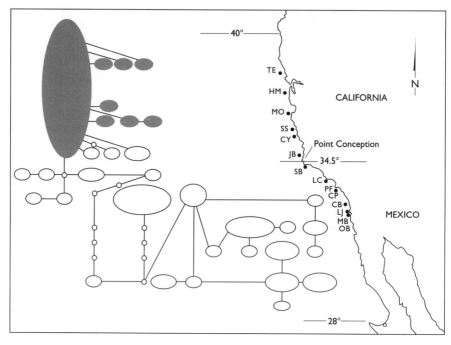

Figure 9.1. Phylogeographic data suggesting a Holocene range expansion in the intertidal snail *Acanthinucella spirata*. Significant differences in measures of allelic diversity for populations north of (shaded) and south of (open) Point Conception, California, are just one measure that indicates a disequilibrium between gene flow and the
geographic range of *A. spirata*. Similar patterns have been noted in other species at
Point Conception (see text; adapted from Hellberg et al. 2001).

Wares and Cunningham 2001). Comparisons of the population structure in the
partially sympatric species *Nucella ostrina* and *N. emarginata* (also distributed
around Point Conception) show that, although these species currently inhabit
adjacent zoogeographic provinces, they partially overlap in geographic range
because of a recent northward range expansion by *N. emarginata* (Marko 1998).
The concordance of both mitochondrial sequence data and nuclear allozyme
markers is evidence that the lower genetic diversity in the region of overlap is
due to the effects of this historical event.

Even this fairly typical response to climatic change may be modified by the
intrinsic traits of particular species. Fossil evidence shows that, in general, large-
bodied molluscan species in the eastern Pacific were more able to shift their
geographic ranges during Pleistocene climatic change than smaller species (Roy
et al. 2001). This suggests that the range limits of certain taxa will be more unstable during climatic change than those of other species. This effect alone may

influence whether phylogeographic patterns among a number of species are concordant.

The correspondence between phylogeographic discontinuities and known areas of biogeographic transition is of particular importance. Studies that have used phylogeographic methods to directly test for the mechanisms responsible for faunal transition (Avise 1992; Burton 1998; Hare and Avise 1998; Wares et al. 2001; Wares 2002) indicate a need to distinguish among transition zones that have been caused by divergent forces of natural selection, historical vicariant events, or other environmental processes that could act as a partial barrier to gene flow.

Another well-known transition zone involves a number of marine taxa that have divergent genetic lineages in the Gulf of Mexico and Atlantic (Avise 1992), including molluscan taxa *Geukensia demissa* (ribbed mussel, Sarver et al. 1992), *Stramonita haemostoma* (southern oyster drill, Liu et al. 1991), the clam *Mercenaria mercenaria* (Ó Foighil et al. 1996), the squid *Loligo pealei* (Herke and Foltz 2002), and the oyster *Crassostrea virginica* (Avise 1992; Hare and Avise 1998; and references therein). Because of the broad range of taxa that appear to be phylogeographically discordant between Gulf and Atlantic populations, there is strong rationale for supporting a late Pleistocene vicariance event as the cause for this common pattern (Avise 1992). Between glacial maxima, sea level was much higher than it is today, and it is believed that many coastal populations that are currently disjunct owing to the projection of the Florida peninsula into tropical waters were continuously distributed at that time. Subsequent isolation of these lineages may be responsible for the common pattern, but the mechanisms that *maintain* this phylogeographic break, which tends to localize to the east coast of Florida near Cape Canaveral (Avise 1992), are less well understood.

The oyster *Crassostrea* has been well studied in this regard. A combination of nuclear and mitochondrial gene trees, and analyses of allele frequency variation in allozymes and nuclear restriction site polymorphisms, have been used to distinguish among the hypotheses that this phylogeographic break represents secondary contact with introgression among the two formerly separate lineages, or that there is an oceanographic or other environmental barrier to dispersal between the two lineages. Because Gulf and Atlantic populations are reproductively compatible but completely distinct at mitochondrial loci and in some nuclear restriction fragment-length polymorphism (RFLP) assays (Karl and Avise 1992), the use of data from additional, independent nuclear loci can be used to illustrate the rate of genetic exchange between these two lineages. In *Crassostrea,* Hare and Avise (1998) show that the combination of data from multiple mitochondrial and nuclear loci are consistent with the idea that barriers to gene flow exist between the two lineages, despite high capacity for gene flow due to planktonic larval dispersal. In contrast to the mitochondrial gene trees that show reciprocal monophyly between Atlantic and Gulf lineages, sequence data from three nuclear loci illustrate a pattern more consistent with incomplete lineage sorting than with partial introgression. This is expected when the isolation event is evolutionarily recent (Avise 1994 2000; Palumbi et al. 2001).

Regional environmental and oceanographic data support the compelling arguments for local ocean currents acting as partial or complete barriers to gene flow at Cape Canaveral (Hare and Avise 1998; but see Engle and Summers 1999). The Gulf Stream diverges from the North American coast at this point, potentially carrying larvae and juveniles from southern populations out to sea, and limiting the spread of northern types southward. In other regions of upwelling or strong oceanographic forces, such patterns have been inferred through phylogeographic analysis (e.g., Rocha-Olivares and Vetter 1999; Wares et al. 2001).

The influence of ocean currents on gene flow is illustrated particularly well in the Australian muricid snail *Bedeva hanleyi*. Populations of *Bedeva* are characterized by low gene flow because their larvae are direct developing; most genetic exchange is accomplished by rafting, although not often enough to overcome the divergent effects of genetic drift (Hoskin 2000). The East Australian Current transports water southward along the outer continental shelf of Australia, impinging on the coast itself from about latitude 26 to 33 degrees South. Where this current separates from the coast, there is also cold coastal upwelling, and so nearshore surface water south of this region is transported offshore. The metapopulation of *B. hanleyi* thus consists of two adjacent regions with very different gene-flow characteristics (Hoskin 2000); in the northern region, relatively higher gene flow occurs among local populations than in the southern region. For marine mollusks, ocean currents appear to be an important influence on the overall population structure of a species and the phylogeographic history of a region.

EXTRINSIC FORCES IN FRESHWATER MOLLUSCAN PHYLOGEOGRAPHY

Freshwaters comprise less than 1% of the total water on earth, yet there is a remarkable diversity of freshwater aquatic forms. Approximately 8% of mollusks occur in freshwater (Pechenik 1995)—more than would be predicted based on a simple species–area relationship alone. The distribution of freshwater habitats is patchy and widely interspersed across a largely terrestrial landscape.

Freshwater habitats are typically connected in a hierarchical, fractal geometric fashion with low-order streams draining into larger streams and rivers. Unlike marine systems, this hierarchical organization permits formulation of a very clear hypothesis of current migration routes, and increases the effective distance separating populations. Freshwater habitats often bear the mark of historical events that have shaped the distribution of organisms. For example, river courses in North America are known to have been dramatically altered by repeated glaciation events (Mayden 1988). Thus, both contemporary and historical extrinsic forces shape the distribution of freshwater mollusks and should strongly affect phylogeographic structure. The patchy nature of freshwater habitats may in some respects account for the high species diversity encountered there because the opportunity for geographic isolation (and presumably allopatric speciation) is greater than in marine habitats.

The phylogeographic approach is expected to be especially fruitful for the study of distributional changes and speciation processes in freshwater mollusks. In turn, an awareness of these distributional patterns will help illuminate and refine our understanding of the timing and scale of extrinsic forces that are likely to affect all freshwater organisms. The reasons are twofold: (1) freshwater mollusks on the whole are poor dispersers (Dillon 2000), and (2) the dynamic and patchy distribution of freshwater environments offers a complex geographic landscape where geographic isolation of populations at small spatial scales is probable.

Nine families of bivalve mollusks and 16 families of gastropods have successfully colonized freshwaters. Phylogeographic history in most of these species remains unstudied. For those already studied, the focus has been the role of geological forces in shaping genetic diversity under an allopatric speciation model. However, many codistributed species differ in intrinsic features that should affect gene flow and response to environmental change, offering enormous potential for uncovering the interactions between intrinsic and extrinsic forces in shaping genetic diversity within and among closely related species. Our review highlights some of these features and indicates where comparative phylogeography is likely to provide insight into the roles of these different factors.

Of the bivalves that have successfully invaded freshwaters, most phylogeographic studies have focused on three families: Unionidae, Corbiculidae, and Dreissenidae. The latter two families have representative species (*Corbicula fluminea*–Asian clam, *Dreissena polymorpha*–zebra mussel, and *Dreissena bugensis*–quagga mussel) that are invasive in western Europe and North America and have rapidly expanded their geographic ranges and regional abundances (McMahon 1983; Strayer 1991). As a result of their potential economic and biotic impact, the life history and ecology of these invasive species is fairly well known (Dillon 2000). Moreover, genetic and distributional data gathered after invasion events have contributed substantial knowledge about the mechanisms of dispersal (Wilson et al. 1999; Muller et al. 2001; Lewis et al. 2000) and extrinsic factors that can limit dispersal or successful colonization in these species (e.g., Allen and Ramcharan 2001). To our knowledge, no detailed phylogeographic studies have been made of these taxa in their native ranges, but several studies have evaluated phylogeography in introduced ranges, seeking to discern the source population(s) of the colonists (Stepien et al. 1999; Renard et al. 2000).

The main focus of freshwater molluscan phylogeographic studies has been the bivalve family Unionidae. Although cosmopolitan, unionids attain highest species diversity in North America. Despite this richness, freshwater bivalves in North America are imperiled. Of roughly 280 described unionid species in the United States, more than 70 species are listed as federally threatened or endangered (Williams et al. 1993). Several authors have recognized that much of the biodiversity of this fauna may be lost before it is fully known (Williams et al. 1993; Lydeard and Mayden 1995; Lydeard and Roe 1998; Neves 1999). This conservation crisis has motivated phylogeographic studies to document genetically distinct lineages within and among closely related species (Mulvey et al. 1997; Lydeard and Roe 1998; Roe and Lydeard 1998; King et al. 1999). Two general observations have arisen: (1) there is often a poor relationship between conchological variation and genetic variation among closely related species (e.g., Mulvey et al. 1997), and (2) over broad spatial scales, genetic divergences appear to conform to expectations based on geological evidence and phylogeographic patterns established in other organisms (e.g., Mulvey et al. 1997; Roe et al. 2001). Attempts to study phylogeography at finer spatial scales have revealed previously unknown population fragmentation events in Atlantic slope drainages in the eastern United States (King et al. 1999) and the Ouachita drainage of southwestern Arkansas (Turner et al. 2000).

Geological processes and climate fluctuations over time have also altered drainage connections in the Colorado River and Rio Grande basins of the American Southwest. Timing and extent of drainage alteration have been hypothesized largely on the basis of zoogeographic trends in fishes coupled with an analysis of the geological histories of the respective basins (Minckley et al. 1986). Hershler et al. (1999) sought to evaluate the concordance of speciation events in the genus *Tryonia* (family Hydrobiidae) with the hypothesized zoogeographic history of the region. They argued that this genus was particularly well suited for studying zoogeography because most species are locally endemic to patchy aquatic habitats and exhibit limited dispersal capability.

Phylogenetic relationships derived from nucleotide sequence variation in the mitochondrial cytochrome *c* oxidase subunit I (COI) gene were consistent with

(1) a freshwater, southeastern United States origin for southwestern *Tryonia,* consistent with well-known vicariant separation of coastal drainages in the Gulf of Mexico (Bermingham and Avise 1986; Avise 1992), and (2) a sequence of speciation events that were largely consistent with geological events that separated drainage basins in the Southwest. Interestingly, two areas of local endemism were identified within a single drainage basin (Amargosa Basin), which is concordant with the hypothesis that this region is a biotic and geological composite attributable to late Miocene and Pliocene tectonic events that rerouted the drainages, with subsequent incursions of the Gulf of California, to separate local fauna (Taylor 1985). The composite nature of the Amargosa drainage is not evident in the phylogeography of similarly distributed desert pupfish species (Echelle and Dowling 1992), suggesting the importance of studying many groups for a biogeographic synthesis of aquatic habitats in the desert southwest.

Two important points emerge from this study. First, the only parthenogenetic southwestern spring snail, *Tryonia protea,* was identified as the most widely distributed species and was characterized by low genetic diversity among populations. Vicariant events apparently had little effect on spatial phylogeographic discontinuities in this species. These results support the idea that mating system (in this case, parthenogenesis) is crucial for determining the distribution of species and genetic diversity on the landscape. Second, shell features could not be distinguished among the relevant clades, rendering the biogeographic interpretation of morphological variation equivocal. There is often poor correspondence of shell features and genetic divergence in freshwater mollusks (e.g., Lydeard et al. 2000). However, a combination of molecular data and shell characteristics may help uncover sources of variation in shell features, whether environmental or a product of local selection on heritable conchological traits (De-Witt et al. 1998).

Vicariant processes can also play an important role in the formation of new parthenogenetic species. The caenogastropod genus *Campeloma* forms a species complex of parthenogenetic and sexual forms in Atlantic and Gulf coastal drainages of the southeastern United States. The distributions of three sexual species, *C. decisum* (west of the Mobile River), *C. geniculum* (between the Mobile and Appalachicola rivers), and *C. limum* (occurs east of the Appalachicola River) are consistent with well-established zoogeographic boundaries (Bermingham and Avise 1986; Wooten et al. 1988; Avise 1992). Phylogeographic analysis of parthenogenetic species, including both spontaneous and allotriploid forms, indicated that the sexual *C. limum* is the maternal ancestor of all parthenogenetic lineages (at least four independent spontaneous parthenogenetic *C. limum* lineages and one allotriploid species, *C. parthenum;* Johnson and Bragg 1999). The male ancestor of the allotriploid form was *C. genicu-*

lum, which suggests that spontaneously parthenogenetic *C. limum* colonized west of the Appalachicola River and hybridized with the local sexual species. Spontaneous and allotriploid parthenogens were more widely distributed than their sexual relatives and were able to transcend zoogeographic boundaries. Johnson and Bragg (1999) speculated that increased probability of successful colonization, coupled with local competition with sexual forms, facilitated the spread of parthenogenetic lineages.

The phylogeography of the pulmonate snail genus *Biomphalaria* provides an important example of the power of a well-documented (and independently derived) vicariance hypothesis coupled with phylogeographic analysis to infer forces that shape genetic diversity within and among closely related species. *Biomphalaria* species, which are the only intermediate hosts for the intensely studied trematode *Schistosoma mansoni,* are widely distributed across South America and Africa. This distributional pattern led early researchers to propose a Gondwanan origin for the genus (Pilsbry 1911; Davis 1980). If correct, then South American and African species should be reciprocally monophyletic with a long branch (indicating a large number of evolutionary changes) separating them. Phylogeographic analysis of the genus indicated an alternative scenario. De-Jong et al. (2001) showed that African species were indeed monophyletic, but were the sister group to *Biomphalaria glabrata,* a widespread (and phylogenetically derived) South American endemic. African species were derived, with South American species at the base of the tree (DeJong et al. 2001). The phylogenetic tree topology supported very strongly the conclusion that *Biomphalaria* originated in South America, and that African species of this genus arose many millions of years after continental separation when colonized by a *B. glabrata*-like ancestor. A probable mechanism for colonization was passive dispersal by birds (see general discussion of passive dispersal by birds in Wesselingh et al. 1999).

INTRINSIC FORCES IN MARINE MOLLUSCAN PHYLOGEOGRAPHY

Phylogeography, at the most fundamental level, considers the interactions between intrinsic effects on both the genetic effective population size and migration. Effective population size may be altered by a number of intrinsic life-history traits of a species, but perhaps most strongly by mating system. Migration ability is also an intrinsic property of species or populations (Peterson and Denno 1997; Collin 2001) and is the basis for gene flow among populations. Acknowledging the importance of these traits permits phylogeographers to go beyond documenting the historical changes in species to establish predictive hypotheses. In each case, generalizations about the effects of mating system or migratory ability are possible, although violations of these expectations are not infrequent. Nevertheless, by generating predictive hypotheses using these traits, the exceptional species will be able to lead us toward identification of other important deterministic traits.

Mating systems in aquatic mollusks are important determinants on genetic effective population size. In marine species, the determinant factors affecting mating system and development and, in turn the genetic effective size of specific populations, include body size, vagility, and latitude (reviewed in Bertness 1999). In general, free-spawning reproduction is common among marine bivalves. This mating system requires a large gametic output and usually results in high prereproductive mortality (type III survivorship) and/or high variance in reproductive success, either of which can reduce the genetic effective population size in marine species (Hedgecock 1994; Grosberg and Cunningham 2001; Turner et al. 2002).

However, many mobile marine species, particularly gastropods, have internal sexual fertilization accompanied by less variation in reproductive success. Genetic effective size will be less directly influenced by this mating system, although the effects of male competition may play an indirect role. Other mating systems exist, of course, including sequential protandry in the slipper limpet (*Crepidula fornicata*). In this species, fertilization is internal, but socially controlled sexual expression reduces the number of individuals that are permitted to reproduce. Larvae settle as males, but in social clusters only the bottom individual is a female and only the largest male is able to reproduce. As the older female dies, the oldest male takes her place as the reproductive female. In this case, the genetic effective population size is not reduced because the generations overlap; eventually each limpet in a social cluster should have the opportunity to reproduce.

Developmental mode in mollusks is strongly related to latitude and the size of the organism (Bertness 1999). Those species that disperse via planktonic larvae are known to have more fluctuations in population densities than organisms with direct development or short-distance dispersal (Thorson 1950), emphasizing the relationship between a number of phylogenetic and other intrinsic factors and the overall genetic effective population size of a species. Because of the importance of the genetic effective population size for the strength of natural selection and localized adaptation within populations (as well as divergence caused by genetic drift among populations), these factors should be considered in phylogeographic studies.

A good example of intrinsic demographic patterns apparently influencing phylogeographic reconstruction is found in the clam *Lasaea* (Park and Ó Foighil

2000a). In comparisons of populations in Florida and Bermuda, levels of allelic diversity and the general polarity of the mitochondrial gene tree suggest an improbable counter-current colonization of eastern Florida from Bermuda. Park and Ó Foighil (2000a) used reconstructions of temporal changes in genetic effective population size, inferred from genealogical data (Kuhner et al. 1998), in both populations to argue that historical population dynamics in this minute direct-developing bivalve are more likely to be the cause of this pattern than actual historical relationships between the two populations.

INTRINSIC FORCES IN FRESHWATER MOLLUSCAN PHYLOGEOGRAPHY

For freshwater bivalves, much of the variation in traits expected to determine demography and effective size is partitioned at the level of family. Table 9.1 shows how these traits are partitioned among four well-studied families that colonized freshwater environments independently (Park and Ó Foighil 2000b). Mating system differences among families may strongly affect phylogeographic differences by affecting genetic effective population sizes. Members of Dreissenidae and Unionidae are mostly gonochoristic (i.e., dioecious), whereas Corbiculidae and Sphaeriidae have a high incidence of self-compatible hermaphroditism (Heard 1965). Mating system differences lead to two predictions: (1) the ratio of genetic effective population size (Ne) to breeding adult census size (N) will be higher in gonochoristic than in hermaphroditic species because overall genetic diversity is expected to be lowered as a result of self-fertilization; and (2) more colonists are necessary to successfully colonize a new population in gonochoristic species than in hermaphroditic species (which require a minimum of one individual colonist to found a new population). Interestingly, the frequency of hermaphroditism is higher in introduced populations of *Corbicula* than in its native range. Plasticity in the mating system of *Corbicula* may be shaped by a trade-off between fitness costs of selfing in the native range, countered by prodigious colonization ability in the introduced range.

Life span also affects the potential for different phylogeographic histories to arise. Longer life span is predicted to increase the ratio of Ne to N, because long-lived species are able to spread their reproductive effort over several years and thus decrease the variance in reproductive success among individuals (Hill 1979). It is important to note that in our discussion, genetic effective size is standardized by adult breeding census size for purposes of comparison among taxa. However, differences in absolute abundance among taxa may counteract any differences arising from mating system or life span. Local abundances vary greatly among freshwater bivalves. For example, Neves and Widlak (1987)

Table 9.1

Life history and demographic variation among four families of freshwater bivalve mollusks, ranked in descending order of predicted levels of gene flow

Taxon	Mating System	LifeSpan (years)	Sex Ratio Male:Female	Larval Form	Maximum Offspring Size (mm)
Dreissenidae	Gonochoristic[a]	>3	1:1	Veliger	0.2
Unionidae	Gonochoristic[a]	>10[b]	1:1	Glochidia/ ectoparasite	0.5
Corbiculidae	Simultaneous hermaphroditic[c]	2–4	Variable	Brooding/ reduced veliger	0.3
Sphaeriidae	Simultaneous hermaphroditic	2–4	—	Brooding/ veliger/juvenile	7.0

Note: Life history information is summarized from accounts in Dillon (2001).
[a]Hermaphroditic individuals occur at low frequency (usually <5% from population surveys).
[b]Heller 1990.
[c]A wide range of sexuality is present, see text.

noted that sphaeriid clams were at least an order of magnitude more abundant than unionids in Big Moccasin Creek, Virginia, and *Corbicula* and *Dreissena* can achieve local densities higher than 1,000/m² (Graney et al. 1980). Comparative phylogeographic studies may uncover very different patterns of genetic diversity because of differences in abundance, rather than historical effects.

As in freshwater bivalves, in gastropods, mating system differences are extremely important for predicting differences in phylogeographic history, and variation in mating systems of freshwater gastropods is largely partitioned among higher taxonomic groups. Members of the subclass Pulmonata are best described as simultaneous hermaphrodites with the capability to self-fertilize (although spermatogenesis precedes oogenesis in many species; Dillon 2000). Genetic data suggest that the incidence of self-fertilization is high in the pulmonates *Biomphalaria, Physa,* and *Bulinus* (Viard et al. 1997; Monsutti and Perrin 1999; Charbonnel et al. 2000; Mavarez et al. 2000). Conversely, most members of the subclass Caenogastropoda are gonochoristic. Apomictic parthenogenesis (eggs produced by mitosis) occurs in three caenogastropod families: Hydrobiidae, Thiaridae, and Viviparidae. Selfing or parthenogenetic species are typically more widely distributed than their sexual relatives, presumably because of increased likelihood of successful colonization. However, mating system has a profound effect on the distribution of genetic diversity on the spatial landscape, as it does in freshwater bivalves. For parthenogenetic species, rapid geographic range expansion usually stems from the introduction of very few clonal genetic lines, resulting in genetic homogeneity over very broad geographic scales (e.g., Jacobsen et al. 1996; Samadi et al. 1999; but see Dybdahl and Lively 1995).

DISPERSAL AND GENE FLOW

We emphasize the relationship between phylogeographic history and diverse life history traits because too much trust is placed in the utility of larval dispersal in marine systems and landscape processes of vicariance in freshwaters to accurately predict gene flow. In general, the idea that larval dispersal is a primary determinant of population structure is not often tested in systems where other factors, such as geographic range and other life history traits, do not vary (but see Emlet 1995; Peterson and Denno 1997; and Collin 2001). The counterexamples to this prediction are abundant enough (reviewed in Grosberg and Cunningham 2001) that it is worth focusing our attention on a broader array of traits.

The relationship between dispersal type and gene flow may be inconsistent because of the influence of historical events and other interactions between the intrinsic traits of a species and its environment. Dispersal is essentially passive in most mollusks, suggesting that these interactions (including, although not exclusively, larval and juvenile dispersal) are worth consideration. Here, we consider a group of studies that describe the relationship between gene flow patterns and the environment of a particular species. Although some generalizations may be made, it should be clear that there are enough idiosyncratic results to direct us toward more refined hypotheses.

Collin (2001) presents a good example of codistributed, closely related species of limpets that vary in dispersal ability. These limpets (*Crepidula*) exhibit not only variation in the extent of larval dispersal among species, but also along the Atlantic and Gulf coasts of North America produce results consistent with typical expectations: high gene flow and spatial genetic homogeneity are associated with species that have planktonic dispersal. However, it is difficult to determine whether population structure in poorly dispersing species of *Crepidula* is due to equilibrium isolation by distance (see Grosberg and Cunningham 2001) or other historical effects; divergent clades separated by large geographic distances can represent either pattern, and further sampling may be necessary to discriminate these two hypotheses for many datasets (Templeton et al. 1995; Wares 2002).

In cases where the geographic range of a species has been disrupted or altered in recent geological history, more examples appear to support historical discontinuities in gene flow as playing a stronger role than ongoing migration in generating phylogeographic signal. For example, the periwinkle *Littorina littorea* has planktonic dispersal, but the direct-developing congener *L. saxatilis* has a broader geographic range in the North Atlantic (Johannesson 1988). Phylogeographic comparisons of the planktonic *L. littorea* (Wares et al. 2002) with the direct-developing *L. obtusata* in the North Atlantic (Wares and Cunningham 2001) show that the difference in larval dispersal is not reflected in differences between the ability of the species to colonize distant lands and, in fact, more phylogeographic differentiation is seen in *L. littorea*. Similarly, the direct-developing dogwhelk *Nucella lapillus* has apparently expanded its range across the North Atlantic since the most recent glacial maximum (Wares and Cunningham 2001), whereas genetic data for the planktonically dispersing mussel *Mytilus edulis* suggests only an ancestral connection between the two coasts (Riginos et al., in review). However, the effects of metapopulation structure can lead to greater broad-scale homogeneity in poorly dispersing species such as *Nucella* (Day et al. 1993), whereas those species that are more demographically stable in the face of rapidly changing environments may be genetically more variable (Wade and McCauley 1988; McCauley 1991).

Some species may maintain intrinsic mechanisms that restrict gene flow and promote diversification in marine settings. Littorinid snails of the subgenus *Neritrema* (Reid et al. 1996), the descendants of an invasion of *Littorina* from the Pacific into the North Atlantic, exhibit such intertidal diversification in a number of independent regions and lineages. The species pair *L. obtusata* and *L. fabalis* diverged during the Pleistocene (Reid et al. 1996; Wares 2000), and speciation was facilitated by differences in habitat and feeding preferences (Watson and Norton 1987; Vermeij 1992). The specialization of *L. obtusata* on fucoid algae is a recent adaptation (Vermeij 1992) and floating populations of *L. obtusata* are commonly observed on drifting algae mats (Ingólfsson 1995). This trait could have been important in the postglacial colonization of North America by European populations of *L. obtusata* (Wares and Cunningham 2001) not accompanied by *L. fabalis*.

This recent, apparently sympatric speciation event would be of interest by itself. Yet it is now apparent that phylogeographic divergence of direct-developing *Littorina* species into high- and low-intertidal forms is common throughout the Atlantic and Pacific. In preliminary stages, this is evidenced by single monomorphic populations of *L. brevicula* that separate into high- and low-intertidal populations during breeding (Takada and Rolán-Alvarez 2000). A more advanced pattern of reproductive isolation has formed between sympatric morphs of *L. saxatilis* on the shores of the Iberian peninsula (Johannesson et al. 1995; Rolán-Alvarez et al. 1997; Erlandsson et al. 1998), with only limited gene flow between high- and low-intertidal morphs. Similar differentiation in this species complex plays out among sheltered versus exposed habitats (Johannesson and Johannesson 1996; Tatarenkov and Johannesson 1998; but see Tatarenkov and Johannesson 1999; Small and Gosling 2000b). Some of this differentiation may be due to isolation in distinct glacial refugia during the Pleistocene (Small and Gosling 2000a; Wilding et al. 2000).

In freshwater systems, an obvious constraint to dispersal is the landscape and other geographic barriers to gene flow. If gene flow is a function of dispersal and the probability of successful mating of a migrant with a member of the recipient population, then any trait that affects dispersal ability should affect phylogeographic patterning on the geographic landscape. For creatures that are mobile as adults, such as freshwater and marine gastropods, dispersal can occur over the life of the organism; thus, dispersal capability can be measured by characteristics of larval, juvenile, and adult life history stages. For species with sedentary adult life stages, adult traits such as body size may play an indirect but crucial role in dispersal capability of larval and juvenile life stages (Dillon 2000).

Important differences in dispersal capability exist among families of freshwater bivalves. Variation in early life history (larval size, behavior, and abundance) should determine dispersal capability because adults are sedentary. Movement of genes across the landscape occurs when sperm is released and transported, and when larvae are released. Juvenile and adult movement is only sporadic and opportunistic, although the passive dispersal of adult forms may be important for explaining the rapid invasions of nonnative freshwater mollusks (Counts 1986; Isom 1986; Strayer 1991).

Dreissenidae, with planktonic veliger larvae (unique among freshwater bivalves), is predicted to have the highest potential for gene flow. Larvae are suspended in the water column for up to three weeks before settling (Dillon 2000). Unionids are predicted to have the next highest potential for gene flow, as glochidial larvae become encysted in vagile fish hosts (Williams et al. 1993). Finally, because corbiculids and sphaeriids often brood their larvae to advanced juvenile stages, they have the lowest potential for gene flow via dispersal of propagules. Juvenile size at birth among and within families is variable (Table 9.1) and appears to co-vary with phylogeny at least in sphaeriids (Cooley and Ó Foighil 2000). Larger, more fully developed juveniles may be more apt to settle on substrates close to their parents, whereas moving water may displace smaller juveniles.

Recent invasions of corbiculoids and dreissenids into Europe and North America offer an opportunity to evaluate how mating system and high-dispersal strategies can lead to the same kinds of geographic distributions but very different genealogical data. *Corbicula fluminea* was introduced into the United States in 1938 in western Washington (Isom 1986). Presently, the species is found in nearly all major drainages of the contiguous 48 states. The primary mechanism of dispersal in *Corbicula* is thought to be human-facilitated introduction of adults into new drainages. Despite their abundance and vast geographic range, there is little genetic variation in the introduced range of *Corbicula* except for a genetic divergence between two widely distributed groups that probably reflect separate colonization events from different source populations (Hillis and Patton 1982; McLeod 1986; Siripattrawan et al. 2000). An identical pattern is observed in Europe (Renard et al. 2000). In *Corbicula* there appears to be a high probability of successful introduction followed by rapid expansion of geographic range, even when very few individuals found a new population. Prodigious colonization ability is undoubtedly related to the ability to self-fertilize. However, this mating system has important consequences for genetic diversity in the introduced range, resulting in virtually no genetic diversity within groups, and spatial partitioning of genetic diversity among groups that reflects colonization history.

Dreissena polymorpha, introduced to the Great Lakes region in 1988, has rapidly expanded its range to encompass much of the lower Mississippi River drainage (USGS nonindigenous mollusk web site, http://nas.er.usgs.gov/images/currzm00.gif). Genetic diversity of the colonizing populations is high and appears to reflect the diversity in the native range (Marsden et al. 1996). This suggests that a large number of colonists were present in initial introductions. Once established, the high dispersal ability of *D. polymorpha* (perhaps aided by human-facilitated movements of large numbers of individuals attached to boat hulls) has resulted in its rapid spread. To our knowledge, no population genetic studies have been conducted over the current range of *Dreissena*. If larval dispersal is the primary mechanism for the spread of the species, then we predict comparable levels of heterozygosity and gene diversity over the entire range and little spatial population structure. However, if local introduction by humans is the primary mechanism, then within-population heterozygosity should be reduced and among-population divergence increased. Genealogical-based analytical methods are well suited to partition the effects of migration and local genetic drift for explaining spatial genetic patterns (Turner et al. 2000; Beerli and Felsenstein 2001).

Perhaps the most fruitful avenue for comparative phylogeography of freshwater bivalves is comparisons within and among sphaeriids and unionids because their distributions often overlap at local and regional scales. Phylogenetic relationships and evolutionary history of demographic and life history characters that should influence phylogeography are well understood within each group (Lydeard et al. 1996; Graf and Ó Foighil 1999; Park and Ó Foighil 2000b) and taxa can be selected that vary in key traits predicted to influence phylogeographic relationships within and among families. Roe et al. (2001) illustrate this in *Lampsilis,* a group that disperses glochidial larvae via fish hosts. There is a strong phylogenetic component to the analysis of this group, as all of the superconglutinate lampsilids form a monophyletic clade. However, designated species are not all reciprocally monophyletic, and neither are distinct river drainages (Roe et al. 2001), suggesting both the recent isolation of these lineages and the need to focus on their interactions with historical events (e.g., Pleistocene changes in sea level) and with other species (their fish hosts; discussed below). Unfortunately, we know of no phylogeographic studies of sphaeriid clams with which to compare these results.

Euryhaline gastropod species have been examined for differences in genetic structure related to dispersal capability and habitat selection. Wilke and Davis (2000) studied two species of gonochoristic mud snails, *Hydrobia ulvae* and *H. ventrosa,* that are codistributed across the Atlantic, Baltic, and North Sea coasts of Europe. Microhabitats differ between the two species. *H. ulvae* is found along shorelines where it is directly exposed to waves and high salinities. *H. ventrosa* is restricted to lower salinity and less energetically stressful backwater environments. These two species also differ in larval characters that can influence dispersal; *H. ulvae* produces free-swimming veliger larvae with planktonic development, and *H. ventrosa* produces offspring that develop directly into juveniles. Based on these differences, Wilke and Davis (2000) predicted that genetic divergence among sample localities would be lower (and gene flow higher) for *H. ulvae* than for *H. ventrosa*.

Indeed, when compared side by side, *H. ulvae* displayed higher gene flow in a pattern consistent with Wright's (1977) island model. Under this model, the probability of migrant exchange is equal among spatially structured populations. Conversely, *H. ventrosa* exhibited lower gene flow and higher divergence across the study area, with population structure consistent with an isolation by distance (IBD) model of gene exchange. Under IBD, the probability of gene flow decreases with increasing geographic distances separating populations (Hellberg 1994; Grosberg and Cunningham 2001). Observed measures of gene flow are consistent with predictions based on dispersal differences among species. Planktonic larvae are known to travel long distances and gene flow from larval movement has been hypothesized to homogenize allele frequencies among geographically distant populations (Scheltema 1986).

Wilke and Davis (2000) noted that this interpretation is confounded by interspecific differences in adult habitat preference, which may relate to probabilities of adult movement. Ecological studies of development and larval behavior of *H. ulvae* suggest that veliger larvae have a rather short life in the plankton (less than three days) stage. In some *H. ulvae* populations, the veliger larval stage was bypassed entirely and offspring developed directly. These observations suggested a diminished role for larval dispersal and implicated adult movement as the most important factor for determining genetic diversity differences between the two species. Adult movement was hypothesized to be more likely for *H. ulvae* because of its proximity to open ocean.

Is it plausible for adult movement to account for differences in genetic patterns observed in these species? A series of studies on *Bembicium vittatum* (a littorinid snail) occurring on the Houtman Abrolhos Islands in Western Australia indicated that the answer is yes, but the magnitude of movement depends on extrinsic environmental factors related to barriers to migration (Johnson and Black 1991, 1998). *Bembicium* populations are distributed on the shoreline of the eastern and western sides of the islands. On eastern shores, the species is also found in isolated tidal pools. Comparison across tidal pool localities on the eastern side of the islands indicated IBD, as did a separate analysis of adjacent shoreline localities. Only the magnitude of the relationships differed and, in general, genetic divergences were pronounced across tidal pool and shoreline localities (Johnson and Black 1998). Conversely, western shoreline localities showed no obvious IBD (based on our examination of the data), and overall levels of genetic divergence were very low across localities (Johnson and Black 1991). Because they face open reefs, western shores are subject to greater flow of water, which increased dispersal among these localities when compared with shore and tidal pool localities on eastern shores.

Passive dispersal by birds (Wesselingh et al. 1999) appears to enhance gene flow in a variety of freshwater gastropod species. For two codistributed *Hydrobia* species, gene flow was greatly increased and of similar magnitude among populations located within a migratory bird flyway across the North Sea (Wilke and Davis 2000), despite differences in larval dispersal and adult microhabitat preferences that had consequences for gene flow in other parts of their range. Wesselingh et al. (1999) showed that species distributions of *Tryonia* and *Planorbarious* in the Caribbean Sea and continental Europe, respectively, were consistent with migratory waterfowl flyways. These authors suggested that the probability of successful colonization depends on intrinsic factors related to mating system, body size, brooding characteristics, and environmental preferences. Hermaphroditic or parthenogenetic mating systems permit a single colonist to give rise to a new population. Sexual species that brood eggs can colonize a new area with a single gravid female.

INTEGRATING THE INTRINSIC AND EXTRINSIC: COMPARATIVE MOLLUSCAN PHYLOGEOGRAPHY

Biogeographic trends are made clear by the concordant distribution patterns of many species (Briggs 1974). By the same reasoning, finding concordance among phylogeographic histories, whether caused by vicariance, range expansion, or similar processes, in a number of species, strengthens the inference for a common mechanism that broadly acted on species dynamics. The trans-Arctic interchange, discussed below, was inferred from the distributional changes in hundreds of molluscan species (Vermeij 1991). Similarly, explicit hypothesis testing of the spatial and temporal concordance of phylogeographic patterns with well-studied geological events ensures a direct link between landscape-level changes and population genetic responses within and among species.

An exemplary study comparing biogeographic and phylogeographic boundaries in the Indo-West Pacific uses most of the extant species of cowries (cypraeid gastropods) to examine the major hypotheses describing the formation of biodiversity (Meyer and Paulay, in review). The isolation patterns of numerous cowrie lineages in this region are concordant with other regional phylogeographic and biogeographic divisions. Meyer and Paulay (in review) illustrate the importance of allopatry-forming mechanisms in producing general distributions of faunal change, although often these are current ecological mechanisms rather than geological or vicariant. The hypothesis that Plio-Pleistocene climatic and oceanographic changes were responsible for diversification across a number of distinct biota is also tested using cowrie phylogeographic data. By carefully establishing the substitution rate for their mitochondrial dataset using a detailed fossil record, Meyer and Paulay were able to reject this hypothesis, instead suggesting that many of the phylogeographic discontinuities in this dataset were not representative of a single set of historical events.

Fossils are one of the few independent means of dating the establishment of new populations or the divergence of older populations. The development of molluscan shells is believed to have promoted the diversification of molluscan species because of protection from predators and desiccation (Vermeij 1987), and shells account for much of the available fossil record. These fossils can be correlated with specific geological events (e.g., Bowen and Sykes 1988), and by comparing this date with the divergence in genes sampled from involved populations or species, we can calculate the rate of substitution for a particular gene. Knowing this substitution rate then allows the estimation of the timing of other events with respect to a phylogeny or genealogy constructed with that gene. Studies that have generated lineage-specific genetic substitution rates have primarily focused on two well-documented historical events: the trans-Arctic interchange and the closure of the Isthmus of Panama.

In fossil beds of Iceland, northern Europe, and the Canadian Maritimes, hundreds of molluscan taxa appear for the first time around 3.5 million years ago (Vermeij 1991; Reid et al. 1996). These species are either found in the Pacific or are closely related to Pacific fauna and arrived after a massive asymmetric exchange of species from the Pacific to the Atlantic when sea levels rose above

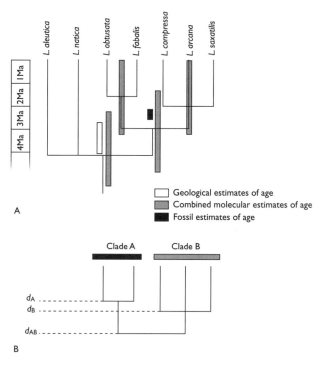

Figure 9.2. (A) Fossil data and geological events used to calibrate the substitution rate for genes used in phylogeographic studies. The intertidal periwinkle genus *Littorina* invaded the North Atlantic via the trans-Arctic Interchange twice; one of the invading lineages has undergone a subsequent radiation, with at least five species found in the North Atlantic. Fossil data for the arrival of *Littorina* and hundreds of other molluscan species in the Atlantic, along with geological evidence regarding the opening of the Bering Strait, have been used to calibrate the substitution rates for a number of mitochondrial genes. (From Reid et al. 1996) (B) Appropriate measures of divergence rates also require estimates of within-population divergence. Measures of *d,* the molecular divergence between two populations or taxa, are modified by the mean divergence within each group. This mean within-group divergence is a correlate of the diversity expected given the effective population size, and must be subtracted from the overall divergence estimate because of the within-population differentiation expected in the ancestral population. This adjusted estimate is usually expressed as $d = d_{AB} - (d_A + d_B)/2$ (Nei and Li 1979; Edwards and Beerli 2000).

the Bering land bridge (Vermeij 1991). Among these invading species were two lineages of the northern hemisphere periwinkle *Littorina*. An excellent fossil record and knowledge of the timing of this trans-Arctic interchange event allowed Reid et al. (1996) to generate robust estimates of the substitution rate at three mitochondrial genes (Figure 9.2), consistent with other widely applied rates.

Collins et al. (1996) followed a similar path in establishing the rate of molecular evolution for the intertidal gastropod *Nucella*. Although a combination of geological and paleontological evidence showed that the North Atlantic populations of trans-Arctic interchange participants could be no older than 3.5 to 5 million years (Vermeij 1991; Collins et al. 1996; Marincovich and Gladenkov 1999), the volcanic activity that formed the Isthmus of Panama ended gene flow between populations in the eastern Pacific and Caribbean. This final closure of the Panamanian isthmus about 3 million years ago (Keigwin 1978; Knowlton and Weigt 1998) indicates a constraint on marine species to have diverged at the same time or before this event. Rate calibrations for *Nucella* using closely related taxa with good fossil records that were separated by each of these geological events are not entirely consistent (Collins et al. 1996), suggesting that rate variation among even closely related lineages has the potential to bias tests of specific phylogeographic hypotheses.

In fact, fossil data suggest that many species separated by the Isthmus of Panama may have diverged long before the final proposed cessation of gene flow through this region. Separation of arcid bivalve geminate pairs may predate the formation of this geological barrier by 10 million years (Marko and Jackson 2001; Marko 2002). Although the appearance of distinct taxa in the fossil record is known to underestimate the true divergence time of two species, the assumption that vicariance among geminate sister species around the Isthmus of Panama occurred at the final closure of this seaway is a potentially much stronger bias in the estimation of substitution rates. Nevertheless, there is generally a strong relationship between phylogenetic evidence, fossil data, and the inferred rates of molecular evolution across a number of studies of different molluscan taxa (e.g., Cunningham and Collins 1994; Rawson and Hilbish 1995; Collins et al. 1996; Reid et al. 1996; Hellberg 1998; Wares and Cunningham 2001; Meyer and Paulay, in review).

The issue of establishing accurate estimates of substitution rates in molecular markers is important for a number of reasons. It is not sufficient for a phylogeography study to show congruence between the topology of genealogical lineages and their geographic origins. If the events causing the separation of genetic lineages and those causing geographic changes do not occur contemporaneously, we may be misled into believing that there is a common geological cause for a phylogeographic event when in fact it is a case of "pseudo-congruence" (Cunningham and Collins 1994). The field of phylogeography is rapidly becoming one based on greater statistical rigor, and the accurate dating of events and their correspondence with competing phylogeographic hypotheses is crucial (Meyer and Paulay, in review).

Direct comparisons between fossil evidence and phylogeographic evidence illustrate the importance of these rate estimates for interpreting the evolutionary processes acting on ancestral populations. In *Acanthinucella spirata*, the combination of fossil evidence for a range expansion in the late Pleistocene, accompanied by genetic data that are spatially and temporally concordant with this hypothesis, confirms the predictions typically cited for genetic data given a range expansion (Hewitt 1996). Additionally, the rate estimates used for the mitochondrial genes in this study were used to reconstruct changes in effective population size (*Ne*) in the ancestral population, and again the results of this analysis coincided with a late Pleistocene population expansion (Hellberg et al. 2001).

These comparisons among genetic and fossil data illustrate the importance of independent lines of support for a hypothesis. In phylogeography, this is typically achieved by comparing the results from multiple genes or loci (Hare and Avise 1998; Wollenberg and Avise 1998). It may also include the use of other physiological or genetic characteristics for a species. We know that speciation involves more than simply neutral processes, and the analysis of the traits involved in speciation is an important line of investigation. The phylogeography of the marine bivalve *Mytilus* has revealed the mechanics of speciation and the dynamics of hybridization among populations and species of the northern hemisphere *M. edulis* group (including *M. edulis*, *M. trossulus*, and *M. galloprovincialis*). The history of *Mytilus* in the Atlantic Ocean can be described by a number of partly independent evolutionary lineages that have been separated for only a brief time (Figure 9.3).

Mytilus offers a unique system for studying adaptive diversification because it is an osmoconformer found in habitats with a broad salinity range. Much early work on *Mytilus* indicated strong clinal variation in some allozyme loci that apparently corresponds to osmolarity gradients in their habitat. For instance, allele frequencies at the *Lap* locus strongly conform to patterns consistent with natural selection (reviewed in Hilbish 1996). Similar allele frequency differences at this locus and other allozyme loci between populations of *M. trossulus* in the Baltic Sea and the North Atlantic *M. edulis* were believed to represent historical differentiation of these populations, but recent work by Riginos et al. (2002) suggests otherwise.

Hybridization between the Baltic populations of *M. trossulus* and *M. edulis* was initially described using data from the mitochondrial genome (Quesada et al. 1995; Rawson et al. 1996a; Rawson and Hilbish 1998). Because the diagnostic allozyme loci for *M. trossulus* were stable in frequency, but the *M. edulis* mitochondrial type is now found throughout the Baltic, it appeared that mitochondrial introgression was more rapid than for nuclear markers. However, Riginos et al. (2002) used a number of additional nuclear genes to test this hy-

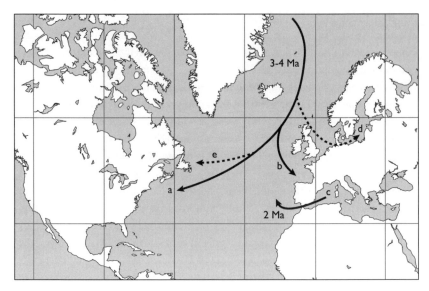

Figure 9.3. The history of *Mytilus* in the North Atlantic. This bivalve genus was not present in the Atlantic before its arrival in the late Pliocene, a range expansion event that involved the ancestral population of the Pacific species *M. trossulus* (Rawson and Hilbish 1995; Beynon and Skibinski 1996). The North Atlantic species *M. edulis* is a direct descendant of this event (a and b), and the combination of Pleistocene vicariance events (c, the brief separation of populations that formed *M. galloprovincialis*; Barsotti and Meluzzi 1968; Rawson and Hilbish 1995) and late Pleistocene or Holocene secondary invasions (d and e, the secondary introduction of *M. trossulus* to the Atlantic during a warm interglacial period; Riginos et al. in review; T. J. Hilbish, personal communication; J. P. Wares, unpublished data) have produced a patchwork of secondary contact zones between putative species of *Mytilus* in the North Atlantic. The distribution of the Mediterranean and eastern Atlantic species *M. galloprovincialis* overlaps with *M. edulis* in and around the British Isles, while recently arrived populations of *M. trossulus* hybridize with *M. edulis* in the Baltic Sea and along the east coast of Canada. Because this system has been intensively studied, it may be useful for comparisons with other molluscan phylogeographic studies for species that have similar geographic distributions.

pothesis. Data from the ribosomal internal transcribed spacer (ITS) region, two protein-coding genes that are responsible for the adhesion of byssal strands (Glu-5'; Rawson et al. 1996a) and sperm packaging (PL-II-a; Heath et al. 1995), and an anonymous nuclear marker (MAL-1; Rawson et al. 1996b) were sampled from Baltic populations of *M. trossulus*. In each case, *M. edulis* alleles are found in high frequency throughout these populations of *M. trossulus*, evidence of strong introgression as well as an indication that the allozyme loci represent an ancestral coadapted gene complex that improves the fitness of *Mytilus* individuals in the low-salinity Baltic (Hilbish 1996; Riginos et al. 2002). The maintenance of such coadapted gene complexes in the face of introgression is probably a general issue in molluscan diversification (e.g., the intertidal whelk *Nucella;* see Kirby 2000). The use of additional genes helps resolve two very different phylogeographic histories involving introgression.

Attempts to elucidate the phylogeographic history of *Mytilus* have illuminated other unique genetic processes that play a role in diversification of this genus. The phenomenon of doubly uniparental inheritance (DUI) of the mitochondrial genome (reviewed in Zouros 2000) complicates our interpretation of the divergence of populations and species of *Mytilus*. Separate male and female mitochondrial genomes that evolve independently at different rates apparently predate the genus itself (Rawson and Hilbish 1995) and may play a role in sex determination in *Mytilus* (Saavedra et al. 1997). The effect that this sexual system has on the reconstruction of phylogeographic history in mussels is unclear, but the two mitochondrial genomes described very different histories. The isolation of European and North American populations of *M. edulis,* for example, is quite ancient in the male mitochondrial lineage, but female lineages are only recently diverged (Wares and Cunningham 2001; Riginos et al., in review). DUI of mtDNA has also been noted in the freshwater unionid bivalve *Anodonta* (Hoeh et al. 1996; Liu et al. 1996a).

In other taxa, phylogeographic patterns may be better explained when additional genetic mechanisms are explored. One good example involves the intertidal snail *Tegula*. Interspecific gene trees using the mitochondrial 12S and COI genes (Hellberg 1998) showed a surprising number of sister species to *Tegula* in sympatry, a relationship that is unexpected if allopatry is believed to be responsible for most speciation. Earlier work using abalone (*Haliotis*) as a model system for reproductive isolation involving gametic interactions (Lee et al. 1995) had shown that gametic recognition proteins might be under strong diversifying selection. Comparisons of these same genes in *Tegula* found that, indeed, the sperm-related lysin protein evolves at a rate that is an order of magnitude faster than the fastest reported mammalian genes (Hellberg and Vacquier 1999). If gametic recognition proteins evolve extremely rapidly, then even short periods of isolation or allopatry among populations could produce lineages and populations that are reproductively incompatible. Clearly, such interactions have strong implications for the generation of diverse lineages in marine settings, which often contain a surprising amount of cryptic species diversity given the lack of obvious geographic barriers to gene flow (Palumbi 1994).

COMPARATIVE BIOGEOGRAPHY

Other comparisons may be made with other marine biogeographic and phylogeographic studies. As detailed above, a tremendous amount of work has been done on the diversification and phylogeographic relationships among different populations of the bivalve *Mytilus* (Figure 9.3). Patterns in other taxa may be due to other mechanisms (e.g., the trans-Atlantic relationships in the ocean quahog, *Arctica islandica,* are indicative of a recent range expansion rather than an equilibrium history; Dahlgren et al. 2000). They may also represent a weaker phylogeographic signal than *Mytilus* (e.g., the pattern of isolation by distance in the cuttlefish *Sepia* that is geographically concordant with the initial isolation of *M. galloprovincialis;* Perez-Losada et al. 1999). Nevertheless, such comparisons should be used when possible. This allows us to find the most general mechanisms that may be responsible for regional diversification, whether they are intrinsic or extrinsic forces.

In the freshwater milieu, little comparative work has been done in phylogeography. However, the potential is great for comparisons among taxa that use different modes of reproduction, dispersal, or other means of interacting with the extrinsic environment. The remarkable reproductive biology of unionid bivalves, featuring an obligate parasitic (glochidial) phase on fish hosts (Coker et al. 1921), could serve as the basis for productive research in comparative phylogeography evaluating coevolutionary relationships of mussels and their fish hosts. We propose a very basic hypothesis for the expected relationship between the spatial distribution of genetic variation in freshwater unionid mussels and their fish hosts based on the following premises: (1) Fish and mussel species richness is positively related, suggesting that dispersal limitation is an important factor structuring freshwater mussel communities (Vaughn and Taylor 2000). (2) Fish and mussel species richness increases with stream order (Sheldon 1968; Gelwick 1990; Kingsolving and Bain 1993; Strayer and Ralley 1993). (3) Maximum body size usually increases as fish communities become more species-rich (Smith 1978). (4) Vagility and gene flow increase in fishes as a function of body size (Roff 1988, 1991; Turner and Trexler 1998).

These premises lead to the hypothesis that unionid mussels with distributions that are limited to low-order streams (e.g., creeks, small tributaries) should exhibit lower gene flow between populations and greater divergence among populations than those found in a broad range of stream orders, or those limited to larger rivers (Berg et al. 1997) if host selection is random. This is because fish hosts, on average, are expected to exhibit lower vagility (as a result of smaller body size) in low-order streams. A compilation of phylogeographic studies of freshwater mussels lends some very weak support to this hypothesis (Table 9.2). Species limited to creeks and small rivers exhibit higher intraspecific genetic distance values and limited dispersion of shared haplotypes (with respect to the maximum distance separating sampling localities) compared with generalist or large-river species. Refinement of this hypothesis will certainly be necessary. If a particular mussel species requires a specific fish host (or a subset of available fish hosts), then the spatial distribution of genetic variation should reflect dispersal dynamics of that particular host, or subset of hosts. Differences among species should be reflected in mean body size differences (weighted by abundance) of available hosts.

Brooding characteristics and the timing of glochidial release may also affect unionid mussel dispersal. Four broad categories describe brooding characteristics in unionids (Dillon 2000): (1) summer tachytictic (short brooding time, release with rising temperatures; (2) winter tachytictic (short brooding time, release with falling temperatures; (3) summer bradytictic (overwinter brooding, release with rising temperatures; and (4) winter bradytictic (overwinter brooding, release with falling temperatures). Adult life cycle differences are expected to influence dispersal of glochidia in at least two ways. First, temperature at release is expected to correspond to developmental rates of glochidia within the fish host (Gillooly et al. 2001), and perhaps net movement of glochidia may increase with time encysted on a fish host. Second, dispersal behavior may differ seasonally in fish hosts (e.g., migration associated with spring reproduction), or local fish assemblage structure may fluctuate seasonally (Gelwick 1990). Brooding characters and other glochidial characteristics (presence/absence of hooks, etc.) are widely studied across unionids and appear to be strongly related to phylogenetic history in the group (Lydeard et al. 1996; Graf and Ó Foighil 1999). This information should permit careful selection of taxa for comparative phylogeographic study.

SUMMARY AND CONCLUSIONS

Molluscan diversification is rapid, especially in the tropics (Flessa and Jablonski 1996). The tremendous amount of diversity found in aquatic mollusks is nevertheless enigmatic. In the oceans, few obvious physical barriers separate populations—a mechanism for allopatric speciation (Mayr 1942)—yet there is a surprising amount of species (and phylogeographic) diversity (Palumbi 1994). In freshwater ecosystems, dispersal from one body of water to the next or even within a drainage is a limiting step that tends to generate increased diversity,

Table 9.2

A compilation of phylogeographic studies of freshwater unionid mussels

Species	Reference	Individuals / Localities Sampled	Genetic Marker(s)	p[a]	Distance Between Samples (km)[b]	Haplotype Distance (km)[c]	Habitat Type[d]
Lampsilis altilis	Roe et al. 2001	5 / 5	16S RNA + CO I	1.01	170	75	1
Lampsilis australis	Roe et al. 2001	5 / 5	16S RNA + CO I	1.89	130	0	1
Lampsilis subangulata	Roe et al. 2001	3 / 3	16S RNA + CO I	0.95	80	0	1
Lampsilis perovalis	Roe et al. 2001	5 / 5	16S RNA + CO I	1.15	140	0	1
Lasmigona subviridis	King et al. 1999	37 / 9	COI	0.23	840	840	1
Lasmigona subviridis	King et al. 1999	40 / 9	ITS-1	0.49	840	375	1
Megalonaias nervosa	King et al. 1999	7 / 6	16S RNA	0.2	1,100	600	2
Lampsilis hydiana	Turner et al. 2000	240 / 14	16S RNA	0.76	500	500	2
Pyganodon (Anodonta) grandis	Liu et al. 1996 Whole mtDNA	72 / 5	RFLP	0.11	335	335	2
Amblema elliotti	Mulvey et al. 1996	11 / 2	16S RNA	0	80	80	3
Amblema plicata	Mulvey et al. 1996	5 / 2	16S RNA	0	115	115	3
Megalonaias nervosa boykiniana	Mulvey et al. 1996	17 / 11	16S RNA	0.5	1,130	1,130	3

[a]p = average pairwise genetic distance between (female) haplotypes identified in each species.

[b]Estimate of the maximum linear distance separating sampling localities.

[c]Estimate of the distance separating populations with a shared haplotype.

[d]Habitat type: 1 = creeks and small rivers, 2 = generalist, 3 = large rivers. Habitat preferences, when not described by the authors of the study, were from descriptions in Cummings et al. (1992), and Illinois Natural History Survey and Florida Museum of Natural History mollusk databases.

but there are predictable ways in which genetic and species diversity will be modified by the life history traits of different molluscan groups. In short, the intrinsic mechanisms used by molluscan species for recognition, development, and dispersal are crucial for producing and maintaining cohesive species groups.

The interactions of these intrinsic features with the extrinsic environment generate the phylogeographic history and patterns of interest. In some cases, we have only begun to realize the potential of molluscan systems for illuminating primary mechanisms of diversification. The marine gastropod *Tegula,* for instance, has undergone a tremendous species radiation in the past 4 million years (Hellberg 1998). Although speciation is typically thought to occur when populations are allopatric, many sister species to *Tegula* in the tropical eastern Pacific are currently sympatric and probably have been for most of their independent evolutionary histories. Landscape processes, including the dimensionality of the habitat, may play a role in species diversification (Hellberg 1998). If different processes are responsible for the divergence of populations and species when they are distributed along one-dimensional habitats (e.g., rivers or coastlines) than when distributed in two-dimensional habitats (e.g., nearby lakes or islands), then comparing these processes will help illustrate how much a species' response to environmental change depends on intrinsic qualities (which affect the genetic effective population size and migration ability of a species), and how much the response depends on extrinsic forces (including geographic forces, oceanography or hydrology, climate, and substrate type).

Essentially, the time has come when we can integrate a broader amount of information about species and their habitat into phylogeographic analysis. A number of life history traits, beyond larval development and dispersal, will contribute to a species' response to changes in the environment. We can go beyond these phenotypic traits to a diverse set of genetic markers that are informative about both the historical and adaptive milieu for diversification. Thanks to the excellent fossil record available for most molluscan taxa, there are few limits on the amount of resolution that can be obtained for phylogeographic studies, including ancestral population dynamics that may be crucial in establishing which of two hypotheses is more accurate. More thorough sampling, including the use of GIS systems to explicitly map genetic variation onto the landscape, will enable discrimination between subtly different but biologically relevant competing histories.

Although our understanding of coevolution in phylogeography is minimal, it is clear that even a single species has the capability to dramatically change the habitat it lives in and these changes may influence the genetic structure of co-occurring taxa. The intertidal snail *Littorina littorea* is a perfect example.

Studied for more than a century as an example of a species that may have been introduced by humans from Europe to North America (Ganong 1886), this snail has had a dramatic impact on the abundance and distribution of other intertidal species (Bertness 1984), including other molluscan species. Examining how this species arrived in North America is an important facet in the study of this system. Recent analysis of *L. littorea* indicates that this species has in fact maintained a limited (and paradoxical) geographic distribution on both the New England and European coasts of the Atlantic (Wares et al., unpublished data). Although *L. littorea* has clearly been extant in Europe for a much longer time (Vermeij 1991), populations in North America were founded long before European colonization of North America (by humans). Phylogeographic exposure of the extraordinary history of trans-Atlantic range expansion, followed by geographic isolation in the Canadian Maritimes until recent times, indicates that we must be cautious in speculating on the sources of diversity in natural populations and that interactions among species in forming phylogeographic patterns may be crucial (Wares 2002; Sotka et al., in press).

The interaction of molluscan species with their community is a vital interaction that must be better understood for precise phylogeographic hypothesis testing to be carried out. Whether the association is based on diet or substrate (Ingólfsson 1992, 1995), parasites, hosts, or commensal interactions, or more general geographic or landscape patterns (e.g., Hellberg 1998), more attention must be paid to these interactions. Furthermore, the appropriate use of null models (Haydon et al. 1994) and comparative studies (Cunningham and Collins 1998; Avise 2000; Wares and Cunningham 2001) must be examined to accurately determine the mechanisms involved. These mechanisms, both extrinsic to the species and part of their natural traits of reproduction, migration, and development, will lead to a better comprehension of the evolutionary forces acting on these traits (Collins et al. 1996).

Most importantly, we believe that a new era in phylogeography is arriving that not only informs us about the history of individual species, but will also detail the interactions that have led to broad evolutionary events in the phylogenetic history of mollusks and other organisms. Many of the comparisons suggested in this chapter are at the family level; to what extent are the distinct features of these higher-level groups a phylogenetic accident, and to what extent are these features deterministic *of* the phylogenetic history of mollusks? Throughout time, diversification in different lineages has accelerated and declined because of the interaction of species with their environment (see Jablonski et al. 1996). These interactions have shaped the extant molluscan fauna in ways that cannot always be isolated by a collection of sequence data.

ACKNOWLEDGMENTS

We would like to thank Megan McPhee, Dominique Aló, Cynthia Riginos, Peter Marko, Mike Hellberg, Randy DeJong, Sam Loker, Jess Morgan, and Jerry Hilbish for organizational and editorial help in writing this chapter. J. Wares would also like to thank Cliff Cunningham for his advice and assistance in this process, and Geerat Vermeij for many helpful discussions about molluscan evolution.

REFERENCES

Allen, Y. C., and C. W. Ramcharan. 2001. *Dreissena* distribution in commercial waterways of the US: Using failed invasions to identify limiting factors. Canadian Journal of Fisheries and Aquatic Sciences 58:898–907.

Avise, J. C. 1992. Molecular population structure and the biogeographic history of a regional fauna: A case history with lessons for conservation biology. Oikos 63: 62–76.

Avise, J. C. 1994. Molecular Markers, Natural History, and Evolution. Chapman and Hall, New York.

Avise, J. C. 2000. Phylogeography: The History and Formation of Species. Harvard University Press, Cambridge, MA.

Barnes, R. S. K., P. Calow, P. J. W. Olive, D. W. Golding, and J. I. Spicer. 2001. The Invertebrates: A Synthesis, 3rd ed. Blackwell, Oxford.

Barsotti, G., and C. Meluzzi. 1968. Observasioni su *Mytilus edulis* L. e *Mytilus galloprovincialis* Lamarck. Conchiglie 4:50–58.

Beerli, P., and J. Felsenstein. 2001. Maximum likelihood estimation of a migration matrix and effective population size in *n* subpopulations by using a coalescent approach. Proceedings of the National Academy of Sciences U.S.A. 98:4563–4568.

Berg, D., S. I. Guttman, and E. G. Cantonwine. 1997. Micro- and macro-genetic differentiation among mussel populations. P. 284 in Conservation and Management of Freshwater Mussels II: Initiatives for the Future (K. S. Cummings, A. C. Buchanan. C. A. Mayer, and T. J. Naimo, eds.). Proceedings of a UMRCC Symposium, St. Louis, MO, Illinois Natural History Survey, Champaign, IL.

Bermingham, E., and J. C. Avise. 1986. Molecular zoogeography of fresh-water fishes in the southeastern United States. Genetics 113:939–965.

Bermingham, E., and C. Moritz. 1998. Comparative phylogeography: Concepts and applications. Molecular Ecology 7:367–369.

Bertness, M. D. 1984. Habitat and community modification by an introduced herbivorous snail. Ecology 65:370–381.

Bertness, M. D. 1999. The Ecology of Atlantic Shorelines. Sinauer Associates, Sunderland, MA.

Beynon, C. M., and D. O. F. Skibinski. 1996. The evolutionary relationships between three species of mussel (*Mytilus*) based on anonymous DNA polymorphism. Journal of Experimental Marine Biology and Ecology 203:1–10.

Bowen, D. Q., and G. A. Sykes. 1988. Correlation of marine events and glaciations on the northeast Atlantic margin. Philosophical Transactions of the Royal Society of London B 318:619–635.

Briggs, J. C. 1974. Marine Zoogeography. McGraw-Hill, New York.

Burton, R. S. 1998. Intraspecific phylogeography across the Point Conception biogeographic boundary. Evolution 52(3):734–745.

Charbonnel, N., B. Angers, R. Razatavonjizay, P. Bremond, and P. Jarne. 2000. Microsatellite variation in the freshwater snail *Biomphalaria pfeifferi*. Molecular Ecology 9:1006–1007.

Coker, R., A. Shira, H. Clark, and A. Howard. 1921. Natural history and propagation of fresh-water mussels. Bulletin of the United States Bureau of Fisheries 37:75–181.

Collin, R. 2001. The effects of mode of development on phylogeography and population structure of North Atlantic *Crepidula* (Gastropoda: Calyptraeidae). Molecular Ecology 10:2249–2262.

Collins, T. M., K. Frazer, A. R. Palmer, G. J. Vermeij, and W. M. Brown. 1996. Evolutionary history of northern hemisphere *Nucella* (Gastropoda, Muricidae): Molecular, morphological, ecological, and paleontological evidence. Evolution 50:2287–2304.

Cooley, L. R., and D. Ó Foighil. 2000. Phylogenetic analysis of the Sphaeriidae (Mollusca: Bivalvia) based on partial mitochondrial 16S rDNA gene sequences. Invertebrate Biology 119:299–308.

Counts, C. L., III. 1986. The zoogeography and history of the invasion of the United States by *Corbicula fluminea* (Bivalvia: Corbiculidae). American Malacological Bulletin, Special Edition 2:7–39.

Cunningham, C. W., and T. M. Collins 1994. Developing model systems for molecular biogeography: Vicariance and interchange in marine invertebrates. Pp. 405–433 in Molecular Ecology and Evolution: Approaches and Applications (B. Schierwater, B. Streit, G. Wagner, and R. DeSalle, eds.). Birkhauser Verlag, Basel.

Cunningham, C. W., and T. M. Collins. 1998. Beyond area relationships: Extinction and recolonization in molecular marine biogeography. Pp. 297–321 in Molecular Approaches to Ecology and Evolution (R. DeSalle and B. Schierwater, eds.). Birkhauser Berlin.

Dahlgren, T. G., J. R. Weinberg, and K. M. Halanych. 2000. Phylogeography of the ocean quahog (*Arctica islandica*): Influences of paleoclimate on genetic diversity and species range. Marine Biology 137:487–495.

Davis, G. M. 1980. Snail hosts of Asian Schistosoma infecting man: Evolution and coevolution. Pp. 195–238 in The Mekong Schistosome (J. Bruce and S. Sornmani, eds.). Malacological Review, MI.

Day, A. J., H. P. Leinas, and M. Austersrud. 1993. Allozyme differentiation of populations of the dogwhelk *Nucella lapillus*, (L.): The relative effects of geographic distance and variation in chromosome number. Biological Journal of the Linnean Society 51:257–277.

DeJong, R. J., J. A. T. Morgan, W. L. Paraense, J. P. Pointer, M. Amarista, and 24 others. 2001. Evolutionary relationships and biogeography of *Biomphalaria* (Gastropoda: Planorbidae) with implications regarding its role as a host of the human bloodfluke. Molecular Biology and Evolution 18:2225–2239.

DeWitt, T. J., A. Sih, and D. S. Wilson. 1998. Costs and limits of phenotypic plasticity. Trends in Ecology and Evolution 13:77–81.

Dillon, R. T. 2000. The Ecology of Freshwater Molluscs. Cambridge University Press, Cambridge.

Dillon, R. T., and J. J. Manzi. 1992. Population genetics of the hard clam, *Mercenaria mercenaria,* at the northern limit of its range. Canadian Journal of Fisheries and Aquatic Sciences 49:2574–2578.

Dybdahl, M. F., and C. M. Lively. 1995. Diverse, endemic and polyphyletic clones in mixed populations of a fresh-water snail (*Potamopyrgus antipodarum*). Journal of Evolutionary Biology 8:385–398.

Echelle, A. A., and T. E. Dowling. 1992. Mitochondrial DNA variation and evolution of the death-valley pupfishes (*Cyprinodon,* Cyprinodontidae). Evolution 46:193–206.

Edwards, S. V., and P. Beerli. 2000. Perspective: Gene divergence, population divergence, and the variance in coalescence time in phylogeographic studies. Evolution 54:1839–1854.

Emlet, R. B. 1995. Developmental mode and species geographic range in regular sea urchins (Echinodermata: Echinoidea). Evolution 49:476–489.

Engle, V. D., and J. K. Summers. 1999. Latitudinal gradients in benthic community composition in Western Atlantic estuaries. Journal of Biogeography 26:1007–1023.

Erlandsson, J., E. Rolán-Alvarez, and K. Johannesson. 1998. Migratory differences between ecotypes of the snail *Littorina saxatilis* on Galician rocky shores. Evolutionary Ecology 12:913–924.

Flessa, K., and D. Jablonski. 1996. The geography of evolutionary turnover: A global analysis of extant bivalves. Pp. 376–397 in Evolutionary Paleobiology (D. Jablonski, D. H. Erwin, and J. H. Lipps, eds.). University of Chicago Press, Chicago.

Ganong, W. F. 1886. Is *Littorina littorea* introduced or indigenous? American Naturalist 20:931–940.

Gelwick, F. P. 1990. Longitudinal and temporal comparisons of riffle and pool fish assemblages in a northeastern Oklahoma Ozark stream. Copeia 1990:1072–1082.

Gillooly J. F., J. H. Brown, G. B. West, V. M. Savage, and E. L. Charnov. 2001. Effects of size and temperature on metabolic rate. Science 293:2248–2251.

Graf, D. L., and D. Ó Foighil. 1999. The evolution of brooding characters among the freshwater pearly mussels (Bivalvia: Unionoidea) of North America. Journal of Molluscan Studies 66:157–170.

Graney, R., D. Cherry, J. Rogers Jr., and J. Cairns. 1980. The influence of thermal discharge and substrate composition on the population structure of the Asiatic clam, *Corbicula fluminea,* in the New River, Virginia. Nautilus 94:130–135.

Grosberg, R. K., and C. W. Cunningham. 2001. Genetic structure in the sea: From populations to communities. Pp. 61–84 in Marine Community Ecology (M. D. Bertness, S. D. Gaines, and M. E. Hay, eds.). Sinauer Associates, Sunderland, MA.

Hare, M. P., and J. C. Avise. 1998. Population structure in the American Oyster as inferred by nuclear gene genealogies. Molecular Biology and Evolution 15:119–128.

Haydon, D. T., B. I. Crother, and E. R. Pianka. 1994. New directions in biogeography? Trends in Ecology and Evolution 9:403–406.

Heard, W. 1965. Comparative life histories of North American pill clams (Sphaeriidae: Pisidium). Malacologia 2:381–411.

Heath, D. D., P. D. Rawson, and T. J. Hilbish. 1995. PCR-based nuclear markers identify introduced *Mytilus edulis* genotypes in British Columbia. Aquaculture 137:51.

Hedgecock, D. 1994. Does variance in reproductive success limit effective population sizes of marine organisms? Pp. 122–134 in Genetics and Evolution of Aquatic Organisms (A. R. Beaumont, ed.). Chapman and Hall, London.

Hellberg, M. E. 1994. Relationships between inferred levels of gene flow and geographic distance in a philopatric coral, *Balanophyllia elegans*. Evolution 48:1829–1854.

Hellberg, M. E. 1998. Sympatric sea shells along the sea's shore: The geography of speciation in the marine gastropod *Tegula*. Evolution 52:1311–1324.

Hellberg, M. E., and V. D. Vacquier. 1999. Rapid evolution of fertilization selectivity and lysin cDNA sequences in Teguline gastropods. Molecular Biology and Evolution 16:839–848.

Hellberg, M. E., D. P. Balch, and K. Roy. 2001. Climate-driven range expansion and morphological evolution in a marine gastropod. Science 292:1707–1710.

Herke, S. W., and D. L. Foltz. 2002. Phylogeography of two squid (*Loligo pealei* and *L. plei*) in the Gulf of Mexico and northwestern Atlantic Ocean. Marine Biology 140:103–115.

Hershler, R., H. P. Liu, and M. Mulvey. 1999. Phylogenetic relationships within the aquatic snail genus *Tryonia:* Implications for biogeography of the North American Southwest. Molecular Phylogenetics and Evolution 13:377–391.

Hewitt, G. M. 1996. Some genetic consequences of ice ages, and their role in divergence and speciation. Biological Journal of the Linnean Society 58:247–276.

Hilbish, T. J. 1996. Population genetics of marine species: The interaction of natural selection and historically differentiated populations. Journal of Experimental Marine Biology and Ecology 200:67–83.

Hill, W. G. 1979. Note on effective population-size with overlapping generations. Genetics 92:317–322.

Hillis, D. M., and J. C. Patton. 1982. Morphological and electrophoretic evidence for two species of *Corbicula* (Bivalvia: Corbiculidae) in North America. American Midland Naturalist 108:74–80.

Hoeh, W. R., D. T. Stewart, G. W. Sutherland, and E. Zouros. 1996. Multiple origins of gender-associated mitochondrial DNA lineages in bivalves (Mollusca: Bivalvia). Evolution 50:2276–2286.

Hoskin, M. G. 2000. Effects of the East Australian Current on the genetic structure of a direct developing muricid snail (*Bedeva hanleyi,* Angas): Variability within and among local populations. Biological Journal of the Linnean Society 69:245–262.

Hudson, R. R. 1990. Gene genealogies and the coalescent process. Pp. 1–44 in Oxford

Surveys in Evolutionary Biology, vol. 7 (D. Futuyma and J. Antonovics, eds.). Oxford University Press, Oxford.

Ingólfsson, A. 1992. The origin of the rocky shore fauna of Iceland and the Canadian Maritimes. Journal of Biogeography 19:705–712.

Ingólfsson, A. 1995. Floating clumps of seaweed around Iceland: Natural microcosms and a means of dispersal for shore fauna. Marine Biology 122:13–21.

Isom, B. G. 1986. Historical review of Asiatic clam (*Corbicula*) invasion and biofouling of waters and industries in the Americas. American Malacological Bulletin, sp. ed. 2:1–5.

Jablonski, D., D. H. Erwin, and J. H. Lipps. 1996. Evolutionary Paleobiology. University of Chicago Press, Chicago.

Jacobsen, R., V. E. Forbes, and O. Skovgaard. 1996. Genetic population structure of the prosobranch snail *Potamopyrgus antipodarum* (Gray) in Denmark using PCR–RAPD fingerprints. Proceedings of the Royal Society of London B. 263:1065–1070.

Johannesson, K. 1988. The paradox of Rockall: Why is a brooding gastropod (*Littorina saxatilis*) more widespread than one having a planktonic larval dispersal stage (*L. littorea*)? Marine Biology 99:507–513.

Johannesson, B., and K. Johannesson. 1996. Population differences in behavior and morphology in the snail *Littorina saxatilis:* Phenotypic plasticity or genetic differentiation? Journal of Zoology 240:475–493.

Johannesson K., E. Rolán-Alvarez, and A. Ekendahl. 1995. Incipient reproductive isolation between two sympatric morphs of the intertidal snail *Littorina saxatilis*. Evolution 49:1180–1190.

Johnson, M. S., and R. Black. 1991. Genetic subdivision of the intertidal snail *Bembicium vittatum* (Gastropoda, Littorinidae) varies with habitat in the Houtman Abrolhos islands, western Australia. Heredity 67:205–213.

Johnson, M. S., and R. Black. 1998. Increased genetic divergence and reduced genetic variation in populations of the snail *Bembicium vittatum* in isolated tidal ponds. Heredity 80:163–172.

Johnson, S. G., and E. Bragg. 1999. Age and polyphyletic origins of hybrid and spontaneous parthenogenetic *Campeloma* (Gastropoda: Viviparidae) from the southeastern United States. Evolution 53:1769–1781.

Karl, S. A., and J. C. Avise. 1992. Balancing selection at allozyme loci in oysters: Implications from nuclear RFLPs. Science 256:100–102.

Keigwin, L. D. 1978. Pliocene closing of the Isthmus of Panama, based on biostratigraphic evidence from nearby Pacific Ocean and Caribbean sea cores. Geology 6:630–634.

King, T. L., M. S. Eackles, B. Gjetvaj, and W. R. Hoeh. 1999. Intraspecific phylogeography of *Lasmigona subviridis* (Bivalvia: Unionidae): Conservation implications of range discontinuity. Molecular Ecology 8:S65–S78.

Kingsolving, A. D., and M. B. Bain. 1993. Fish assemblage recovery along a riverine disturbance gradient. Ecological Applications 3:531–544.

Kirby, R. R. 2000. An ancient transpecific polymorphism shows extreme divergence in a multitrait cline in an intertidal snail (*Nucella lapillus* (L.)). Molecular Biology and Evolution 17:1816–1825.

Knowlton, N. 2000. Molecular genetic analyses of species boundaries in the sea. Hydrobiologia 420:73–90.

Knowlton, N., and L. A. Weigt. 1998. New dates and new rates for divergence across the Isthmus of Panama. Proceedings of the Royal Society of London B 265:2257–2263.

Kuhner, M. K., J. Yamato, and J. Felsenstein. 1998. Maximum likelihood estimation of population growth rates based on the coalescent. Genetics 149:429–434.

Lee, Y.-H., T. Ota, and V. D. Vacquier. 1995. Positive selection is a general phenomenon in the evolution of abalone sperm lysin. Molecular Biology and Evolution 12:231–238.

Lewis, K. M., J. L. Feder, and G. A. Lamberti. 2000. Population genetics of the zebra mussel, *Dreissena polymorpha* (Pallas): Local allozyme differentiation within midwestern lakes and streams. Canadian Journal of Fisheries and Aquatic Science 57:637–643.

Liu, L. L., D. W. Foltz, and W. B. Stickle. 1991. Genetic population structure of the southern oyster drill *Stramonita* (=*Thais*) *haemostoma*. Marine Biology 111:71–79.

Liu, H. P., J. B. Mitton, and S. K. Wu. 1996a. Paternal mitochondrial DNA differentiation far exceeds maternal mitochondrial DNA and allozyme differentiation in the freshwater mussel, *Anodonta grandis grandis*. Evolution 50:952–957.

Liu, H. P., J. B. Mitton, and S. J. Herrmann. 1996b. Genetic differentiation in and management recommendations for the freshwater mussel, *Pyganodon grandis* (Say, 1829). American Malacological Bulletin 13:177–124.

Lydeard, C., and R. L. Mayden. 1995. A diverse and endangered aquatic ecosystem of the southeast United States. Conservation Biology 9:800–805.

Lydeard, C., R. L. Minton, and J. D. Williams. 2000. Prodigious polyphyly in imperiled freshwater pearly-mussels (Bivalvia: Unionidae): A phylogenetic test of species and generic designations. Pp. 145–158 in The Evolutionary Biology of the Bivalvia (E. M. Harper, J. A. Crame, and J. D. Taylor, eds.). Geological Society of London, Special Publication 177.

Lydeard, C., M. Mulvey, G. M. Davis. 1996. Molecular systematics and evolution of reproductive traits of North American freshwater unionacean mussels (Mollusca: Bivalvia) as inferred from 16S rRNA gene sequences. Philosophical Transactions of the Royal Society of London B 351:1593–1603.

Lydeard, C, and K. J. Roe. 1998. Phylogenetic systematics: The missing ingredient in the conservation of freshwater unionid bivalves. Fisheries 23:16–17.

Marincovich, L., and A. Y. Gladenkov. 1999. Evidence for an early opening of the Bering Strait. Nature 397:149–151.

Marko, P. B. 1998. Historical allopatry and the biogeography of speciation in the Prosobranch snail genus *Nucella*. Evolution 52:757–774.

Marko, P. B. 2002. Fossil calibration of the COI molecular clock reveals ancient divergence times for geminate species pairs separated by the Isthmus of Panama. Molecular Biology and Evolution 19:2005–2021.

Marko, P. B., and J. B. C. Jackson. 2001. Patterns of morphological diversity among and within arcid bivalve species pairs separated by the isthmus of Panama. Journal of Paleontology 75:590–606.

Marsden, J. E., A. P. Spidle, and B. May. 1996. Review of genetic studies of *Dreissena* spp. American Zoologist 36:259–270.

Mavarez, J., M. Amarista, J. P. Pointier, and P. Jarne. 2000. Microsatellite variation in the freshwater schistosome transmitting snail *Biomphalaria glabrata*. Molecular Ecology 9:1009–1011.

Mayden, R. L. 1988. Vicariance biogeography, parsimony, and evolution in North American freshwater fishes. Systematic Zoology 37:329–355.

Mayr, E. 1942. Systematics and the Origin of Species. New York: Columbia University Press.

McCauley, D. E. 1991. Genetic consequences of local population extinction and recolonization. Trends in Ecology and Evolution 6:5–8.

McLeod, M. 1986. Electrophoretic variation in North American *Corbicula*. Malacological Bulletin, sp. ed. 2:125–132.

McMahon, R. 1983. Ecology of an invasive pest bivalve, *Corbicula*. Pp. 505–561 in The Mollusca, Vol. 6 (W. Russell-Hunter, ed.). Academic Press, New York.

Meyer, C. and G. Paulay. In review. Recurrent speciation at biogeographic boundaries drives diversification of reef fauna.

Minckley, W. L., D. A. Hendrickson, and C. E. Bond. 1986. Geography of western North American freshwater fishes: Descriptions and relationships to intracontinental tectonism. Pp. 519–613 in The Zoogeography of North American Freshwater Fishes (C. E. Hocutt and E. O. Wiley, eds.). Wiley, New York.

Monsutti, A., and N. Perrin. 1999. Dinucleotide microsatellite loci reveal a high selfing rate in the freshwater snail *Physa acuta*. Molecular Ecology 8:1076–1078.

Muller, J., S. Woll, U. Fuchs, A. Seitz. 2001. Genetic interchange *of Dreissena polymorpha* populations across a canal. Heredity 86:103–109.

Mulvey, M., C. Lydeard, D. L. Pyer, K. M. Hicks, J. Brim-Box, J. D. Williams, and R. S. Butler. 1997. Conservation genetics of North American freshwater mussels *Amblema* and *Megalonaias*. Conservation Biology 11:868–878.

Nei, M., and W.-H. Li. 1979. Mathematical model for studying genetic variation in terms of restriction endonucleases. Proceedings of the National Academy of Sciences U.S.A. 76:5269–5273.

Neves, R. J. 1999. Conservation and commerce: Management of freshwater mussel (Bivalvia: Unionoidea) resources in the United States. Malacologia 41:461–474.

Neves, R., and J. Widlak. 1987. Habitat ecology of juvenile freshwater mussels (Bivalvia: Unionidae) in a headwater stream in Virginia. American Malacological Bulletin 5:1–7.

Nielsen, R., and J. Wakeley. 2001. Distinguishing migration from isolation: A Markov Chain Monte Carlo approach. Genetics 158:885–896.

Ó Foighil, D., T. J. Hilbish, and R. M. Showman. 1996. Mitochondrial gene variation in *Mercenaria* clam sibling species reveals a relict secondary contact zone in the western Gulf of Mexico. Marine Biology 126:675–683.

Palumbi, S. R. 1992. Marine speciation on a small planet. Trends in Ecology and Evolution 7:114–118.

Palumbi, S. R. 1994. Genetic divergence, reproductive isolation, and marine speciation. Annual Review of Ecology and Systematics 25:547–572.

Palumbi, S. R., F. Cipriano, and M. P. Hare. 2001. Predicting nuclear gene coalescence from mitochondrial data: The three-times rule. Evolution 55:859–868.

Park, J.-K., and D. Ó Foighil. 2000a. Genetic diversity of oceanic island *Lasaea* (Mollusca: Bivalvia) lineages exceeds that of continental populations in the northwestern Atlantic. Biological Bulletin 198:396–403.

Park, J.-K., and D. Ó Foighil. 2000b. Sphaeriid and corbiculid clams represent separate heterodont bivalve radiations into freshwater environments. Molecular Phylogenetics and Evolution 14:75–88.

Peterson, M. A., and R. F. Denno. 1997. The influence of intraspecific variation in dispersal strategies on the genetic structure of planthopper populations. Evolution 51:1189–1206.

Perez-Losada, M., A. Guerra, and A. Sanjuan. 1999. Allozyme differentiation in the cuttlefish *Sepia officinalis* (Mollusca: Cephalopoda) from the NE Atlantic and Mediterranean. Heredity 83:280–289.

Pilsbry, H. A. 1911. Non-marine mollusca of Patagonia. Report of the Princeton University expedition to Patagonia, 1896–1899.

Quesada, H., R. Wenne, and D. O. F. Skibinski. 1995. Differential introgression of mitochondrial DNA across species boundaries within the marine mussel genus *Mytilus*. Proceedings of the Royal Society of London B 262:51–56.

Rawson, P. D., and T. J. Hilbish. 1995. Evolutionary relationships among the male and female mitochondrial DNA lineages in the *Mytilus edulis* species complex. Molecular Biology and Evolution 12:893–901.

Rawson, P. D., and T. J. Hilbish. 1998. Asymmetric introgression of mitochondrial DNA among European populations of blue mussels (*Mytilus* spp.). Evolution 52:100–108.

Rawson, P. D., K. Joyner, and T. J. Hilbish. 1996a. Evidence for intragenic recombination with a novel genetic marker that distinguishes mussels in the *Mytilus edulis* species complex. Heredity 77:599–607.

Rawson, P. D., C. L. Secor, and T. J. Hilbish. 1996b. The effects of natural hybridization of on the regulation of doubly uniparental mtDNA inheritance in blue mussels (*Mytilus* spp.). Genetics 144:241–248.

Reid, D. G., E. Rumbak, and R. H. Thomas. 1996. DNA, morphology and fossils: Phylogeny and evolutionary rates of the gastropod genus *Littorina*. Philosophical Transactions of the Royal Society of London B 351:877–895.

Renard, E., V. Bachmann, M. L. Cariou, and J. C. Moreteau. 2000. Morphological and molecular differentiation of invasive freshwater species of the genus *Corbicula* (Bivalvia, Corbiculidea) suggest the presence of three taxa in French rivers. Molecular Ecology 9:2009–2016.

Riginos, C., M. J. Hickerson, C. M. Henzler, and C. W. Cunningham. In review. Differential patterns of male and female trans-Atlantic gene flow in the blue mussel, *Mytilus edulis*.

Riginos, C., K. Sukhdeo, and C. W. Cunningham. 2002. Extreme discordance of allozyme and non-allozyme introgression across a mussel hybrid zone. Molecular Biology and Evolution 19:347–351.

Rocha-Olivares, A., and R. D. Vetter. 1999. Effects of oceanographic circulation on the gene flow, genetic structure, and phylogeography of the rosethorn rockfish (*Sebastes helvomaculatus*). Canadian Journal of Fisheries and Aquatic Sciences 56:803–813.

Roe, K. J., P. D. Hartfield, and C. Lydeard. 2001. Phylogeographic analysis of the threatened and endangered superconglutinate-producing mussels of the genus *Lampsilis* (Bivalvia: Unionidae). Molecular Ecology 10:225–2234.

Roe, K. J., and C. Lydeard. 1998. Molecular systematics of the freshwater mussel genus *Potamilus* (Bivalvia: Unionidae). Malacologia 39:195–205.

Roff, D. A. 1988. The evolution of migration and some life-history parameters in marine fishes. Environmental Biology of Fishes 22:133–146.

Roff, D. A. 1991. Life-history consequences of bioenergetic and biomechanical constraints on migration. American Zoologist 31:205–215.

Rolán-Alvarez, E., K. Johannesson, and J. Erlandsson. 1997. The maintenance of a cline in the marine snail *Littorina saxatilis:* The role of home site advantage and hybrid fitness. Evolution 51:1838–1847.

Roy, K., D. Jablonski, and J. W. Valentine. 2001. Climate change, species range limits and body size in marine bivalves. Ecology Letters 4:366–370.

Saavedra, C., M.-I. Reyero, and E. Zouros. 1997. Male-dependent doubly uniparental inheritance of mitochondrial DNA and female-dependent sex-ratio in the mussel *Mytilus galloprovincialis*. Genetics 145:1073–1082.

Samadi, S., E. Artiguebielle, A. Estoup, J. P. Pointier, J. F. Silvain, J. Heller, M. L. Cariou, and P. Jarne. 1999. Density and variability of dinucleotide microsatellites in the parthenogenetic polyploid snail *Melanoides tuberculata*. Molecular Ecology 7:1233–1236.

Sarver, S. K., M. C. Landrum, and D. W. Foltz. 1992. Genetics and taxonomy of ribbed mussels (*Geukensia* spp.). Marine Biology 113:385–390.

Scheltema, R. S. 1986. On dispersal and planktonic larvae of benthic invertebrates: An eclectic overview and summary of problems. Bulletin Marine Science 39:290–322.

Sheldon, A. L., 1968. Species diversity and longitudinal succession in stream fishes. Ecology 49:193–198.

Siripattrawan, S., J. K. Park, and D. Ó Foighil. 2000. Two lineages of the introduced Asian freshwater clam *Corbicula* occur in North America. Journal of Molluscan Studies 66:423–429.

Small, M. P., and E. M. Gosling. 2000a. Genetic structure and relationships in the snail species complex *Littorina arcana* Hannaford Ellis, *L. compressa* Jeffreys and *L. saxatilis* (Olivi) in the British Isles using SSCPs of cytochrome-*b* fragments. Heredity 84:692–701.

Small, M. P., and E. M. Gosling. 2000b. Species relationships and population structure of *Littorina saxatilis* Olivi and *L. tenebrosa* Montagu in Ireland using single-strand conformational polymorphisms. Molecular Ecology 9:39–52.

Smith, G. R., 1978. Biogeography of intermountain fishes. Pp. 17–42 in Intermountain Biogeography: A Symposium. Great Basin Naturalist Memoirs. Vol. 2. Brigham Young University Press, Salt Lake City, UT.

Sotka, E. E., J. P. Wares, and M. E. Hay. (in press). Geographic and genetic variation in feeding preference for chemically-defended seaweeds. Evolution.

Stepien, C. A., A. N. Hubers, and J. L. Skidmore. 1999. Diagnostic genetic markers and evolutionary relationships among invasive dreissenoid and corbiculoid bivalves in North America: Phylogenetic signal from mitochondrial 16S rDNA. Molecular Phylogenetics and Evolution 13:31–49.

Strayer, D. L. 1991. Projected distribution of the zebra mussel, Dreissena polymorpha, in North America. Canadian Journal of Fisheries and Aquatic Science 48:1389–1395.

Strayer, D. L., and J. Ralley. 1993. Microhabitat use by an assemblage of stream-dwelling unionaceans (Bivalvia), including two rare species of Alasmidonta. Journal of the North American Benthological Society 12:247–258.

Takada, Y., and E. Rolán-Alvarez. 2000. Assortative mating between phenotypes of the intertidal snail Littorina brevicula: A putative case of incipient speciation? Ophelia 52:1–8.

Taylor, D. W. 1985. Evolution of freshwater drainages and mollusks in western North America. Pp. 256–321 in Late Cenozoic History of the Pacific Northwest (A. E. Levinton, ed.). American Association for the Advancement of Science, San Francisco, CA.

Tatarenkov, A., and K. Johannesson. 1998. Evidence of a reproductive barrier between two forms of the marine periwinkle Littorina fabalis (Gastropoda). Biological Journal of the Linnean Society 63:349–365.

Tatarenkov, A., and K. Johannesson. 1999. Micro- and macrogeographic allozyme variation in Littorina fabalis; Do sheltered and exposed forms hybridize? Biological Journal of the Linnean Society 67:199–212.

Templeton, A. R., E. Routman, and C. A. Phillips 1995. Separating population structure from population history: A cladistic analysis of the geographical distribution of mitochondrial DNA haplotypes in the tiger salamander, Ambystoma tigrinum. Genetics 140:767–782.

Thorson, G. 1950. Reproductive and larval ecology of marine bottom invertebrates. Biological Review 25:1–45.

Turner, T. F., and J. C. Trexler. 1998. Ecological and historical associations of gene flow in darters (Teleostei: Percidae). Evolution 53:1781–1801.

Turner, T. F., J. C. Trexler, J. L. Harris, and J. L. Haynes. 2000. Nested cladistic analysis indicates population fragmentation shapes genetic diversity in a freshwater mussel. Genetics 154:777–785.

Turner, T. F., J. P. Wares, and J. R. Gold. 2002. Genetic effective size is three orders of magnitude smaller than adult census size in an abundant, estuarine-dependent marine fish (Sciaenops ocellatus) Genetics 162:1329–1339.

Vaughn, C. C., and C. M. Taylor. 2000. Macroecology of a host–parasite relationship. Ecography 23:11–20.

Vermeij, G. J. 1987. Evolution and Escalation, Princeton University Press.

Vermeij, G. J. 1991. Anatomy of an invasion: The trans-Arctic interchange. Paleobiology 17:281–307.

Vermeij, G. J. 1992. Time of origin and biogeographical history of specialized relationships between northern marine plants and herbivorous molluscs. Evolution 46:657–664.

Viard, F., F. Justy, and P. Jarne. 1997. The influence of self-fertilization and population dynamics on the genetic structure of subdivided populations: A case study using microsatellite markers in the freshwater snail *Bulinus trucatus*. Evolution 51:1518–1528.

Wade, M. J., and D. E. McCauley. 1988. Extinction and recolonization: Their effects on the genetic differentiation of local populations. Evolution 42:995–1005.

Wares, J. P. 2000. Abiotic influences on the population dynamics of marine invertebrates. Ph.D. thesis, Duke University.

Wares, J. P. 2002. Community genetics in the Northwestern Atlantic intertidal. Molecular Ecology 11:1131–1144.

Wares, J. P., and C. W. Cunningham. 2001. Comparative phylogeography and historical ecology of the North Atlantic intertidal. Evolution 55:2455–2469.

Wares, J. P., S. D. Gaines, and C. W. Cunningham. 2001. A comparative study of asymmetric migration events across a marine biogeographic boundary. Evolution 55:295–306.

Wares, J. P., D. S. Goldwater, B. Y. Kong, and C. W. Cunningham. 2002. Refuting a controversial case of a human-mediated marine species introduction. Ecology Letters 5:577–584.

Watson, D. C., and T. A. Norton. 1987. The habitat and feeding preferences of *Littorina obtusata* (L.) and *L. mariae* Sacchi et Rastelli. Journal of Experimental Marine Biology and Ecology 112:61–72.

Wesselingh, F. P., G. C. Cadee, and W. Renema. 1999. Flying high: On the airborne dispersal of aquatic organisms as illustrated by the distribution histories of the gastropod genera *Tryonia* and *Planorbarius*. Geologie en Mijnbouw 78:165–174.

Wilke, T., and G. M. Davis. 2000. Intraspecific mitochondrial sequence diversity in *Hydrobia ulvae* and *Hydrobia ventrosa* (Hydrobiidae: Rissooidea: Gastropoda): Do their different life histories affect biogeographic patterns and gene flow? Biological Journal of the Linnean Society 70:89–105.

Wilding, C. S., J. Grahame, and P. J. Mill. 2000. Mitochondrial DNA COI haplotype variation in sibling species of rough periwinkles. Heredity 85:62–74.

Williams, J. D., M. L. Warren, K. S. Cummings, J. L. Harris, and R. J. Neves. 1993. Conservation status of freshwater mussels of the United States and Canada. Fisheries 18:6–22.

Wilson, A. B., K. A. Naish, and E. G. Boulding. 1999. Multiple dispersal strategies of the invasive quagga mussel (*Dreissena bugensis*) as revealed by microsatellite analysis. Canadian Journal of Fisheries and Aquatic Science 56:2248–2261.

Wollenberg, K., and J. C. Avise. 1998. Sampling properties of genealogical pathways underlying population pedigrees. Evolution 52:957–966.

Wooten, M. C., K. T. Scribner, and M. H. Smith. 1988. Genetic variability and systematics of *Gambusia* in the southeastern United States. Copeia 1988:283–289.

Wright, S. 1977. Evolution and the Genetics of Populations: A Treatise. Vol. 4. Variability Within and Among Natural Populations. University of Chicago Press, Chicago.

Zouros, E. 2000. The exceptional mitochondrial DNA system of the mussel family Mytilidae. Genes and Genetic Systems 75:313–318.

ANDREW HUGALL, JOHN STANISIC, AND CRAIG MORITZ

10

PHYLOGEOGRAPHY OF TERRESTRIAL GASTROPODS

The Case of the *Sphaerospira* Lineage and History of Queensland Rainforests

Phylogeography attempts to identify the mechanisms governing the geographic distributions of genealogical lineages (phylogenies) among and within closely related species (Avise 2000; Chapter 9, this volume), particularly the role of historical landscape processes. As such, it mirrors aspects of biogeography using species-level phylogenies (Nelson and Platnick 1981). Both disciplines require some way of describing the congruence or match to the historical landscape, typically using a comparative approach of identifying concordance among un-related, codistributed taxa to infer common extrinsic historical forces. A next step is to link the spatial and temporal scales of intraspecific phylogeography to those of the underlying interspecific phylogeny. This can give insights into the geography and tempo of speciation (Barraclough and Nee 2001) and the connection between spatial patterns of phylogenetic diversity and species richness. For example, comparison of the age of phylogeographic groups with that of species has been taken as a measure of the tempo of allopatric speciation (Avise and Walker 1998). In this study, we go back to the case of a single phylogenetic series of taxa and endeavor to assess congruence or fit to landscape via bioclimatic distribution modeling, using a trans-species mtDNA phylogeny.

mtDNA PHYLOGEOGRAPHY IN LAND SNAILS

Terrestrial snails and slugs have strong potential to retain more historical signal of population structure—to be good "tracers of history"—because the combination of high persistence and low vagility results in numerous differentiated populations at various levels of geographic and genetic isolation. What is the role of this isolation in the evolution of local and of regional species diversity? Here a phylogeographic framework spanning multiple species and including dense sampling of geographic diversity within species may enable partitioning of the relative importance of incidental divergence in allopatry, environmentally driven divergent selection, and parapatric processes in the development of regional and local diversity (Woodruff 1978; Schilthuizen and Gittenberger 1996; Moritz et al. 2000; Barraclough and Nee 2001; Pfenniger and Magnin 2001).

Numerous studies of presumed neutral genetic diversity in land snails have confirmed a strong link between geography and genetic diversity at very fine scales; examples include analysis of allozymes in *Albinaria corrugata* (Schilthuizen and Lombaerts 1994) and microsatellites in *Helix aspersa* (Arnaud et al. 1999, 2001) or *Cepaea nemoralis* (Davison and Clarke 2000). Such studies have also frequently encountered high levels of mtDNA diversity, often structured into several deeply divergent lineages (starting with Thomaz et al. 1996). Various hypotheses have been proposed about the source of this deeper structure; most of them focus on the role of historical vicariance processes in generating the structures and on the metapopulation structure necessary to retain them.

Untangling the more recent processes of the spread and sorting of ancestral diversity from those that created the diversity requires some reasonable alternative information about historical landscape processes. Therefore, more recent intensive regional-scale phylogeographic studies have used historical landscape frameworks for interpreting the distribution of the deeper level aspects of land snail mtDNA diversity. Studies include secondary contact after dispersal from centers of refuge, such as the postglacial expansion of European *Cepaea* (Davison 2000); river drainage patterns of *Discus* (Ross 1999); complex geological histories of the Kanto region of Japanese *Euhadra* (Hayashi and Chiba 2000; Watanabe and Chiba 2001); secondary contact between nascent *Helix* species in North Africa (Guiller et al. 2001); and habitat specialization in *Gyliotrachela* (Schilthuizen et al. 1999). However, some cases involving high levels of mixing, introgression, and possible cryptic species underscore the potential complexity of interpreting the diversity of land snail mtDNA (Ross 1999; Shimizu and Ueshima 2000; Pfenninger and Magnin 2001; Watanabe and Chiba 2001).

Scaling-up taxonomically, and looking for the link between geography and speciation, a lack of concordance between mtDNA phylogeny and species boundaries in *Partula* (Clarke et al. 1996; Goodacre and Wade 2001) and *Albinaria* (Schilthuizen and Gittenberger 1996; Douris et al. 1998) has been attributed variously to peripheral isolation, retention of ancestral polymorphism, and introgression among closely related species. Although these phylogenetic studies have confirmed a geographic component in the gene phylogeny above the species level, they also show the folly of assuming the monophyly of island radiations (e.g., Crete: *Albinaria*, Douris et al. 1998; Society Island: *Partula*, Goodacre and Wade 2001; Hawaii: achatinellid tree snails, Thacker and Hadfield 2000).

VALUE OF BIOCLIMATIC MODELING FOR PHYLOGEOGRAPHY

Lest the genetic data be left to do all of the work—the gene tree versus species tree problem—it is best to reinforce molecular phylogeographies with independent, spatially explicit hypotheses about population structure; bioclimatic distribution modeling has some claim to providing this (Hugall et al. 2002). Both phylogeographic analysis and distribution modeling will work best in a situation where the climate governs the habitat, which is strongly structured by geography, and together these factors essentially govern the distribution of the organism. Bioclimatic distribution modeling also introduces the possibility of predicting the effects of historical climate change on the potential distribution—paleoclimate modeling (Nix and Kalma 1972; Nix 1991; Hilbert and Ostendorf 2001).

This combination of modeling and phylogeography is strongest in groups with a dense geographic distribution and a gradual dispersal, so that gaps in spatial and temporal distributions of lineages can be used to infer extinction. Because they can persist in small areas, phylogeographic structuring in land snails should also be informative about the spatial and temporal patterns of speciation. With phylogeography coupled to distribution modeling providing the spatial context, the molecular clock implicit in mtDNA phylogenies can provide the temporal aspect. This can be explored through analyses of lineage diversification models (Nee et al. 1994; Nee 2001) that can provide insights into the long-term balance between speciation and extinction. Altogether, these may allow identification of the speciation and extinction components in current species diversity.

BRIEF SUMMARY OF BIOGEOGRAPHIC STUDIES OF EAST AUSTRALIAN FORESTS

Hooker (1859) was the first to recognize the deep historical origins of the Australian biota as a mix of "Antarctic and tropical elements"—the Gondwanan element overlain by a more recent Southeast Asian component. Spencer (1896) began the organization of biogeographic regions, recognizing the northeastern "Torresian" region from the southeastern "Bassian." Snails were first applied to these bioregions by Iredale (1937), who identified a subtropical "Oxlyean" zone between the Torresian and the Bassian. Burbidge's (1960) landmark phytogeography recognized the special diversity of Wet Tropics and Border Ranges (the Iredale "Oxlyean"). In summarizing the distribution of birds, Keast (1961, 1981) identified four regions: Cape York (CY), Wet Tropics (WT), Mid-East Queensland (MEQ), and South East Queensland-Border Ranges (SEQ-NNSW) as centers of endemism and possible historical refuges (see Figures 10.1 and 10.2). Kikkawa and Pearse (1969) introduced analytical methods, identifying the Burdekin Gap separating the WT from southeastern Australia at species and genus levels. Schodde and Calaby's (1972) exegesis of birds cemented many of the names for the developing biogeography: Laura Gap, Burdekin Gap, Fitzroy Gap (also known as the St. Lawrence Gap), and the "Tumbunan" biota; the last recognizing the connection between eastern Australian and mid-montaine New Guinea rainforest avifauna. These pioneering works have been followed by cladistic approaches using various taxa (birds, lizards, frogs: Cracraft 1986; Phoracanthine beetles: Wang et al. 1996; plants: Crisp et al. 1995), but have also taken on a new perspective in the work of Henry Nix and colleagues in demonstrating the method and importance of a bioclimatic approach to the natural history of Australia. In 1972, Nix and Kalma introduced paleoclimate analysis to indicate the tenuous link between northeastern Australia and New Guinea rainforests. Nix (1982) formed a climate-based plant growth index classification and developed the concept of the mesothermal "archipelago" of closed forest along eastern Australia, culminating in the bioclimatic modeling of Pleistocene patterns of refugia in Wet Tropics rainforests (Nix 1991).

Joseph et al. (1993) and Joseph and Moritz (1994) provided the first molecular phylogeographic tests of these biogeographic scenarios, showing they were not just ecological, but also historical refuges. Intraspecific structuring in birds across the Burdekin Gap (the foremost biogeographic barrier) was dependent on ecology: rainforest birds showed a 5 to 7% divergence in the mtDNA cytochrome *b* gene, open forest birds did not. Subsequent studies of frog and lizard taxa introduced bioclimatic modeling (McGuigan et al. 1998) and comparative studies to phylogeography across parts of the region (James and Moritz 2000; Moritz 2000; Schäuble and Moritz 2001). These show a trend of increasing distinction among areas from south to north but with little overarching phylogenetic congruence, perhaps in part because of substantial differences in ecology (Schäuble and Moritz 2001). Interspecific phylogenetic studies of vertebrate taxa show old species with little regional diversification, which is interpreted as taxa dominated by extinction (reviewed in Moritz et al. 1997). An

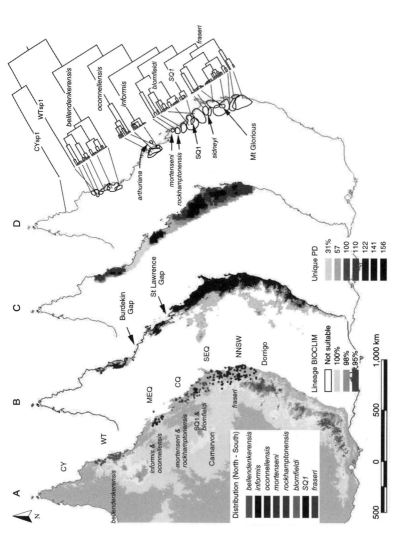

Figure 10.1. Species distribution points, bioclimatic model, and major phylogeographic units mapped onto eastern Australia. In this and other figures of eastern Australia, named biogeographic regions and barriers are indicated (for key, see text). (A) *Sphaerospira* lineage species known distribution points over DEM map. (B) *Sphaerospira* lineage bioclimatic model (current climate). (C) interpolated phylogenetic diversity (PD) from ultrametric tree of 24 major phylogeographic units (expressed as a percentage of tree height, 20-minute grid scale). (D) mtDNA haplotype tree linked to the approximate geographic distribution of the 24 phylogeographic units delimited by the marked 3% thresh-

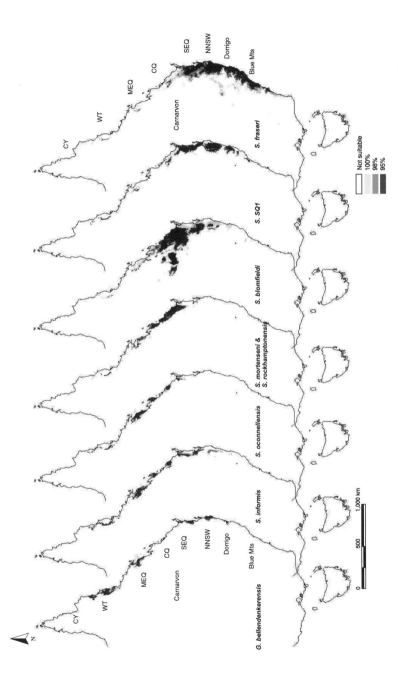

Figure 10.2. Bioclimatic models for seven species or groups based on three climate parameters (annual mean temperature, annual mean precipitation, and precipitation of the driest quarter) using minimum and maximum from the 100%, 98%, and 95% set of observed distribution points (see text). Minimum/maximum (100%) only for *S.* SQ1 and combined *S. mortenseni* and *S. rockhamptonensis* models. The models are arranged left to right for species from north to south.

exception to this pattern is the diversification of Carphodactyline geckos in the MEQ region (Couper et al. 2000); the comparative study of codistributed skinks by Stuart-Fox et al. (2001) strongly supports this pattern as regional endemicity driven by vicariance.

In summary, the history of the mosaic of forests along the east coast of Australia has been dominated by the long-term drying of the climate from the late Miocene on, leading to the decline and fragmentation of a formerly vast and rich rainforest habitat (Archer et al. 1991; Adam 1992; Truswell 1993). As they currently stand, major rainforest domains are restricted to the eastern areas of the Great Dividing Range, centered on a disjunct series of upland areas—an "archipelago" of cooler mesothermal climate. During the Pleistocene, these forests waxed and waned to an uncertain but considerable extent. For mammals and other large vertebrates, extinction caused by habitat contraction is a major theme, with most extant species being quite old (Moritz et al. 1997); however, to the land snail, this mosaic of forests may harbor multiple ongoing processes of vicariant and environmentally driven diversification.

The Australian Camaenidae are probably Miocene/post-Miocene immigrants from a large pool of Laurasian taxa—one of the Malesian elements (Solem 1979, 1997; but see Scott 1997), with the east coast dominated by the Camaeninae, including the Hadroid subgroup containing the "Sphaerospira facies." These large, often striped snails are distributed in a broad band down eastern Australia (see Figure 10.1). Here we apply biogeographic analyses to the trans-species mtDNA phylogeny of a snail lineage identified as mapping to eastern Australian rainforests. Specific questions for this *Sphaerospira* lineage include: Is the sympatric diversity due to filtering of ancestral diversity or in situ diversification? How does geographic population structure relate to speciation? What does this say about the biogeography of Queensland rainforests?

GNAROSOPHIA BELLENDENKERENSIS AS BENCHMARK

Gnarosophia bellendenkerensis is the Wet Tropics member of the *Sphaerospira* lineage. Previous detailed study of *G. bellendenkerensis* in the light of comparative phylogeography studies with vertebrates and bioclimatic modeling (Hugall et al. 2002) demonstrates that the distribution of mtDNA diversity maps a population structure formed by an interaction of climate and geography into patterns of contraction to and expansion from discrete refuges. The bioclimatic distribution modeling, by providing independent information on historical population structure, enables multiple processes that complicate the genetic data to be teased apart. Thus *G. bellendenkerensis* becomes our benchmark for understanding levels of genetic diversity down to a fine scale, and by reasonable extension, guides our interpretation of the larger-scale patterns observed across the entire *Sphaerospira* lineage.

TAXA INCLUDED

The taxa included in our discussion (including outgroups) are distributed across mesothermal, mesic forests in coastal eastern Australia, from Cape York in the north to the rainforests of northeastern New South Wales in the south. To describe the areas, we use a series of regional names established from previous biogeographic studies: Cape York (CY), Wet Tropics (WT), Mideast Queensland (MEQ), Central Queensland (CQ; Byfield and Shoalwater Bay area), Southeast Queensland (SEQ; Bulburin and Kroombit Tops to Main Range), and the McPherson (Border) Ranges and northeastern New South Wales (NNSW).

The lineage we studied, which included eight species and two subspecies, is defined as a monophyletic group in a larger molecular phylogenetic analysis of camaenid snails that encompasses 800 individuals from more than 200 taxa (A. Hugall, unpublished). This same analysis identified two undescribed taxa as outgroups: one taxa endemic to the isolated Iron Range rainforests of Cape York (CYsp1), and one taxa endemic to Thornton Peak within the Wet Tropics rainforests of northeastern Queensland (WTsp1). Sampling of this "*Sphaerospira*" lineage is complete in that all living species and subspecies are represented and geographic sampling is extensive for most taxa. Conversely, some species traditionally included within *Sphaerospira* are excluded because they lie phylogenetically outside the lineage. The taxonomy is from the latest zoological catalogue (Smith 1992) except where noted. All included taxa are terrestrial leaf litter/understory generalists.

The species concerned are *Gnarosophia bellendenkerensis* endemic to the WT rainforests; *Sphaerospira informis,* endemic to MEQ rainforest (the largest camaenid in Australia: 51.7 mm high; 52.1 mm wide; $n = 52$); *S. oconnellensis,* overlapping distribution and microsympatric with *S. informis,* and subspecies *S. o. arthuriana* on the adjacent Whitsunday Islands; *S. mortenseni* and *S. rockhamptonensis,* narrow endemics in central Queensland (CQ); *S. blomfieldi,* broadly distributed in vine scrub, wet sclerophyll, and rainforest across the CQ-SEQ region, and subspecies *S. b. sidneyi* at the southern limit; *Sphaerospira* sp., SQ1 endemic to the Dawes Range (Bulburin SF) in the SEQ region, microsympatric with *S. blomfieldi,* until recently considered as *S. fraseri* (Scott 1996); and *S. fraseri,* broadly distributed in mesic forest and rainforest across

SEQ-NNSW regions. *Sphaerospira fraseri* (Griffith and Pidgeon, 1833) is the type species of *Sphaerospira* (Mörch, 1867).

DISTRIBUTION MODELING

Locality records for bioclimatic modeling were obtained from the Queensland Museum database. The number of locations per species ranged from 315 distribution records for *S. fraseri* to 7 for the narrowly endemic *S. mortenseni* (Table 10.1). Because the number of records was small, *S. mortenseni* was combined with the geographically adjacent sister species, *S. rockhamptonensis,* for modeling (combined $n = 22$). The number of records per 20-minute grid was more consistent, ranging from 4.4 to 12.4. For modeling the total *Sphaerospira* lineage, 728 location records were used. Each species range was measured as the area within 0.25 degrees of each grid cell datapoint, falling within the coastline and within the 100% *Sphaerospira* lineage bioclimatic model.

Climate-based distribution modeling followed the BIOCLIM procedure (Nix 1986), which uses spatially interpolated estimates of annual mean temperature, annual mean precipitation, and precipitation of the driest quarter on a 36-second (approximately 1-km^2) resolution digital elevation model for the whole of Australia (ANUCLIM; Houlder et al. 2000). Bioclimatic models and map analyses were made using ARCVIEW GIS 3.2a ESRI, using only one record per 36-second grid cell per species.

The BIOCLIM procedure sets upper and lower limits for each climate layer based on the upper and lower limits of the observed distribution points, allowing for trimming of outliers. Distributions show absolute minimum and maximum (100%) and limits leaving out 2% and 5% of the observed distribution points at the upper and lower climate limits (98% and 95% models). This approach, although simple and based on few climate parameters, has proven highly effective for modeling current and paleodistributions of species and their rainforest habitats in eastern Australia (Nix 1991; McGuigan et al. 1998), including the Wet Tropics *G. bellendenkerensis* (Hugall et al. 2002).

Models were constructed for each species to estimate the extent of overlap of potential distributions and to visualize the biogeographic implication of physiological evolution. In addition, we combined models across species in two ways: (1) by summing the spatial overlap of individual models (±98% limits); areas present in all models representing a conserved distribution for the lineage; and (2) by modeling the distribution for all species simultaneously using the climate limits of the entire lineage. Quantitative comparisons of models and other map layers used map calculation functions in ARCVIEW.

Table 10.1

Genetic and distribution data for the *Sphaerospira* lineage bioclimatic modeling study

Species	Specimens	Locations /Species	Distribution Points	Max Dxy[a]	Π Species	Π Pops	Records per 20' Cell	98% Model Area[b] (km²)	100% Model Area[c] (km²)	Area[d] (km²)
G. bellendenkerensis	121	53	102	0.152/0.222	0.129	0.010	9.1	21,826	30,084	11,007
S. informis	17	9	75	0.037	0.024	0.008	6.3	27,268	27,268	7,391
S. oconnellensis	15	7	26	0.165/0.211	0.110	0.007	4.6	14,732	14,732	6,619
S. mortenseni	2	1	7	NA	NA	NA	12.4	24,270	24,270	2,898
S. rockhamptonensis	4	2	15	0.02	0.013	0.005	6.7	–	–	–
S. blomfieldi	27	10	116	0.143/0.249	0.091	0.011	5.4	86,790	108,828	27,436
S. SQ1	9	3	13	0.036	0.022	0.013	5.8	36,926	36,926	1,331
S. fraseri	43	22	315	0.162/0.233	0.113	0.016	7.2	86,955	121,293	55,283
S. b. sidneyi	5	2	50	0.11/	NA	NA	4.4	NA	NA	NA
S. o. arthuriana	1	1	9	NA	NA	NA	8.0	NA	NA	NA
Sister group	2	2	6	NA	NA	NA	NA	NA	NA	NA
Mean					0.072	0.011	6.3			
Total[e]	246	112	734					136,080	239,037	109,521

Note: Π Species = nucleotide diversity of entire species; Π Pops = average nucleotide diversity of single site populations within a species.

[a] Maximum divergence, by the COII Tamura-Nei model / alldata GTR-Γ MLK model.

[b] Modeled area using 98% percentile limits.

[c] Modeled area using 100% percentile limits, interpolated from points, 36-second cell = 1 km², *rockhamptonensis* and *mortenseni* combined.

[d] Best estimate of observed range, measured as area within 0.25 degrees from each datapoint, bounded by the coast and the full *Sphaerospira* lineage bioclimatic model.

[e] Total areas may not add because of overlap among models and ranges.

To predict potential distributions under paleoclimates, we used climate shifts estimated for the Wet Tropics region (Nix 1991; see also Hugall et al. 2002), and a modification of this (Nix, personal communication in Williams 1991) that allows for estimated latitudinal variation in paleoclimate parameters along the east Australian coast. To represent the climate ranges of the Pleistocene, we modeled two extremes: the restrictive cool and dry conditions of the last glacial maximum (18,000 years ago, annual mean temperature –3.0°C; annual mean precipitation 67% and precipitation of the driest quarter 75% of current monthly averages), and the more favorable cool-wet period (8,000 to 6,000 years ago: annual mean temperature –2.0°C; annual mean precipitation 120% and precipitation of the driest quarter 170% of current averages). The Williams model has the function –0.05°C per degree of latitude and precipitation layers reduced to 55 to 85% of current monthly averages weighted against summer rainfall. As in previous analyses, the modeled distributions are restricted to the current coastline.

MOLECULAR ANALYSES

DNA was extracted from either freshly collected snails or ethanol-preserved specimens from the Queensland Museum using the Chelex (Bio-Rad) method. All specimens were sequenced by ABI automated sequencing, most with both strands. The final ingroup dataset used here comprises 244 individuals from 110 localities (Table 10.1), for a total of 120 kb of sequence comprising COII, 16S rRNA, and 12S rRNA mtDNA. Sequencing was structured hierarchically. All individuals were sequenced for the most rapidly evolving gene, COII (480 sites), to identify tightly knit phylogeographic lineages. Nineteen individuals representing these phylogeographic lineages were also sequenced for 16S rRNA (450 sites), and a subset of nine of these was sequenced for 12S rRNA (520 sites) to further resolve branch lengths and relationships among the deepest lineages. Sequences were aligned in ClustalX (Thompson et al. 1994). Sequences are lodged with GenBank (accession numbers AY151055 to 151082 and AY151291 to AY151354).

We performed phylogenetic analyses using PAUP* 4.0b8 (Swofford 2000), lineage-through-time analyses using End-Epi v1.0 (Rambaut et al. 1997), tree manipulation using TreeEdit v1.0-a4.61 (Rambaut and Charleston 2000), divergences using the maximum likelihood (ML) model with ultrametric tree, and calculated diversities and divergences using REAP (McElroy et al. 1992). The ML model we used is derived from a larger tree of 66 taxa with all genes (1,450 sites), including the nine *Sphaerospira* lineage specimens with all genes. The model was chosen with reference to the gain in lnL as parameters were added, leading to an empirical base content, four substitution rate category, five median site rate category GTR-Γ model. The topology of this larger tree and the model framework provides a backbone for adding the further subset of taxa delimiting the major lineages within the group (19 taxa with COII and 16S), and then again for all haplotypes within the study lineage (164 haplotypes from 244 individuals, all with COII). With so many taxa, searching the ML tree space was not directly feasible. We therefore determined relationships among closely related haplotypes (within populations) with neighbor-joining and maximum parsimony. We then fixed these relationships and confined ML analyses to relationships between these phylogeographic groups. The wide divergence scale poses difficulties in developing an ultrametric tree. We used the complex site rate and substitution rate ML model to provide a best estimate of branch length for the all-taxa tree.

For lineage-through-time analyses, we generated an ultrametric tree using the ML tree and model, with clock assumption (MLK). This process was checked by comparing branch lengths of pruned trees with ML-modeled subset trees and comparing all gene trees with COII-only trees. The cost of imposing the clock is 88.9 ΔlnL — not considered significant given the number of nodes (using df = taxa – 2; Felsenstein 1981). Compared with COII-only trees, the all-data trees showed negligible difference at the tips of the branches and relative stretching of the tree length across the deeper nodes defining the position of the nearest outgroups and the primary splits within the ingroup.

We conducted lineage-through-time analyses both for all lineages and for recognized taxa only, including the subspecies *arthuriana* and *sidneyi*. These subset trees were pruned from the larger tree using PAUP* 4.0b8 and TreeEdit v1.0-a4.61. Log-linear plots were analyzed for deviation from linear with the Wilcoxon test in End-Epi v1.0, considering uncertainty in branch lengths estimated from nonparametric bootstrapping (Baldwin and Sanderson 1998).

PAUP* handles missing data in ML by calculating the lnL for all possible states. A state close to one for which there is data (a reasonably close relative with the additional genes) will score a higher likelihood, therefore contributing more than a state for which there is no close pattern — PAUP* essentially fills in the missing data with the most likely pattern. Therefore, such apparently large amounts of missing data — deadly for neighbor-joining distance methods — has little effect on likelihood estimations for both topology and branch length, unlike maximum parsimony analysis, which fails for the latter. The likelihood analysis of nested sequence data (incorporating several genes) provides a way of rationalizing the amount of sequence needed to be gathered for large trees with enough sites for accurate branch length estimation.

MOLECULAR DIVERSITY AND PHYLOGENY

Figure 10.1 provides an overview of the region, distribution, and three depictions of the genetic data: phylogeny, spatial phylogenetic diversity, and phylogeographic units. The phylogeny is shown again in Figure 10.3. The *Sphaerospira* lineage forms a discrete well-supported, well-resolved clade, within which each species is monophyletic (Figures 10.1 and 10.3), with the more widely distributed species composed of multiple, deeply divided phylogeographic lineages (see below for more detail). Nucleotide diversity within populations ranges from 0.001 to 0.032 (mean, 0.011; 24 populations with $n \geq$ 4 individuals within a distance of 0.25 degrees), whereas nucleotide diversity within species ranges from 0.013 to 0.129 (mean 0.072), with maximum intraspecific mtDNA divergence up to 0.25 using the MLK model (Table 10.1).

Broadly, the *Sphaerospira* lineage forms a phylogenetically nested series from north to south spanning some 2,000 km and 15 degrees of latitude. Each of the few instances of sympatry combines one lineage with the northernmost representative of a more widely distributed sister clade; examples are WTsp1 (an outgroup) and *G. bellendenkerensis* in WT rainforests; *Sphaerospira oconnellensis* and *S. informis* in MEQ; and *S. blomfieldi* and *S.* SQ1 in SEQ. The northernmost member, *G. bellendenkerensis,* comprises a series of discrete phylogeographic lineages distributed across upland areas of the WT rainforests (see Hugall et al. 2002 for detail). Next to the south is MEQ in which *S. informis* and *S. oconnellensis* each show phylogeographic structure separating the Clarke Ranges (including Eungella) from the southern end of the MEQ region, the Connors Range. Although the divergence is shallow in *S. informis* (<4%), *S. oconnellensis* has a deeply divergent lineage (16%). The two widely distributed species from the south, *S. blomfieldi* (northern SEQ) and *S. fraseri* (southern SEQ to NNSW), consist of multiple, mostly discrete phylogeographic lineages, but in the latter we start to see more geographic overlap between lineages (Figure 10.1). Each of these species has allopatric sister taxa located to the north: *S.*

Figure 10.3. mtDNA haplotype molecular clock tree with lineage-through-time-plot. (A) Combined data likelihood model ultrametric tree of 139 ingroup haplotypes, with taxonomic designations. Heavy lines denote branches inferred from all three genes (1,450 sites); lighter lines denote two genes (930 sites); and black light lines denote COII only (480 sites). For clarity, only 139 of 164 haplotypes are shown, the remainder all fall within the most recent nodes. Range of bootstrap values from various methods and combinations of taxa are shown for some nodes. This is the same tree as in Figure 10.1. (B) Log lineage-through-time plot showing both all lineages (triangles, up to a ceiling of 100) and 10 recognized taxa (diamonds, pruned from lineage tree). Dashed lines are 95% CI bounds from bootstrap resampling; bars are $N(t,T)$, the ex-

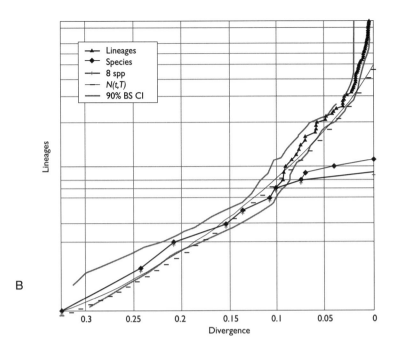

B

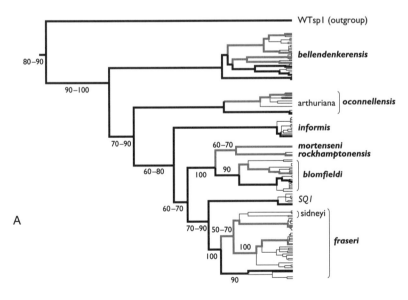

A

pected theoretical *b/d* equilibrium plot from estimate *b* = 0.28 and *d* = 0.16 (from Equation 2 in Harvey et al. 1994). The bar demarks 3% divergence phylogeographic threshold.

mortenseni and *S. rockhamptonensis* adjacent to *S. blomfieldi,* and *S.* SQ1 to the north of *S. fraseri.*

This intraspecific mtDNA diversity shows considerable geographic structuring down to small scales (e.g., *G. bellendenkerensis*); however, we will limit discussion of geographic patterns of mtDNA lineages to those above the level encountered within populations (samples from within a few grid cells), using phylogeographic structure in *G. bellendenkerensis* as a benchmark. Thus in these qualitative descriptions of the phylogeography, a threshold of about 3% divergence (along the molecular clock tree) delimits 24 geographically coherent groups, indicated on Figures 10.1 and 10.3. This phylogeographic scale is appropriate to the sampling we used and to the biogeographic scope of our investigations; study of finer-scale patterns below this level requires more targeted sampling.

BIOCLIMATIC MODELING

Not surprisingly, the potential ranges predicted by paleoclimatic modeling closely reflect the distribution of mesic mesothermal forests on the east coast (individual species in Figure 10.2, combined analyses in Figures 10.1 and 10.4). The model for each species predicts discrete zones, typically coastal wet forests, where other species are found. Most individual species models predict relatively small areas (Table 10.1), except those for *S. blomfieldi* and *S. fraseri,* which predict much larger ranges that include extensive areas not actually occupied to the south and the west (south to central NSW and west to the Carnarvon Ranges). The ratio of actual to modeled ranges for species ranges from 12 to 63% (within 98% bioclimatic limits) and is 80% for the whole lineage model. Quantitatively, model overlap ranges from zero (*S. blomfieldi* and *G. bellendenkerensis*) to 48% (mean 13%; Table 10.2). However, qualitatively, each species model predicts suitable habitat across all of the major biogeographic regions: WT, MEQ, CQ, and SEQ.

These models are spatial manifestations of subtle shifts in environmental envelopes spanning observed current distribution points, presumably reflecting a combination of physiological tolerances and available environmental space, throughout the evolutionary history of this complex. The general picture is one of conserved environmental limits, but with the southern species, *S. blomfieldi* and *S. fraseri,* occupying somewhat drier and cooler habitats, respectively. This overall conservatism, and the departures from it, is reflected in the combined analyses. The total lineage analysis encompasses the individual models and connects the coastal regions from CQ to NNSW, but the Burdekin and

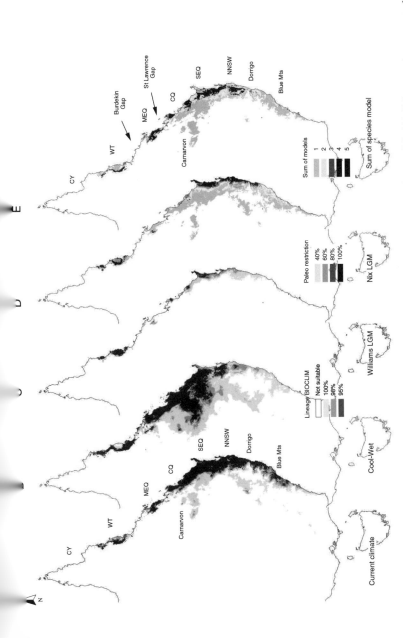

Figure 10.4. Combined lineage bioclimatic paleodistribution models, with sum of individual species model. (A) *Sphaerospira* current climate bioclimatic distribution model. (B) Cool wet scenario model (annual mean temperature [AMT] −2.0°C, annual mean precipitation [AMP] 120%, precipitation of the driest quarter [PDQ] 170%). (C) The Williams (1991) last glacial maximum (LGM) model (for SEQ this equates to AMT −3.0°C, AMP 67%, PDQ 75%). (D) The Nix (1991) wet tropics last glacial maximum model indicated as percentage of the full (100%) last glacial maximum scenario (AMT −3.5°C, AMP 50%, PDQ 60%). (E) Sum of the seven individual species current climate models shown in Figure 10.2.

Table 10.2

Model and range overlap for the *Sphaerospira* lineage bioclimatic modeling study

	bellendenkerensis	informis	oconnellensis	CQ taxa	blomfieldi	SQ1	fraseri
bellendenkerensis	**0.50**	—	—	—	—	—	—
informis	0.20	**0.27**	—	—	—	—	—
oconnellensis	0.14	0.48	**0.45**	—	—	—	—
CQ taxa[a]	0.04	0.21	0.25	**0.12**	—	—	—
blomfieldi	0.00	0.05	0.04	0.21	**0.32**	—	—
SQ1	0.11	0.23	0.03	0.07	0.07	**0.04**	—
fraseri	0.06	0.09	0.00	0.02	0.00	0.33	**0.64**

Note: Overlap between models is expressed as overlap-to-sum ratio. Actual-to-modeled area on diagonal (≥98%). Ranges measured as area within 0.25 degrees of each datapoint, bounded by the coast and the full *Sphaerospira* lineage bioclimatic model.

[a] *S. rockhamptonensis* and *mortenseni* combined.

St. Lawrence gaps remain. The sum of the models (Figures 10.1 and 10.4) highlights the conserved component of the bioclimatic profiles across the whole lineage, these being focused on the wet forests of WT, MEQ, CQ, and SEQ-NNSW, but excluding the wettest areas (eastern WT). When compared with the total lineage model, the sum model serves to highlight the major geographic disjunctions within the range of the *Sphaerospira* lineage, these being the Burdekin Gap between WT and MEQ and the St. Lawrence Gap between MEQ and CQ, in contrast to the relative continuity across SEQ-NNSW.

The last glacial maximum paleoclimatic models for the total lineage predict highly restricted distributions within each of WT, MEQ, CQ, and SEQ-NNSW, with separate predicted refugia in the north and south of WT and northern and central regions in SEQ-NNSW. The most obvious feature of the paleomodels is the elimination of the large extensions to the west and south, these representing substantial overpredictions in the current climate models of *S. blomfieldi* and *S. fraseri,* respectively. Relative to the Wet Tropics last glacial maximum paleoclimate scenario (Nix 1991), the model based on the Williams (1991) paleoclimate parameters predicts more extensive areas in WT and CQ, but greater restriction in SEQ-NNSW. The difference between the actual and modeled ranges for the lineage model corresponds to range contractions predicted in the last glacial maximum paleomodel; the modeled area is reduced by 79%, accounting for 94% of the difference between actual and (current) modeled ranges. At the other extreme, the cool-wet scenario (Figure 10.4) predicts increased connectivity between MEQ and southern areas and indicates that drier areas to the west became suitable, but does not extend the predicted range south of NNSW. Even with these permissive conditions, the Burdekin Gap remains. Preliminary explorations (not shown) suggest that it takes a further 2°C temperature decline and a marked reduction in seasonality to bridge this gap, and even more to connect the present CY rainforests.

The sum model and the last glacial maximum paleomodel are similar in picking out centers for WT, MEQ, two within CQ (Shoalwater and Bulburin/ Kroombit), SEQ, and NNSW. Taken together, the lineage, sum, and paleomodels provide a map of biogeographic regions, their size and stability, and a type of hierarchy of the degree of connection among them: (WT, (MEQ, (SW, (CQ, (SEQ, NNSW)))))). These correspond to the general pattern observed in the phylogeny.

RELATIONSHIP OF BIOCLIMATIC MODELS TO PHYLOGENY

Bioclimatic models of the whole clade, and the sum across taxa, highlight the major disjunctions in the phylogeny. The primary split between the observed ranges of *G. bellendenkerensis* and the other taxa corresponds to the Burdekin Gap and the secondary split between MEQ and south—the St. Lawrence or Fitzroy gap—separating the MEQ taxa *S. informis* from the southern group. This latter region, from Shoalwater south, is more or less continuous in the lineage bioclimatic model, but shows some structuring in the individual species, the sum model, and also in the more restrictive last glacial maximum paleoclimate scenario. The northern extremity of this region forms a narrowly connected bioclimatic domain containing the restricted endemics *S. mortenseni* and *S. rockhamptonensis,* sister group to the widely distributed *S. blomfieldi,* which itself is split into several phylogeographic units. The remainder is split into the CQ (Bulburin) endemic species *S.* SQ1, and its sister species *S. fraseri,* which is widespread further to the south in SEQ-NNSW, and is again dissected into multiple mtDNA sublineages. The bioclimatic models of *S. fraseri* and *S.* SQ1 are similar (33% overlap), with *S.* SQ1 occupying an isolated zone to the north of the main area of models correlating with the range of *S. fraseri* (Figure 10.2). Of the subspecies, *S. o. arthuriana* is distributed on the offshore island of the MEQ region (the Whitsunday Islands) and can be described as a sublineage within the phylogeographic structure of *S. oconnellensis* (but note the relatively limited sampling); and *S. b. sidneyi* forms two related sublineages within the *S. fraseri* mtDNA phylogeographic structure, distributed at the northern extremity of this range and at the edge of the core bioclimatic model.

DISTRIBUTION OF PHYLOGENETIC DIVERSITY

For a quantitative approach to the biogeographic distribution of genetic diversity encompassing the whole phylogeny, we turn to the analyses of the distribution of phylogenetic diversity (Faith 1994), shown in Figures 10.1 and 10.5. The phylogenetic diversity of individual areas is estimated as the sum of branch lengths of lineages within that area, expressed as the percentage of the total height of the tree (rooted with WTsp1 = 0.325). This is represented spatially in two ways: (1) a geographic map of the phylogenetic diversity interpolated from the 244 genetic data points onto 20-minute grids (Figure 10.1; here we consider only the phylogeographic units, i.e., lineages with less than 3% sequence divergence), and (2) as a latitudinal transect (Figure 10.5). At this scale, most grids have genetic data; however, where there is no genetic sampling but a species is (probably) present (known distribution and/or predicted in the bioclimatic model), the value of the surrounding points is interpolated. Where the lineage is absent (never found and/or outside modeled range), only the ancestral internodes can be interpolated (this assumes the presence of ancestors, i.e., steady fine-scale dispersal, some time in the past, a not unreasonable assumption for land snails at this spatial scale). A strict geographic pattern would have only one tip lineage at each point and so the phylogenetic diversity value is the height of the tree. Elevations above this level are due to (1) overlap of tip lineages (e.g., another species and secondary contact), and/or (2) internodes from the overlap of clades—this can be thought of as a measure of ancestral dispersal. Phylogenetic diversity values below the height of the tree imply extinction.

Although 44% of the actual geographic range has phylogenetic diversity greater than the basal value (tree height), only 10% of the total phylogenetic diversity is due to overlap (sympatry) of lineages. The tip component of overlap is due to sympatric species in MEQ and CQ (Bulburin); secondary contact (partial overlap) in WT, SEQ, and NNSW; and a Mount Glorious lineage of *S. fraseri*. Little overlap is due to internodes; across the whole, the ordering of relationships is such that only 3% can be ascribed to ancestral dispersal. In the more complex CQ-NNSW region (among some 14 phylogeographic clades), only 2% of phylogenetic diversity is due to historical dispersal. The principal components of this are due to relationships among *S. blomfieldi* and *S. fraseri* phylogeographic units (maximum overlap phylogenetic diversity 8% to 13%) and implied presence of SQ1 lineage (maximum 10%), but this is slight compared with the component due to tip lineages (presence of SQ1 at Bulburin; overlap of lineages in *S. fraseri* in SEQ, maximum increase 41, or 22%). With the caveat of limited sampling at the western edge of *S. fraseri*, the complete

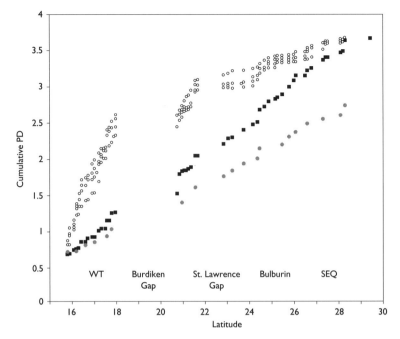

Figure 10.5. mtDNA phylogenetic diversity (PD) accumulation across a latitudinal transect from north to south. All haplotypes (black squares) and 24 phylogeographic groups (grey dots) are from haplotype MLK tree (rooted with WTsp1), per 0.1 degree of latitude. Major biogeographic areas and barriers are indicated. The randomized distribution plot is also indicated (white dots). See text for details.

overlap of lineages on the D'Aguilar Range (in particular Mount Glorious, see Figure 10.1) is less easily dismissed as secondary contact.

If vicariance fragments an otherwise fairly uniform distribution, uniform accumulation of divergence through time (geographically evenly spread) leads to a constant phylogenetic diversity-to-area relationship. If gaps in current geographic distribution represent local extinction of preexisting phylogenetic diversity, we should see departures from linearity, whereas if lineages dispersed to adjacent refugia (overlap), this phylogenetic diversity is maintained, preserving the underlying historical pattern. To examine the consequence of potential extinction of lineages in the climatically unsuitable areas between WT and MEQ (Burdekin Gap) and between MEQ and CQ (St. Lawrence Gap), we plotted cumulative phylogenetic diversity from north to south (Figure 10.5). Nominally the line can take any monotone, and random distributions should give hyperbolic curves. The observed pattern is close to linear (slope = 0.23

phylogenetic diversity per degree of latitude, $R^2 = 0.99$). Furthermore, using only the 24 phylogeographic clades produces a similarly linear relationship. This linearity has two aspects: (1) the phylogenetic diversity needed to span the large Burdekin Gap is found in MEQ as two sympatric lineages, and the phylogenetic diversity gained in the sympatric species of northern SEQ corrects for a slight slump crossing the St. Lawrence Gap (from regression residuals not shown). (2) The within-region intraspecific phylogeographic component of the phylogenetic diversity has the same function as that between regions (= species). This linearity reflects the uniform accumulation of diversity across an essentially intact ancestral geographic system; the implication here is that there has been no large-scale removal of preexisting phylogenetic diversity (i.e., extinction of local clades within the current large gaps in the range).

TEMPO OF DIVERSIFICATION: LINEAGE-THROUGH-TIME ANALYSES

The modeling provides a perspective on climatic limitations on distributions under current conditions and across the climatic extremes of the Quaternary, but does not directly address the longer time scale of speciation implied by the levels of sequence divergence in the molecular phylogeny. Here we further explore this temporal aspect of diversification, in particular, the relationship of phylogeographic groups to speciation, using lineage-through-time analyses (Harvey et al. 1994; Nee et al. 1994). This method uses an ultrametric tree to plot the log of lineage diversity through time and compares the observed trends with those expected with a constant birth-death (b–d, speciation-extinction) process or a constant birth (speciation) process (Nee 2001). This approach requires dense sampling across lineages. Our mtDNA phylogeny traverses the species level. Among species and major phylogeographic groups, sampling is complete; among local populations, the sampling becomes more sparse. Molecular phylogenies only show extant lineages. A key consequence is that lineage accumulation patterns reflect both birth *and* extinction rates. Put simply, the slope from the base of the phylogeny represents net diversification (b–d), whereas the most recent lineages can have a steeper slope dominated by birth rates alone. It is in this framework that changes in speciation/extinction rates (e.g., episodic speciation or mass extinction) must be evaluated.

In the following analysis, we contrast the temporal pattern of lineage accumulation considering only recognized taxa (species only, $n = 8$; or species + subspecies, $n = 10$) versus lineage accumulation that includes the 24 phylogeographic lineages as well as recognized taxa. Inclusion of the latter is prompted by the hypothesis (Avise and Walker 1998) that phylogeographic units represent "species-in-waiting." Fine-scale analysis in *G. bellendenkerensis* indicates that Holocene scales of population expansion from refugia affect only the extreme tips of the tree (Hugall et al. 2002), well within the 3% threshold, and do not contribute to the shape of the lineage-through-time plot.

The tempo of reconstructed cladogenesis along our ultrametric tree is shown in Figure 10.3B. Two levels of diversity have been plotted: all of the observed lineages (the "haplotype tree"), and a phylogeny of recognized taxa only, pruned from the haplotype tree. For the taxon tree, we included both 8- and 10-taxa plots, the latter includes the two subspecies. The 3% divergence threshold is shown. Qualitatively, the plot appears triphasic. From left (basal) to right (recent) the following elements are represented: (1) a region from the root up to around the split between phylogroup and taxon plots (corresponding to maximum intraspecific mtDNA divergence), which is linear and corresponds to a net cladogenesis rate of one lineage formed per 8% divergence (or 0.12 births per lineage per percentage divergence); (2) an intermediate section, between the point of maximum intraspecific divergence and approximately the 3% threshold broadly separating phylogeographic units and within-lineage diversity, which has a steeper slope (Wilcoxon sign test, $p < 0.05$, including bootstrap variance) corresponding to a rate of one lineage formed per 3.4% sequence divergence (or 0.28 births per lineage per percentage divergence); and (3) the most recent segment of less than 3% divergence corresponding mostly to within-population diversity and which is expected to conform to coalescent expectations of population demographics—a birth-minus-death (b–d) process, in this case consistent with a constant population size (endemic transform Wilcoxon test, End-Epi v1.0).

Using the concomitant species/lineage line as a measure of b–d and estimating b from the slope of the midsection (phylogroups) yields the following estimates of lineage birth and death rates: $b = 0.28$, $d = 0.16$ (per lineage per percentage divergence), the birth-to-death (b/d) ratio is 1.7 (56% extinction rate), and a net b–d rate of one birth per lineage per 0.08 divergence (±10% from bootstrap variance in slope). At levels of sequence divergence <3%, the slope (b) trebles and gives a b/d ratio approaching 1, corresponding to intraspecific genetic drift/mutation equilibrium.

The divergence of lineage-through-time plots around the boundary between species and phylogroups indicates a change in tempo. The species curve is linear up to 8 or 9 lineages, then tapers somewhat. Given the hypothesis that "speciation" will take time, we do not necessarily expect new taxa to arise within a short time period relative to this net "speciation" rate. Therefore, including the subspecies *S. o. arthuriana* and *S. b. sidneyi,* the taxon curve is very close to the mtDNA phylogeographic unit *b/d* expectations. Excluding the drift/fixation process within phylogroups (i.e., below the 3% threshold), we have 24 lineages, 10 species or subspecies, with equilibrium expectation of 11 taxa. In this model interpretation, the phylogenetic structure of species diversity (including sympatric species) can be maintained by 20% of phylogeographic units becoming fully independent lineages or recognized species, at a tempo of one per lineage per 0.08 mtDNA divergence time.

IMPLICATIONS OF MOLECULAR CLOCK RATES ON THE AGE OF PHYLOGEOGRAPHIC BREAKS

Although we do not have a reference for absolute time, the relative ages of the different process levels in the tree can be reconciled with interpretations of rainforest history. This in turn gives us some limits to clock rates for the MLK tree. If the ultrametric transform is effective, the depth of the tree, and depth at various levels within it (e.g., the local phylogeographic groups, species), constrains relative times of these levels and forms a temporal and biogeographic reference. For the group to have been shaped by the biogeographic forces outlined in the beginning of this chapter, the tree is bounded by Pleistocene/Holocene ages toward the tips and, more approximately, by a mid-Miocene limit for the deeper nodes, this last representing the time at which the major east coast rainforest blocks are thought to have become isolated (Adams 1992).

Applying recently postulated molecular rates of 10% per million years (Chiba 1999; Hayashi and Chiba 2000; Thacker and Hadfield 2000) places the 3% phylogroup at 0.3 million years, the deeper phylogeographic units (the speciation rate gauge of 0.08 divergence) at 0.8 million years, and the base of the group at 2.4 million years. Given the persistence of the Burdekin Gap to climatic ranges postulated for the Quaternary, and its presence in supraspecific biogeographic analyses, these dates strain credibility. Alternatively, a rate of 2% per million years (per lineage) corresponds to 2, 4, and 12 million years. The latter scenario is more consistent with the time scales usually considered for major events in the biogeography of these wet forest communities.

In biogeography, the issues of how to define regions—the geographic scale at which speciation and extinction shape biodiversity (Rosenzweig 1995; Hubbell 2001)—and how to account for these being different for different taxa (Nelson and Platnick 1981; Crisp et al. 2001)—may be quantitatively addressed by integrating distributional modeling and molecular phylogeny. The modeling and the spatial analysis of phylogenetic diversity (spatial phylogenetic diversity) provide readily integrated perspectives on ecogeographic structure and endemicity. For lineages with high levels of allopatric diversity (e.g., terrestrial snails), different spatial and temporal scales of diversification can be linked via a trans-species phylogenetic analysis.

What are the spatial and temporal scales of extinction and speciation in this system? Here the spatial scales are derived from modeled paleoclimate and actual ranges combined with intraspecific phylogeography, and the temporal scales are derived from lineage-through-time analyses and consilience of phylogeny with preexisting biogeographic inference. Spatial phylogenetic diversity is an attempt to integrate the spatial and temporal information. For this group, the phylogenetic diversity map is relatively simple. With more lineages, more complex phylogenetic diversity landscape patterns can be expected in this temporal endemicity index, thus allowing identification of associated landscape characteristics via quantitative comparisons of phylogenetic diversity and actual or modeled species diversities using GIS. For example, a "museum" of ancestral diversity could have a large phylogenetic diversity–to–species diversity ratio, whereas regions of geographic or adaptive radiations with many young endemics would have a low phylogenetic diversity–to–species diversity ratio. In the current analysis, CQ (Shoalwater/Byfield area) and Kroombit/Bulburin (northern SEQ) are both bioclimatic (ecological) refuges (Figure 10.4) and both contain two species (CQ, *S. mortenseni* and *S. rockhamptonensis;* Kroombit/Bulburin, *S.* SQ1 and *S. blomfieldi*). However, CQ has a much lower phylogenetic diversity (Figure 10.1) and, correspondingly, is less clearly identified as a historical refuge (Figure 10.4).

SHAPE OF VICARIANCE: PHYLOGENY AS BIOGEOGRAPHY

The monophyletic *Sphaerospira* lineage maps to and elaborates on the previously proposed biogeography of the mesic forests of the east Australian mesothermal archipelago. Distribution modeling highlights the discrete nature of these environmental zones and identifies core areas and connections, encapsulated in the lineage and sum species models. The spatial phylogenetic diversity analysis highlights the match of this environmental domain with the phylogeny. The distribution of sister lineages and taxa, in context with the distribution models, suggests that the principal mode of development of these sister lineages is via peripheral vicariant isolation. In each case of the more recently diverged taxa (i.e., *S. rockhamptonensis* and *S. mortenseni, S.* SQ1, *S. o. arthuriana*, and *S. b. sidneyi*), narrowly endemic taxa represent geographically and bioclimatically peripheral isolates of a more widely distributed sister lineage.

Allopatric divergence and vicariance are most clearly evident at the intraspecific phylogeographic level, but what of the entire phylogeny? Broadly, there are two interpretations of the phylogeny: (1) static, comprising a series of historically unique species with essentially ephemeral phylogeographic diversity, or (2) an emergent property of a continuous process of diversification. The first would interpret the lineage-through-time dynamic as different birth rates for events associated with the formation and distribution of the primary lineages (= species resident in the core refugia) and for phylogeographic structuring within these dating from a later era of fragmentation. The second treats it as indicating the balance of a continuous environmentally driven vicariant birth/death process. It would be difficult to argue either way were it not for the additional information in the form of the spatial context.

The lineage-through-time model can be taken as either high birth rates *or* high extinction rates for phylogeography relative to species. The *b/d* model unifies these in one process: gradual allopatric divergence with recognizable species emerging from phylogeographic lineages at a rate of approximately one per 8% mtDNA sequence divergence. Not all phylogeographic units of a species survive to form separate species (only 20%); most fade away either through physical extinction or genetic swamping, dwindling down over time to only the one lineage. *G. bellendenkerensis*, *S. oconnellensis*, and *S. informis* are good examples; each has a very long internode. Most if not all extinction is at the tips, but this does not mean that there was no extinction in the past, just as it does not mean there were no phylogeographic groups in the past.

In lineage-through-time methods, the more clade-specific the extinction, the less visible it is in the reconstructed phylogeny; the more evenly distributed across the phylogeny extinction is, the more visible (Harvey et al. 1994). In *Sphaerospira* this would mean, respectively, the geographically localized elimination of an old lineage versus global contraction and elimination of sublineages within regions. In turn, these scenarios could equate to (1) the deep divisions between lineages separated by the localized Burdekin and St. Lawrence gaps, and (2) strong phylogeographic structure within regions. Although the lineage-through-time analyses may not detect the extinction processes in the first, localized loss of basal lineages should cause sharp reductions in phylogenetic diversity in regions away from historical refugia. However, in the present case, the consistency of phylogenetic diversity down the east coast transect (Figure 10.5) suggests that there has been little loss of ancestral diversity, and that the underlying pattern of phylogenetic diversity is similar to the within-region phylogeographic pattern.

Although the distribution of diversity is the result of a series of specific events—the actual history—the feature of the *b/d* process is that it accounts not just for the recent phylogeography but also for the deeper lineages and the spatial pattern of phylogenetic diversity. Our conclusion is that the *Sphaerospira* lineage is as near to being a perfect vicariant system as nature can devise. In this view, the formation of the phylogeny is contemporaneous with the environmental history; the geography, as embodied in the distribution models, continuously prunes the phylogeny so that it cannot have anything other than a precise fit. This dominating *b/d* process includes the sympatric species: historically *S. oconnellensis* and *S. informis* were sister taxa and *S.* SQ1 and *S. blomfieldi* were sister taxa—that is, their ancestral manifestations were; and to account for filling the gaps in the spatial phylogenetic diversity analysis, *S. oconnellensis* probably started as a phylogeographic unit somewhat to the north of its current distribution, as did *S. blomfieldi*.

BIOGEOGRAPHIC IMPLICATIONS

The potential of snails to provide a fine-grained perspective on the historical biogeography of their habitats is well realized in this study. The combination of spatial phylogenetic diversity and bioclimatic models produces an integrated assessment of historical biogeography, emphasizing more than most previous studies have the endemicity contained within MEQ and historical subdivisions of the wet forests to the south. Of particular interest is the Bulburin/Kroombit area, the separation of NNSW, and the phylogenetic diversity richness of the central SEQ region. The phylogeny and lineage-through-time patterns highlight the considerable age of this fine-grained biogeography.

The phylogeny has spatial and temporal consilience with the proposed biogeography of the region (e.g., the influence of the Burdekin Gap) and provides the first substantial phylogenetic evidence for an effective St. Lawrence (also known as the Fitzroy) Gap, thus the distinction between MEQ and wet forests to the south. Indeed MEQ has never been recognized in any analytical biogeography (cf. Cracraft 1986; Crisp et al. 1995). The MEQ rainforests are supposed to have suffered severe reduction with substantial extinction of taxa through the Pleistocene, as evidenced by the absence of several rainforest-specialist vertebrates that are found to the north and south (WT and SEQ/NNSW; reviewed in Joseph et al. 1993; Low 1993). For the *Sphaerospira* lineage, the high phylogenetic diversity and phylogeographic congruence with endemic leaftail geckos (*Phyllurus;* Couper et al. 2000) suggests that long-term persistence of multiple (albeit small) refugia within MEQ has promoted, as well as preserved, diversity (Stuart-Fox et al. 2001).

Although effectively no previous phylogeographic analysis addresses historical subdivision across the St. Lawrence Gap, some studies traverse the extended SEQ bioregion (including CQ to NNSW as defined here). Of these, the consistent feature is of deepening structure/isolation northward, with three of five species showing a major break between Kroombit/Bulburin and the rest

(Moritz 2000). The pattern in the *Sphaerospira* lineage shows that, like the MEQ region to the north, the CQ (Shoalwater, Bulburin/Kroombit) region is not just a refuge, but is also a source of endemic diversification, and so can be argued to be a biogeographic region in its own right. Further south, the combination of overlapping tips and internodes gives a substantial peak in PD in the D'Aguilar Range. Could this be a sign of nascent speciation and/or admixture in the *S. fraseri* complex, a parallel to the processes (sympatric species) causing the PD peaks in CQ and MEQ?

SPECIATION

The phylogeny and distribution patterns (both actual and bioclimatic models) provide spatial and temporal support for the proposal that local peripheral isolation is the primary process driving diversification. The current intraspecific phylogeography provides a calibration of this ongoing process that has shaped virtually the entire phylogeny. Combined with limited evolution in ecology (inferred from climatic profiles), this indicates that ecological divergence comes after allopatric divergence, with limited local (alpha) diversity obtained secondarily. Although subsequent evolutionary processes leading to speciation, ecological divergence, and lineage overlap (alpha diversity) are not analyzed here, this framework provides avenues for further investigation. For example, we can ask what governs the tempo of morphological change: in particular, the relative importance of age of a refuge, its degree of geographic isolation, and stability of environment.

In addition to drift in allopatry, two other possible mechanisms for the evolution of ecomorphological divergence and sympatric species diversity are (1) environmentally driven divergent selection among isolated units, and (2) reinforcement or displacement processes leading up to secondarily derived sympatry (Rice and Hostert 1993). These processes may be assessed by a comparative analysis of morphological environmental variables across the phylogeny in the context of genetic divergence, spatial overlap, and modeled environmental change, thus encompassing population and species categories available in the *Sphaerospira* trans-species phylogeography.

ACKNOWLEDGMENTS

Thanks to Adnan Moussalli and Catherine Graham for modeling advice, Stuart Baird for discussion of likelihood analyses, the Museum of Vertebrate Zoology for GIS resources and general hospitality, and Mike Lee for comments on the manuscript as well as financial support. This research was supported by a grant from the Australian Research Council to C. Moritz and by Rainforest CRC.

REFERENCES

Adam, P. 1992. Australian Rainforests. Clarendon Press, Oxford.

Archer, M., S. J. Hand, and H. Godthelp. 1991. Riversleigh: The Story of Animals in Ancient Rainforest of Inland Australia. Reed Books, Balgowlah, NSW, Australia.

Arnaud, J. F., L. Madec, A. Bellido, and A. Guiller. 1999. Microspatial genetic structure in the land snail *Helix aspersa* (Gastropoda: Helicidae). Heredity 93:110–119.

Arnaud, J. F., L. Madec, A. Guiller, and A. Bellido. 2001. Spatial analysis of allozyme and microsatellite DNA polymorphisms in the land snail *Helix aspersa* (Gastropoda: Helicidae). Molecular Ecology 10:1563–1576.

Avise, J. C. 2000. Phylogeography: The History and Formation of Species. Harvard University Press, Cambridge, MA.

Avise, J. C., J. Arnold, R. M. Ball, E. Bermingham, T. Lamb, J. E. Neigel, C. A. Reeb, and N. C. Saunders. 1987. Intraspecific phylogeography: The mitochondrial DNA bridge between population genetics and systematics. Annual Review of Ecology and Systematics 18:489–522.

Avise, J. C., and D. Walker. 1998. Pleistocene phylogeographic effects on avian populations and the speciation process. Proceedings of the Royal Society of London B 265:457–463.

Baldwin, B. G., and M. J. Sanderson. 1998. Age and rate of diversification of the Hawaiian silversword alliance (Compositae). Proceedings of the National Academy of Sciences, U.S.A. 95:9402–9406.

Barraclough, T. G., and S. Nee. 2001. Phylogenetics and speciation. Trends in ecology and evolution 16:391–399.

Burbidge, N. T. 1960. The phytogeography of the Australian region. Australian Journal of Botany 8:75–210.

Chiba, S. 1999. Accelerated evolution of land snails *Mandarina* in oceanic Bonin islands: Evidence from mitochondria DNA sequences. Evolution 53:460–471.

Clarke, B., M. S. Johnson, and J. Murray. 1996. Clines in the genetic distance between two species of island land snails: How "molecular leakage" can mislead us about speciation. Philosophical Transactions of the Royal Society of London 351:773–784.

Couper, P. J., C. J. Schneider, C. J. Hoskin, and J. A. Covacevich. 2000. Australian leaf-tail geckos endemic to eastern Australia: A new genus, two new species, and other new data. Memoirs of the Queensland Museum 45:253–265.

Cracraft, J. 1986. Origin and evolution of continental biotas: Speciation and historical congruence within the Australian avifauna. Evolution 40:977–996.

Crisp, M. D., S. Laffan, H. P. Linder, and A. Monro. 2001. Endemism in the Australian flora. Journal of Biogeography 28:183–198.

Crisp, M. D., H. P. Linder, and P. H. Weston. 1995. Cladistic biogeography of plants in Australia and New Guinea: Congruent pattern reveals two endemic tropical tracks. Systematic Biology 44:457–473.

Davison, A. 2000. An East-west distribution of divergent mitochondrial haplotypes in British populations of the land snail, *Cepaea nemoralis* (Pulmonata). Biological Journal of the Linnean Society 70:697–706.

Davison, A., and B. Clarke. 2000. History or current selection? A molecular analysis of "area effects" in the land snail *Cepaea nemoralis*. Proceedings of the Royal Society of London. B 267:1399–1405.

Douris, V., R. A. D. Cameron, G. C. Rodakis, and R. Lecanidou. 1998. Mitochondrial phylogeography of the land snail *Albinaria* in Crete: Long-term geological and short-term vicariance effects. Evolution 52:116–125.

Faith, D. P. 1994. Phylogenetic pattern and the quantification of organismal biodiversity. Philosophical Transactions of the Royal Society of London 345:45–58.

Felsenstein, J. 1981. Evolutionary trees from DNA sequences: A maximum likelihood approach. Journal of Molecular Evolution 17:368–376.

Goodacre, S. L., and C. M. Wade. 2001. Patterns of genetic variation in Pacific island land snails: The distribution of cytochrome *b* lineages among Society Island *Partula*. Biological Journal of the Linnean Society 73:131–138.

Guiller, A., M. Coutellec-Vreto, L. Madec, and J. Deunff. 2001. Evolutionary history of the land snail *Helix aspersa* in the Western Mediterranean: Preliminary results inferred from mitochondrial DNA sequences. Molecular Ecology 10:81–87.

Harvey, P. H., R. M. May, and S. Nee. 1994. Phylogenies without fossils. Evolution 48:523–529.

Hayashi, M., and S. Chiba. 2000. Intraspecific diversity of mitochondrial DNA in the land snail *Euhadra peliomphala* (Bradybaenidae). Biological Journal of the Linnean Society 70:391–401.

Hilbert, D. W., and B. Ostendorf. 2001. The utility of artificial neural networks for modeling the distribution of vegetation in past, present and future climates. Ecological Modeling 146:311–327.

Hooker, J. D. 1859. Introductory essay. Pp. i–cxviii in Botany of the Antarctic Voyage of H.M. Discovery Ships *Erebus* and *Terror* in the Years 1839–1843. Vol. 3, Flora Tasmaniae. Reeve, London.

Houlder, D. J., M. F. Hutchinson, H. A. Nix, and J. P. McMahon. 2000. ANUCLIM Users Guide v. 5.1, CRES, ANU, Canberra.

Hubbell, S. P. 2001. The Unified Neutral Theory of Biodiversity and Biogeography. Monographs in Population Biology 32, Princeton University Press, Princeton and Oxford.

Hugall, A. F., C. Moritz, A. Moussalli, and J. Stanisic. 2002. Reconciling paleodistribution models and comparative phylogeography in the wet tropics rainforest land snail *Gnarosophia bellendenkerensis* (Brazier 1875). Proceedings of the National Academy of Sciences U.S.A. 99:6112–6117.

Iredale, T. 1937. A basic list of the land Mollusca of Australia. Australian Journal of Zoology 8:287–333.

James, C. H., and C. Moritz. 2000. Intraspecific phylogeography in the sedge frog *Litoria fallax* (Hylidae) indicates pre-Pleistocene vicariance of an open forest species from eastern Australia. Molecular Ecology 9:349–358.

Joseph, L., and C. Moritz. 1994. Mitochondrial DNA phylogeography of birds in eastern Australian rainforests: First fragments. Australian Journal of Zoology 42:385–403.

Joseph, L., C. Moritz, and A. Hugall. 1993. A mitochondrial perspective on the historical biogeography of mideastern Queensland rainforest birds. Memoirs of the Queensland Museum 34:201–214.

Keast, J. A. 1961. Bird speciation on the Australian continent. Bulletin of the Museum of Comparative Zoology Harvard 123:303–495.

Keast, J. A. 1981. The evolutionary biogeography of Australian birds. Pp. 1585–1635 in Ecological Biogeography of Australia (J. A. Keast, ed.). W. Junk, The Hague.

Kikkawa, J., and K. Pearse. 1969. Geographical distribution of land birds in Australia: A numerical analysis. Australian Journal of Zoology 17:821–840.

Low, T. 1993. Last of the rainforests: Rainforest refuges of the Mackay region. Wildlife Australia 4:18–21.

McElroy, D., P. Moran, E. Bermingham, and I. Kornfield. 1992. REAP: The restriction enzyme analysis package. Journal of Heredity 83:157–158.

McGuigan, K., K. McDonald, K. Parris, and C. Moritz. 1998. Mitochondrial DNA diversity and historical biogeography of a wet forest-restricted frog (*Litoria pearsoniana*) from mid-east Australia. Molecular Ecology 7:175–186.

Moritz, C. 2000. A molecular perspective on the conservation of diversity. Pp. 21–34 in The Biology of Biodiversity (M. Kato, ed.). Springer-Verlag.

Moritz, C., L. Joseph, M. Cunningham, and C. Schneider. 1997. Molecular perspectives on historical fragmentation of Australian tropical and subtropical rainforest: Implications for conservation. Pp. 442–454 in Tropical Rainforest Remnants: Ecology, Management, and Conservation of Fragmented Communities (W. F. Laurance and R. O. Bierregard, eds.). University of Chicago Press, Chicago.

Moritz, C., J. L. Patton, C. J. Schneider, and T. B. Smith. 2000. Diversification of rainforest faunas: An integrated molecular approach. Annual Review of Ecology and Systematics 31:533–563.

Nee, S. 2001. Inferring speciation rates from phylogenies. Evolution 55:661–668.

Nee, S., E. C. Holmes, R. M. May, and P. H. Harvey. 1994. Extinction rates can be estimated from molecular phylogenies. Philosophical Transactions of the Royal Society of London B 344:77–82.

Nelson, G., and N. Platnick. 1981. Systematics and Biogeography: Cladistics and Vicariance. Columbia University Press, New York.

Nix, H. A. 1982. Environmental determinates of biogeography and evolution in Terra Australis. Pp. 47–66 in Evolution of the Flora and Fauna of Arid Australia (W. R. Barker and P. J. M. Greenslade, eds.). Peacock Publications, Frewville, South Australia.

Nix, H. A. 1986. A biogeographic analysis of Australian elapid snakes. Pp. 4–15 in Atlas of Elapid Snakes of Australia (R. Longmore, ed.). AGPS, Canberra.

Nix, H. A. 1991. Biogeography: Patterns and process. Pp. 11–39 in Rainforest Animals: Atlas of Vertebrates Endemic to Australia's Wet Tropics (H. A. Nix and M. Switzer, eds.). ANPWS, Canberra.

Nix, H. A., and J. D. Kalma. 1972. Climate as the dominant control in the biogeography of northern Australia and New Guinea. Pp. 61–92 in Bridge and Barrier: The Natural and Cultural History of Torres Strait (D. Walker, ed.). Australian National University, Canberra.

Pfenninger, M., and F. Magnin. 2001. Phenotypic evolution and hidden speciation in *Candidula Unifasciata* spp. (Helicellinae, Gastropoda) inferred by 16S variation and quantitative shell traits. Molecular Ecology 10:2541–2554.

Rice, W. R., and E. E. Hostert. 1993. Laboratory experiments on speciation: What have we learned in 40 years? Evolution 47:1637–1653.

Rambaut, A., and M. Charleston. 2000. TreeEdit: Phylogenetic Tree Editor, v. 1.0-a4.61, University of Oxford.

Rambaut, A., P. H. Harvey, and S. Nee. 1997. End-Epi: An application for inferring phylogenetic and population dynamical processes from molecular sequences. CABIOS 13:303–306.

Rosenzweig, M. L. 1995. Species diversity in space and time. Cambridge University Press, Cambridge.

Ross, T. K. 1999. Phylogeography and conservation genetics of the Iowa Pleistocene snail. Molecular Ecology 8:1363–1373.

Schäuble, C. S., and C. Moritz. 2001. Comparative phylogeography of two open forest frogs from eastern Australia. Biological Journal of the Linnean Society 74:157–170.

Schilthuizen, M., and E. Gittenberger. 1996. Allozyme variation in some Cretan *Albinaria* (Gastropoda): Paraphyletic species as natural phenomena. Pp. 301–311 in Origin and Evolutionary Radiation of the Mollusca (J. D. Taylor, ed.). Oxford University Press, Oxford.

Schilthuizen, M., and M. Lombaerts. 1994. Population structure and levels of gene flow in the Mediterranean land snail *Albinaria corrugata* (Pulmonata: Clausiliidae). Evolution 48:577–586.

Schilthuizen, M., J. J. Vermeulen, G. W. H. Davison, and E. Gittenberger. 1999. Population structure in a snail species from isolated Malaysian limestone hills, inferred from ribosomal DNA sequences. Malacologia 41:283–296.

Schodde, R., and J. H. Calaby. 1972. Climate as the dominant control in the biogeography of northern Australia and New Guinea. Pp. 257–300 in Bridge and Barrier: The Natural and Cultural History of Torres Strait (D. Walker, ed.). Australian National University, Canberra.

Scott, B. J. 1996. Systematics, phylogeny and biogeography of the larger camaenid land snails of eastern Queensland rainforests (Pulmonata: Stylommatophora: Camaenidae). Ph.D. thesis, Department of Zoology, James Cook University, Townsville, Australia.

Scott, B. J. 1997. Biogeography of the Helicoidea (Mollusca: Gastropoda: Pulmonata): Land snails with Pangean distribution. Journal of Biogeography 24:399–407.

Shimizu, Y., and R. Ueshima. 2000. Historical biogeography and interspecific mtDNA introgression in *Euhadra peliomphala* (the Japanese land snail). Heredity 85:84–96.

Smith, B. J. 1992. Non-marine Mollusca. Pp. 110–174 in Zoological Catalogue of Australia (W. W. K. Houston, ed.). Australian Government Printing Service, Canberra.

Solem, A. 1979. Land-snail biogeography: A true snail's pace of change. Pp. 197–221 in Vicariance Biogeography: A Critique (G. Nelson and D. E. Rosen, eds.). Columbia University Press, New York.

Solem, A. 1997. Camaenid land snails from Western and Central Australia (Mollusca: Pulmonata: Camaenidae) VII. Taxa from Dampierland through the Nullabor. Records of the Western Australian Museum Supplement 50:1461–1899.

Spencer, W. B. 1896. Summary. Pp. 137–199 in Report on the Work of the Horn Scientific Expedition to Central Australia Part 1 (W. B. Spencer, ed.). Dulau, London.

Stuart-Fox, D. M., C. J. Schneider, C. Moritz, and P. J. Couper. 2001. Comparative phylogeography of three rainforest-restricted lizards from mid-east Queensland. Australian Journal of Zoology 49:119–127.

Swofford, D. L. 2000. PAUP*. Phylogenetic Analysis Using Parsimony (*and other methods). Version 4. Sinauer Associates, Sunderland, MA.

Thacker, R. W., and M. G. Hadfield. 2000. Mitochondrial phylogeny of extant Hawaiian tree snails (Achatinellinae). Molecular Phylogenetic and Evolution 16:263–270.

Thomaz, D., A. Guiller, and B. Clarke. 1996. Extreme divergence of mitochondrial DNA within species of pulmonate land snails. Proceedings of the Royal Society of London B 263:363–368.

Thompson, J. D., D. G. Higgins, and T. J. Gibson. 1994. CLUSTAL W: Improving the sensitivity of progressive multiple sequence alignment through sequence weighting, positions-specific gap penalties and weight matrix choice. Nucleic Acids Research 22:4673–4680.

Truswell, E. 1993. Vegetation changes in the Australian Tertiary in response to climatic and phytogeographic forcing factors. Australian Systematic Botany 6:533–557.

Wang, Q., I. W. B. Thornton, and T. R. New. 1996. Biogeography of the Phoracanthine beetles (Coleoptera: Cerambycidae). Journal of Biogeography 23:75–94.

Watanabe, Y., and S. Chiba. 2001. High within-population mitochondrial DNA variation due to microvicariance and population mixing in the land snail *Euhadra quaesita* (Pulmonata: Bradybaenidae). Molecular Ecology 10:2635–2646.

Williams, J. E. 1991. Biogeographic patterns of three sub-alpine eucalypts in south-east Australia with special reference to *Eucalyptus pauciflora* Sieb. Ex Spreng. Journal of Biogeography 18:223–230.

Woodruff, D. S. 1978. Evolutionary and adaptive radiation of *Cerion:* A remarkably diverse group of West Indian land snails. Malacologia 17:223–239.

CONTRIBUTORS

Allen G. Collins
Ecology, Behavior and Evolution
   Section
Division of Biology, Muir Biology
   Building
University of California, San Diego
La Jolla, CA 92093-0116

Benoît Dayrat
Invertebrate Zoology and Geology
California Academy of Sciences
San Francisco, CA 94118

Daniel L. Distel
Biochemistry, Microbiology, and
   Molecular Biology
School of Marine Science
University of Maine
Orono, ME 04469-5735

Gonzalo Giribet
Department of Organismic and
   Evolutionary Biology
Museum of Comparative Zoology

Harvard University
Cambridge, MA 02138

M. G. Harasewych
Department of Invertebrate Zoology
National Museum of Natural History
Smithsonian Institution
Washington, DC 20560

Walter R. Hoeh
Ecology and Evolutionary Biology
   Program
Department of Biological Sciences
Kent State University
Kent, OH 44242

Andrew Hugall
Cooperative Research Centre for
   Rainforest Ecology and
   Management
Department of Zoology and
   Entomology
University of Queensland
Brisbane 4072 Australia

David R. Lindberg
Department of Integrative Biology
University of California
Berkeley, CA 94720

Charles Lydeard
Biodiversity and Systematics
Department of Biological Sciences
University of Alabama
Tuscaloosa, AL 35487

Andrew G. McArthur
Josephine Bay Paul Center for
  Comparative Molecular Biology
  and Evolution
Marine Biological Laboratory
Woods Hole, MA 02543-1015

Mónica Medina
Genomic Diversity
Joint Genome Institute
Walnut Creek, CA 94598

Craig Moritz
Museum of Vertebrate Zoology
Department of Integrative Biology
University of California
Berkeley, CA 94720-3160

Patrick D. Reynolds
Hamilton College
Clinton, NY 13323

Kevin J. Roe
Delaware Museum of Natural
  History
Wilmington, DE 19807-0937

John Stanisic
Queensland Centre for Biodiversity

Queensland Museum
Brisbane 4000 Australia

Gerhard Steiner
Institute of Zoology
University of Vienna
A-1090 Vienna, Austria

Simon Tillier
Muséum National d'Histoire
  Naturelle
Service de Systématique Moléculaire
75231 Paris, France

Thomas F. Turner
University of New Mexico
Castetter Hall
Albuquerque, NM 87131

Verena Vonnemann
Spezielle Zoologie
Ruhr-Universität Bochum
44780 Bochum, Germany

Heike Wägele
Spezielle Zoologie
Ruhr-Universität Bochum
44780 Bochum, Germany

Wolfgang Wägele
Spezielle Zoologie
Ruhr-Universität Bochum
44780 Bochum, Germany

John P. Wares
University of California
Davis, CA 95616

INDEX